T0212984

Lecture Notes in Computer Science   **10456**

Commenced Publication in 1973
Founding and Former Series Editors:
Gerhard Goos, Juris Hartmanis, and Jan van Leeuwen

More information about this series at http://www.springer.com/series/7409

Gareth J.F. Jones · Séamus Lawless
Julio Gonzalo · Liadh Kelly
Lorraine Goeuriot · Thomas Mandl
Linda Cappellato · Nicola Ferro (Eds.)

Experimental IR Meets Multilinguality, Multimodality, and Interaction

8th International Conference
of the CLEF Association, CLEF 2017
Dublin, Ireland, September 11–14, 2017
Proceedings

 Springer

Editors

Gareth J.F. Jones
Dublin City University
Dublin
Ireland

Séamus Lawless
Trinity College Dublin
Dublin
Ireland

Julio Gonzalo
National University of Distance Education
Madrid
Spain

Liadh Kelly
Dublin City University
Dublin
Ireland

Lorraine Goeuriot
Université Grenoble Alpes
Grenoble
France

Thomas Mandl
University of Hildesheim
Hildesheim
Germany

Linda Cappellato
University of Padua
Padua
Italy

Nicola Ferro
University of Padua
Padua
Italy

ISSN 0302-9743   ISSN 1611-3349   (electronic)
Lecture Notes in Computer Science
ISBN 978-3-319-65812-4   ISBN 978-3-319-65813-1   (eBook)
DOI 10.1007/978-3-319-65813-1

Library of Congress Control Number: 2017949173

LNCS Sublibrary: SL3 – Information Systems and Applications, incl. Internet/Web, and HCI

Printed on acid-free paper

This Springer imprint is published by Springer Nature
The registered company is Springer International Publishing AG
The registered company address is: Gewerbestrasse 11, 6330 Cham, Switzerland

Preface

Since 2000, the Conference and Labs of the Evaluation Forum (CLEF) has played a leading role in stimulating research and innovation in the domain of multimodal and multilingual information access. Initially founded as the Cross-Language Evaluation Forum and running in conjunction with the European Conference on Digital Libraries (ECDL/TPDL), CLEF became a standalone event in 2010 combining a peer-reviewed conference with a multi-track evaluation forum. The combination of the scientific program and the track-based evaluations at the CLEF conference creates a unique platform to explore information access from different perspectives, in any modality and language.

The CLEF conference has a clear focus on experimental information retrieval (IR) as seen in evaluation forums (CLEF Labs, TREC, NTCIR, FIRE, MediaEval, RomIP, TAC) with special attention to the challenges of multimodality, multilinguality, and interactive search ranging from unstructured, to semi-structured and structured data. CLEF invites submissions on significant new insights demonstrated by the use of innovative IR evaluation tasks or in the analysis of IR test collections and evaluation measures, as well as on concrete proposals to push the boundaries of the Cranfield/TREC/CLEF paradigm.

CLEF 2017[1] was hosted by the ADAPT Centre[2], Dublin City University and Trinity College Dublin during September 11–14, 2017. The conference format consisted of keynotes, contributed papers, lab sessions, and poster sessions, including reports from other benchmarking initiatives from around the world. This year's conference was also co-located with MediaEval[3] and the program included joint sessions between both MediaEval and CLEF to allow for cross fertilization.

CLEF 2017 received 38 submissions, of which a total of 22 papers were accepted. Each submission was reviewed by Program Committee (PC) members, and the program chairs oversaw the reviewing and follow-up discussions. CLEF 2017 continued a novel track introduced at CLEF 2015, i.e., inviting CLEF lab organizers to nominate a "best of the labs" paper that was reviewed as a full paper submission to the CLEF 2017 conference according to the same review criteria and PC. In total, 15 long papers were received, of which seven were accepted; 17 short papers were received, of which nine were accepted; six Best of Labs track papers were received, all of which were accepted.

The conference integrated a series of workshops presenting the results of lab-based comparative evaluations. CLEF 2017 was the 8th year of the CLEF Conference and the 18th year of the CLEF initiative as a forum for IR Evaluation. The labs were selected after peer review based on their innovation potential and the quality of the resources created. The labs represented scientific challenges based on new data sets and

---

[1] http://clef2017.clef-initiative.eu/.

[2] http://adaptcentre.ie/.

[3] http://www.multimediaeval.org/.

real-world problems in multimodal and multilingual information access. These data sets provide unique opportunities for scientists to explore collections, to develop solutions for these problems, to receive feedback on the performance of their solutions, and to discuss the issues with peers at the workshops.

In addition to these workshops, the ten benchmarking labs reported results of their year-long activities in overview talks and lab sessions. Overview papers describing each of these labs are provided in this volume. The full details for each lab are contained in a separate publication, the Working Notes, which are available online[4].

The eight labs and two workshops running as part of CLEF 2017 were as follows:

News Recommendation Evaluation Lab (NEWSREEL)[5] provides a vehicle for the IR/recommender system communities to move from conventional offline evaluation to online evaluation. We address the following information access challenge: Whenever a visitor of an online news portal reads a news article on their side, the task is to recommend other news articles that the user might be interested in.

LifeCLEF[6] aims at boosting research on the identification of living organisms and on the production of biodiversity data in general. Through its biodiversity informatics related challenges, LifeCLEF aims to push the boundaries of the state of the art in several research directions at the frontier of multimedia information retrieval, machine learning, and knowledge engineering.

Uncovering Plagiarism, Authorship, and Social Software Misuse (PAN)[7] provides evaluation of uncovering plagiarism, authorship, and social software misuse. PAN offered three tasks at CLEF 2017 with new evaluation resources consisting of large-scale corpora, performance measures, and web services that allow for meaningful evaluations. The main goal is to provide for sustainable and reproducible evaluations, to get a clear view of the capabilities of state-of-the-art-algorithms. The tasks are: author identification; author profiling; and, author obfuscation.

CLEFeHealth[8] provides scenarios which aim to ease patients' and nurses' understanding and accessing of eHealth information. The goals of the lab are to develop processing methods and resources in a multilingual setting to enrich difficult-to-understand eHealth texts, and provide valuable documentation. The tasks are: multilingual information extraction; technologically assisted reviews in empirical medicine; and, patient-centered information retrieval.

Cultural Microblog Contextualization (CMC) Workshop[9] deals with how cultural context of a microblog affects its social impact at large. This involves microblog search, classification, filtering, language recognition, localization, entity extraction, linking open data and summarization. Regular Lab participants have access to the private massive multilingual microblog stream of The festival galleries project.

---

[4] http://ceur-ws.org/Vol-1866.

[5] http://clef-newsreel.org/.

[6] http://www.lifeclef.org/.

[7] http://pan.webis.de/.

[8] https://sites.google.com/site/clefehealth2017/.

[9] https://mc2.talne.eu/.

ImageCLEF[10] organizes three main tasks with a global objective of benchmarking lifelogging retrieval and summarization, tuberculosis type prediction from CT images, and bio-medical image caption prediction; and a pilot task on remote sensing image analysis.

Early risk prediction on the Internet (eRisk)[11] explores issues of evaluation methodology, effectiveness metrics, and other processes related to early risk detection. Early detection technologies can be employed in different areas, particularly those related to health and safety. For instance, early alerts could be sent when a predator starts interacting with a child for sexual purposes, or when a potential offender starts publishing antisocial threats on a blog, forum or social network. Our main goal is to pioneer a new interdisciplinary research area that would be potentially applicable to a wide variety of situations and to many different personal profiles.

Personalized Information Retrieval at CLEF (PIR-CLEF)[12] provides a framework for evaluation of Personalized Information Retrieval (PIR). Current approaches to the evaluation of PIR are user-centered, i.e., they rely on experiments that involve real users in a supervised environment. PiR-CLEF aims to develop and demonstrate a methodology for evaluation PIR which enables repeatable experiments to enable the detailed exploration of personal models and their exploitation in IR.

Dynamic Search for Complex Tasks[13] Information Retrieval research has traditionally focused on serving the best results for a single query – so-called ad hoc retrieval. However, users typically search iteratively, refining and reformulating their queries during a session. A key challenge in the study of this interaction is the creation of suitable evaluation resources to assess the effectiveness of IR systems over sessions. The goal of the CLEF Dynamic Search lab is to propose and standardize an evaluation methodology that can lead to reusable resources and evaluation metrics able to assess retrieval performance over an entire session, keeping the "user" in the loop.

Multimodal Spatial Role Labeling[14] explores the extraction of spatial information from two information resources that is image and text. This is important for various applications such as semantic search, question answering, geographical information systems, and even in robotics for machine understanding of navigational instructions or instructions for grabbing and manipulating objects.

CLEF 2017 was accompanied by a social program encompassing some of Dublin's most popular locations. The Welcome Reception took place at the Guinness Storehouse, Ireland's most popular tourist attraction, including a introduction to the brewing of Guinness, an exhibition of the famous cartoon advertising campaigns, and the main reception in the Gravity Bar with panoramic views across the city. The conference dinner was held jointly with MediaEval in the Dining Hall at Trinity College Dublin. Participants were also able to join a Literary Pub Crawl exploring Dublin's historic literary tradition and its social settings.

---

[10] http://imageclef.org/2017.

[11] http://early.irlab.org/.

[12] http://www.ir.disco.unimib.it/pirclef2017/.

[13] https://ekanou.github.io/dynamicsearch/.

[14] http://www.cs.tulane.edu/~pkordjam/mSpRL_CLEF_lab.htm/.

The success of CLEF 2017 would not have been possible without the huge effort of several people and organizations, including the CLEF Association[15] and the ADAPT Centre, Ireland, the Program Committee, the Lab Organizing Committee, Martin Braschler, Donna Harman, and Maarten de Rijke, the local Organizing Committee in Dublin, Conference Partners International, the reviewers, and the many students and volunteers who contributed.

July 2017

Gareth J.F. Jones
Séamus Lawless
Julio Gonzalo
Liadh Kelly
Lorraine Goeuriot
Thomas Mandl
Linda Cappellato
Nicola Ferro

---

[15] http://www.clef-initiative.eu/association.

Organization

CLEF 2017, Conference and Labs of the Evaluation Forum – Experimental IR meets Multilinguality, Multimodality, and Interaction, was organized by the ADAPT Centre in Dublin City University and Trinity College, Dublin, Ireland.

General Chairs

Gareth J.F. Jones   Dublin City University, Ireland
Séamus Lawless   Trinity College Dublin, Ireland

Program Chairs

Julio Gonzalo   UNED, Spain
Liadh Kelly   Dublin City University, Ireland

Lab Chairs

Lorraine Goeuriot   Université Grenoble Alpes, France
Thomas Mandl   University of Hildesheim, Germany

Proceedings Chairs

Linda Cappellato   University of Padua, Italy
Nicola Ferro   University of Padua, Italy

Local Organization

Piyush Arora   Dublin City University, Ireland
Mostafa Bayomi   Trinity College Dublin, Ireland
Annalina Caputo   Trinity College Dublin, Ireland
Joris Vreeke (Webmaster)   ADAPT Centre, Ireland

Program Committee

Helbert Arenas   IRIT - CNRS, France
Leif Azzopardi   University of Strathclyde, UK
Kevin B. Cohen   University of Colorado School of Medicine, France
Alvaro Barreiro   IRLab, University of A Coruña, Spain
Md Bayzidul Islam   Institute of Geodesy, TU Darmstadt, Germany
Patrice Bellot   Aix-Marseille Université - CNRS (LSIS), France
Giulia Boato   University of Trento, Italy

Vitali Liauchuk	United Institute of Informatics Problems, Republic of Belarus
Yiqun Liu	Tsinghua University, China
Andreas Lommatzsch	Technische Universität Berlin, Germany
Titouan Lorieul	Inria, France
David Losada	Universidade de Santiago de Compostela, Spain
Mihai Lupu	Vienna University of Technology, Austria
Umar Manzoor	The University of Salford, Manchester, UK
Stefania Marrara	Bicocca University of Milan, Italy
Paul McNamee	Johns Hopkins University, USA
Marie-Francine Moens	KULeuven, Belgium
Boughanem Mohand	IRIT University Paul Sabatier Toulouse, France
Manuel Montes	Instituto Nacional de Astrofísica, Óptica y Electrónica, Mexico
Josiane Mothe	Institut de Recherche en Informatique de Toulouse, France
Henning Müller	HES-SO, Switzerland
Philippe Mulhem	LIG-CNRS, France
Fionn Murtagh	University of Derby; Goldsmiths University of London, UK
Jian-Yun Nie	Université de Montréal, Canada
Aurélie Névéol	LIMSI, CNRS, Université Paris-Saclay, France
Simone Palazzo	University of Catania, Italy
Joao Palotti	Vienna University of Technology, Austria
Javier Parapar	University of A Coruña, Spain
Gabriella Pasi	Università degli Studi di Milano Bicocca, Italy
Pavel Pecina	Charles University in Prague, Czech Republic
Raffaele Perego	ISTI - CNR, Italy
Bonnet Pierre	CIRAD, France
Karen Pinel-Sauvagnat	IRIT, France
Luca Piras	University of Cagliari, Italy
Martin Potthast	Bauhaus-Universität Weimar, Germany
James Pustejovsky	Computer Science Department, Brandeis University, USA
Paulo Quaresma	Universidade de Evora, Portugal
Taher Rahgooy	Bu-Ali Sina University of Hamedan, Iran
Francisco Manuel Rangel Pardo	Autoritas Consulting, Spain
Philip Resnik	University of Maryland, USA
Grégoire Rey	Inserm, France
Michael Riegler	Simula Research Laboratory, Norway
Aude Robert	CépiDc-Inserm, France
Kirk Roberts	Human Language Technology Research Institute, University of Texas at Dallas, USA

Paolo Rosso	Technical University of Valencia, Spain
Eric Sanjuan	Laboratoire Informatique d'Avignon- Université d'Avignon, France
Ralf Schenkel	Trier University, Germany
Immanuel Schwall	ETH Zurich, Switzerland
Dimitrios Soudris	National Technical University of Athens, Greece
Concetto Spampinato	University of Catania, Italy
Rene Spijker	Cochran Netherlands, The Netherlands
Efstathios Stamatatos	University of the Aegean, Greece
Benno Stein	Bauhaus-Universität Weimar, Germany
Hanna Suominen	The ANU, Australia
Lynda Tamine	IRIT, France
Xavier Tannier	LIMSI, CNRS, Univ. Paris-Sud, Université Paris-Saclay, France
Juan-Manuel Torres-Moreno	Laboratoire Informatique d'Avignon/UAPV, France
Theodora Tsikrika	Information Technologies Institute, CERTH, Greece
Wp Vellinga	xeno-canto foundation for nature sounds, The Netherlands
Mauricio Villegas	UPV, Spain
Christa Womser-Hacker	Universität Hildesheim, Germany
Guido Zuccon	Queensland University of Technology, Australia
Pierre Zweigenbaum	LIMSI, CNRS, Université Paris-Saclay, France

Sponsors

CLEF Steering Committee

Contents

Best of the Labs

Labs Overviews

Full Papers

A Pinch of Humor for Short-Text Conversation: An Information Retrieval Approach

Vladislav Blinov[✉], Kirill Mishchenko, Valeria Bolotova, and Pavel Braslavski

Ural Federal University, Yekaterinburg, Russia
vladislav.blinov@urfu.ru, ki.mishchenko@gmail.com,
lurunchik@gmail.com, pbras@yandex.ru

Abstract. The paper describes a work in progress on humorous response generation for short-text conversation using information retrieval approach. We gathered a large collection of funny tweets and implemented three baseline retrieval models: BM25, the query term reweighting model based on syntactic parsing and named entity recognition, and the *doc2vec* similarity model. We evaluated these models in two ways: *in situ* on a popular community question answering platform and in laboratory settings. The approach proved to be promising: even simple search techniques demonstrated satisfactory performance. The collection, test questions, evaluation protocol, and assessors' judgments create a ground for future research towards more sophisticated models.

Keywords: Natural language interfaces · Humor generation · Short-text conversation · Information retrieval · Evaluation · Community question answering

1 Introduction

Humor is an essential aspect of human communication. Therefore, sense of humor is a desirable trait of chatbots and conversational agents aspiring to act like humans. Injection of humor makes human-computer conversations more engaging, contributes to the agent's personality, and enhances the user experience with the system [10,18]. Moreover, a humorous response is a good option for out-of-domain requests [3] and can soften the negative impact of inadequacies in the system's performance [4]. However, if we look at existing mobile personal assistants (for example, Apple Siri[1]), it can be noticed that their humorous answers work on a limited set of stimuli and are far from being diverse.

In this study, we approached the problem of generating a humorous response to the user's utterance as an information retrieval (IR) task over a large collection of presumably funny content. Our approach is exploratory and is not based on a certain theory of humor or a concrete type of jokes. The aim of our study is to implement several retrieval baselines and experiment with different methods

---

[1] https://www.apple.com/ios/siri/.

© Springer International Publishing AG 2017
G.J.F. Jones et al. (Eds.): CLEF 2017, LNCS 10456, pp. 3–15, 2017.
DOI: 10.1007/978-3-319-65813-1_1

of evaluation. IR is a promising approach in conversational systems [28, 29] that can significantly improve quality and diversity of responses.

First, we gathered about 300,000 funny tweets. After that, we implemented three baselines for tweet retrieval: (1) BM25 – a classical IR model based on term statistics; (2) a query term reweighting model based on syntactic parsing and NER; and (3) a retrieval model based on document embeddings. Finally, we collected user questions and evaluated three baselines *in situ* on a community question answering (CQA) platform and in laboratory settings.

To the best of our knowledge, IR has not been applied to humorous response generation in short-text conversation scenario and no formal evaluation has been conducted on the task before. We have made the tweet collection (as a list of tweet IDs), test questions and assessors' judgments freely available[2] for research. The data creates a solid ground for future research in the field.

2 Related Work

There are two main directions in computational humor research: humor recognition and humor generation.

Humor recognition is usually formulated as a classification task with a wide variety of features – syntactic parsing, alliteration and rhyme, antonymy and other WordNet relations, dictionaries of slang and sexually explicit words, polarity and subjectivity lexicons, distances between words in terms of *word2vec* representations, etc. [11, 16, 24, 30, 31]. A cognate task is detection of other forms of figurative language such as irony and sarcasm [19, 20, 25]. Several recent studies dealing with humor and irony detection are focused on the analysis of tweets, see [19, 20, 31].

Most humor generation approaches focus on puns, as puns have relatively simple surface structure [6, 21, 26]. Stock and Strapparava [23] developed *HAHAcronym*, a system that generates funny decipers for existing acronyms or produces new ones starting from concepts provided by the user. Valitutti et al. [26] proposed a method for 'adult' puns made from short text messages by lexical replacement. A related study [6] addresses the task of automatic template extraction for pun generation.

Mihalcea and Strapparava [17] proposed a method for adding a joke to an email message or a lecture note from a collection of 16,000 one-liners using latent semantic analysis (LSA). A small-scale user study showed a good reception of the proposed solution. This study is the closest to ours; however, we use an order of magnitude larger collection, implement several retrieval models and place emphasis on evaluation methodology.

Wen et al. [27] explore a scenario, when a system suggests the user funny images to be added to a chat. The work also employs an IR technique among others: candidate images are partly retrieved through Bing search API using query "funny *keywords*", where *keywords* are *tf-idf* weighted terms from the last three utterances.

---

[2] https://github.com/micyril/humor.

Shahaf et al. [22] investigate the task of ranking cartoon captions provided by the readers of New Yorker magazine. They employ a wide range of linguistic features as well as features from manually crafted textual descriptions of the cartoons. Jokes comparison/ranking task is close to ours, however, the settings and data are quite different.

Augello et al. [2] described a chatbot nicknamed *Humorist Bot*. The emphasis was made on humor recognition in the humans' utterances following the approach proposed in [16]; the bot reacted to jokes with appropriate responses and emoticons. Humorous response generation was restricted to a limited collection of jokes that was triggered when the user asked the bot to tell one.

The information retrieval approach to short-text conversations became popular recently [8, 28, 29]. The method benefits from the availability of massive conversational data, uses a rich set of features and learning-to-rank methods. Our approach follows the same general idea; however our exploratory study employs simpler retrieval models with a weak supervision.

3 Data

3.1 Joke Collection

To gather a collection of humorous tweets, we started with several "top funny Twitter accounts" lists that can be easily searched online[3]. We filtered out accounts with less than 20,000 followers, which resulted in 103 accounts. Table 1 lists top 10 most popular accounts in the collection. Then, we downloaded all available text-only tweets (i.e. without images, video, and URLs) and retained those with at least 30 likes or retweets (366,969 total). After that, we removed duplicates with a Jaccard similarity threshold of 0.45 using a Minhash index implementation[4] and ended up with a collection of 300,876 tweets. Here is an example from our collection (359 likes, 864 retweets)[5]:

Life is a weekend when you're unemployed.

To validate the proposed data harvesting approach, we implemented a humor recognizer based on a dataset consisting of 16,000 one-liners and 16,000 non-humorous sentences from news titles, proverbs, British National Corpus, and Open Mind Common Sense collection [16]. We employed a concatenation of *tf-idf* weights of unigrams and bigrams with document frequency above 2 and 300-dimensional *doc2vec* representations (see Sect. 4.3 for details) as a feature vector. A logistic regression classifier achieved 10-fold cross-validation accuracy of 0.887 (which exceeds previously reported results [16, 30]). We applied this classifier to our collection as well as to a sample of tweets from popular media accounts, see Table 2. The results confirm that the approach is sound; however, we did not filter the collection based on the classification results since the training data

---

[3] See for example http://www.hongkiat.com/blog/funny-twitter-accounts/.

[4] https://github.com/ekzhu/datasketch.

[5] https://twitter.com/MensHumor/status/360113491937472513.

Table 1. Accounts with highest numbers of followers in the collection

Account	# of followers
ConanOBrien	21,983,968
StephenAtHome	11,954,015
TheOnion	9,128,284
SteveMartinToGo	7,828,907
Lmao	5,261,116
AlYankovic	4,355,144
MensHumor	3,543,398
TheTweetOfGod	2,285,307
Lord_Voldemort7	2,026,292
michaelianblack	2,006,624

Table 2. Humor recognition in tweets

Collection/account	Classified as humorous	# of tweets
Funny Accounts	258,466/85.9%	300,876
The Wall Street Journal (wsj)	142/9.7%	1,464
The Washington Post (washingtonpost)	195/21.5%	907
The New York Times (nytimes)	240/19.8%	1,210
Donald J. Trump (realDonaldTrump)	7,653/59.1%	12,939

is quite different from the tweets. For example, many emotional and sentiment rich tweets from the *realDonaldTrump* account are considered to be funny by our classifier.

3.2 Yahoo!Answers

We used *Jokes & Riddles* category of Yahoo!Answers[6] for *in situ* evaluation: as a source of users' questions and measuring reactions of community members to automatically retrieved answers.

Yahoo!Answers is a popular CQA platform where users can ask questions on virtually any subject and vote for answers with 'thumb up' and 'thumb down'; the asker can also nominate the 'best answer' [1]. Figure 1 represents Yahoo!Answers' user interface with a question and two answers provided by the community in the *Jokes & Riddles* category. Each question has a title and an optional longer description, which we disregarded. Approximately 20 questions are posted in *Jokes & Riddles* daily. The category has an obvious topical bias: there are noticeably many ironic questions on atheism, faith and theory of evolution (see for instance the second question in Table 3). Apart from using

---

[6] https://answers.yahoo.com/dir/index?sid=396546041.

Fig. 1. Yahoo!Answers interface

Yahoo!Answers to evaluate the retrieved responses, we also gathered historical data to weakly supervise our Query Term Rewighting model (see Sect. 4.2). We collected 1,371 questions asked during two months in the *Jokes & Riddles* category along with submitted answers; 856 of the threads contain 'best answer' nominations.

4 Retrieval Models

We implemented three joke retrieval baselines: (1) a classical BM25 model based on term statistics; (2) a query term reweighting model based on structural properties of the stimulus (dependency tree and the presence of named entities); and (3) a model based on document embeddings that retrieves semantically similar 'answers' and does not require word overlap with the 'query'. Table 3 shows examples of top-ranked responses by these models.

4.1 BM25

BM25 is a well-known ranking formula [9], a variant of the *tf-idf* approach. It combines term frequency within document (*tf*) and collection-wide frequency (*idf*) to rank documents that match query terms. We did not perform stop-word removal: in has been shown that personal pronouns are important features for humorous content [15,22]. Since documents (tweets) in our case are rather short, ranking is dominated by the *idf* weights, i.e. rare words. It can potentially be harmful for humor retrieval, since many popular jokes seem to contain mostly common words [15,22].

4.2 Query Term Reweighting Model (QTR)

This approach is inspired by the notion of *humor anchors* introduced in [30]. According to the authors, a humor anchor is a set of word spans in a sentence that enables humorous effect. To detect humor anchors, they firstly extracted a set of candidates based on syntactic parsing, and then searched for a subset that caused a significant drop in humor score when removed from the original sentence. In our study, we followed the idea of humor anchors to modify term weighting scheme: BM25 scores in a query are adjusted corresponding to the syntactic roles of matched terms.

In order to calculate weight adjustments, we used 573 question-'best answer' pairs that have word overlap (out of 856, see Sect. 3.2).

We applied dependency parsing and named entity recognition (NER) from Stanford CoreNLP [14] to the questions and counted dependency and NER labels of the overlapping words (see Figs. 2 and 3, respectively). Then, we set adjustment coefficients accordingly to the counts. As expected, the *nominal subject* and *main verb* (root) roles and the *person* NE are the most frequent.

Question:

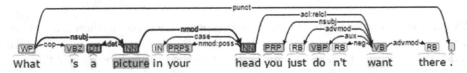

What   's a picture in your   head you just do n't   want   there .

Best Answer:

The picture of Clinton ... it can be either one of them ... :)

Fig. 2. Learning term reweighting based on dependency parsing: accounting for syntactic tags of overlapping words (*picture, nsubj*)

Question: LOCATION
Why is the   Earth   flat ?

Best answer: LOCATION
I have never understood how the   Earth   is

flat and other planets are reportedly round .

Fig. 3. Learning term reweighting based on NER: accounting for NE types of overlapping words (*Earth, LOCATION*)

4.3 doc2vec

doc2vec model [13] generates vector representations of sentences and paragraphs, thus extending *word2vec* model from word to document embeddings. In contrast to two previous models, *doc2vec* is capable of finding jokes semantically close to the 'query' even when there is no word overlap between them (recall, about one third of 'best answers' in *Jokes & Riddles* category have no word overlap with the corresponding questions). See for example the *doc2vec* response to the second question in Table 3. We used cosine similarity between the question's and documents' vector representations to find semantically closest jokes.

We followed the same pipeline as described in [12]: tokenizing, lowercasing, and inferring embeddings with the same initial learning rate and number of epochs. In our experiments, we used the Distributed Bag of Words (DBOW) model pre-trained on the English Wikipedia[7].

Table 3. Examples of top-ranked responses by the three models

BM25	QTR	doc2vec
Is it true Hilary[a] Clinton is secretly Donald Trump's mom?		
Is it true eminem thanked his mom's spaghetti	At the very least, I'm far less concerned about Hillary Clinton's physical ailments than I am about Donald Trump's mental ones	I'm not convinced Donald Trump knows what sex is
Why do you atheist not apply the same standards of evidence on your own "theories" as you do to challenge the existence of God?		
The existence of conspiracy theories is a myth	I have a mosquito bite on the inside of the arch of my foot thus disproving the existence of God	Science is true whether or not you believe it, but religion is true whether or not it's true

[a]Original spelling.

5 Evaluation

Humor evaluation is challenging, since the perception of humor is highly subjective and is conditioned on the situation and the socio-cultural background. We evaluated joke retrieval in two modes: (1) *in situ* – top-1 ranked responses of each model were presented to the CQA users in a 'humorous' category and (2) top-3 responses for a subset of the same CQA questions were assessed by three judges in lab settings. The former approach allows evaluation in real-life environment, however it scales poorly and is harder to interpret. The latter one is more controllable, but it is not clear how well few judges represent an 'average' user.

---

[7] https://github.com/jhlau/doc2vec.

5.1   Yahoo!Answers

For six consecutive days, we manually posted top-1 ranked responses by the models to questions asked during the day. We have submitted responses to 101 questions in total; in five threads there was no voting activity (i.e. neither a 'best answer' selected, nor any votes submitted), so we excluded them from the analysis. Each question received 22 answers on average (including three from our models). The CQA users' votes were collected two weeks later. Evaluation on Yahoo!Answers allows potentially for evaluation models against each other, as well as comparison of automatic responses with those by the users. Table 6 summarizes the users' reaction to retrieved answers by model: upvotes (+), downvotes (−), 'best answer' nominations (BA), and number of times the model outperformed the other two ('best model'). If we rank all answers by their votes (the 'best answer' on the first position if present), there are 5.72 rank positions on average; mean position of an oracle (the best variant out of three models' answers) is 3.92. The last column in Table 6 presents an average percentage of the users, whose answers had lower rank positions than the model's responses. Table 4 shows similar statistics for seven most active answerers in the same 96 question threads. Obviously, users are much more selective, they do not answer all questions in a row and thus have higher average scores per answer. Nevertheless, automatically retrieved answers do not look completely hopeless when compared to these users' performance (except for several good scorers, e.g. User6).

Table 4. The most active CQA users (96 questions)

User	# answers	+	−	BA	Users below
User1	23	25	0	1	32.1%
User2	20	33	1	0	33.1%
User3	20	13	1	2	15.9%
User4	17	20	1	0	16.6%
User5	15	24	0	0	45.6%
User6	13	28	0	7	71.1%
User7	13	8	0	0	19.5%

5.2   Lab Evaluation

For lab evaluation we sampled 50 questions from the ones we answered on Yahoo!Answers. Top-3 results for each model were collected for evaluation, yielding 433 unique question-answer pairs. The question-response pairs were presented to three assessors in a dedicated evaluation interface in a random order, three at a time. The assessors were asked to judge responses with the context in mind, i.e. an out-of-context joke, even when it is funny by itself, is expected

Question:

Are you secretly planing to buy a bottle of soy sauce?

Answers:

I buy soy milk because I can't drink regular milk before it goes bad.

Who called it soy sauce instead of MSG in a bottle

I just put Worcestershire sauce into fried rice instead of soy sauce. Hey ladies.

Fig. 4. The annotation tool for laboratory evaluation

to be scored low. The responses were judged on a four-point scale (from 0 to 3), with corresponding emoticons in the evaluation interface (see Fig. 4).

The relevance score for a question–response pair is an average over three assessors' labels (see Table 5 for some examples). Table 7 shows the averaged scores of the top-ranked responses and DCG@3 scores [7] for the three models and the oracle that composes the best output from the nine pooled results. The averaged pairwise weighted Cohen's kappa [5] is 0.13, which indicates a low agreement among assessors. Here is an example of a question–answer pair that received three different labels (☺/☻/☹) from three assessors:

    Q: Do you scream with excitement when you walk into a clothes shop?
    A: Do hipsters in the Middle East shop at Turban Outfitters?

Table 5. Example question–response pairs and their averaged relevance scores

Score	Question	Response
3.00	Does evolution being a theory make it subjective?	There is no theory of evolution, just a list of creatures Chuck Norris allows to live
2.67	Can you find oil by digging holes in your backyard?	Things to do today: 1. Dig a hole 2. Name it love 3. Watch people fall in it
1.33	Why don't they put zippers on car doors?	Sick of doors that aren't trap doors
0.67	What if you're just allergic to working hard?	You're not allergic to gluten
0.33	What test do all mosquitoes pass?	My internal monologue doesn't pass the Bechdel test :(

Table 6. CQA users' reaction

Model	+	−	BA	Best model	Users below
BM25	19	3	0	3	16.5%
QTR	14	1	2	6	16.1%
doc2vec	15	2	2	3	14.3%
Oracle	23	1	4	–	19.4%

Table 7. Lab evaluation results (50 questions)

Model	Avg. score @1	*DCG@3*
BM25	1.34	2.78
QTR	1.15	2.38
doc2vec	1.25	2.63
Oracle	1.91	3.61

6   Conclusion and Future Work

The most important outcome of the conducted experiment is that a combination of a simple approach to harvesting a joke collection and uncomplicated retrieval models delivers satisfactory performance for humorous response generation task. On the one hand, we may hypothesize that the size of the collection does matter – even simple methods can yield reasonable results when they have a lot of variants to choose from. On the other hand, it seems that when a situation implies a whimsical response, an unexpected, illogical or even inconsistent answer can still be considered funny.

The evaluation on the CQA platform showed that automatic methods for humorous response generation have at least some promise compared to humans. At the same time this evaluation does not reveal an absolute winner among three models. Keeping in mind short-text conversation scenario, best answer nominations seem to be the most appropriate quality measure that proves the advantage of the QTR and *doc2vec* models. However, best answer selection is very competitive in contrast to one-to-one conversation scenario (the asker receives about 20 answers on average); 'thumb up' and 'thumb down' scores from community members seem to be less subjective and biased. In terms of these two scores, BM25 slightly outperforms two other models.

If we look at CQA users' up– and downvotes only, lab evaluation confirms the advantage of BM25 over the other two models to some extent. What seems to be more important in case of top-3 results evaluation is that the models deliver quite diverse responses – the oracle's scores are significantly higher. The average score of the oracle's top response is close to *funny* ☺ , which is promising. The results suggest that a deeper question analysis, humor-specific features and advanced ranking methods can potentially deliver higher-quality responses.

Although a low agreement among assessors in laboratory settings is expected, it constitutes a serious obstacle for future work. Lab evaluation, successfully used in various information retrieval tasks, in our case proves that humor is a highly subjective and contextualized area. Additional efforts must be undertaken to ensure a higher inter-annotator agreement and reliability of judgments. We will explore the opportunity to account for assessors' personality traits (such as Big Five[8]), socio-demographic characteristics, language proficiency, and humor-specific profiling (cf. Jester project[9]) that can potentially help interpret and reconcile divergent assessments. We will also consider crowdsourcing humor evaluation, as several recent studies suggest. In addition, we plan to conduct user studies to better understand the perception and role of humor in short-text conversations.

We also plan to build a sizable collection of dialog jokes, which will allow us to harness advanced features already explored in humor recognition and combine them using learning-to-rank methods. State-of-the-art humor recognition methods can also be applied to improve the quality of the joke corpus.

To sum up, the study demonstrates that the information retrieval approach to humorous response generation is a promising direction of research. The current collection of tweets, test questions, evaluation protocol and assessors' judgments create a solid ground for further investigations of the IR-based humorous response generation.

References

1. Adamic, L.A., Zhang, J., Bakshy, E., Ackerman, M.S.: Knowledge sharing and yahoo answers: everyone knows something. In: Proceedings of WWW, pp. 665–674 (2008)
2. Augello, A., Saccone, G., Gaglio, S., Pilato, G.: Humorist bot: bringing computational humour in a chat-bot system. In: Proceedings of CISIS, pp. 703–708 (2008)
3. Bellegarda, J.R.: Spoken language understanding for natural interaction: the Siri experience. In: Mariani, J., Rosset, S., Garnier-Rizet, M., Devillers, L. (eds.) Natural Interaction with Robots, Knowbots and Smartphones, pp. 3–14. Springer, New York (2014). doi:10.1007/978-1-4614-8280-2_1
4. Binsted, K.: Using humour to make natural language interfaces more friendly. In: Proceedings of the AI, ALife and Entertainment Workshop (1995)
5. Carletta, J.: Assessing agreement on classification tasks: the kappa statistic. Comput. Linguist. 22(2), 249–254 (1996)
6. Hong, B.A., Ong, E.: Automatically extracting word relationships as templates for pun generation. In: Proceedings of CALC, pp. 24–31 (2009)
7. Järvelin, K., Kekäläinen, J.: Cumulated gain-based evaluation of IR techniques. TOIS 20(4), 422–446 (2002)
8. Ji, Z., Lu, Z., Li, H.: An information retrieval approach to short text conversation. arXiv preprint arXiv:1408.6988 (2014)

---

[8] https://en.wikipedia.org/wiki/Big_Five_personality_traits.
[9] http://eigentaste.berkeley.edu/about.html.

9. Jones, K.S., Walker, S., Robertson, S.E.: A probabilistic model of information retrieval: development and comparative experiments. Inf. Process. Manag. **36**(6), 779–840 (2000)
10. Khooshabeh, P., McCall, C., Gandhe, S., Gratch, J., Blascovich, J.: Does it matter if a computer jokes? In: Proceedings of CHI, pp. 77–86 (2011)
11. Kiddon, C., Brun, Y.: That's what she said: double entendre identification. In: Proceedings of ACL-HLT, vol. 2, pp. 89–94 (2011)
12. Lau, J.H., Baldwin, T.: An empirical evaluation of doc2vec with practical insights into document embedding generation. In: Proceedings of the 1st Workshop on Representation Learning for NLP, pp. 78–86 (2016)
13. Le, Q.V., Mikolov, T.: Distributed representations of sentences and documents. In: Proceedings of ICML, pp. 1188–1196 (2014)
14. Manning, C.D., Surdeanu, M., Bauer, J., Finkel, J., Bethard, S.J., McClosky, D.: The Stanford CoreNLP natural language processing toolkit. In: ACL System Demonstrations, pp. 55–60 (2014)
15. Mihalcea, R., Pulman, S.: Characterizing humour: an exploration of features in humorous texts. In: Gelbukh, A. (ed.) CICLing 2007. LNCS, vol. 4394, pp. 337–347. Springer, Heidelberg (2007). doi:10.1007/978-3-540-70939-8_30
16. Mihalcea, R., Strapparava, C.: Learning to laugh (automatically): computational models for humor recognition. Comput. Intell. **22**(2), 126–142 (2006)
17. Mihalcea, R., Strapparava, C.: Technologies that make you smile: adding humor to text-based applications. IEEE Intell. Syst. **21**(5), 33–39 (2006)
18. Niculescu, A., van Dijk, B., Nijholt, A., Li, H., See, S.L.: Making social robots more attractive: the effects of voice pitch, humor and empathy. Int. J. Soc. Robot. **5**(2), 171–191 (2013)
19. Rajadesingan, A., Zafarani, R., Liu, H.: Sarcasm detection on Twitter: a behavioral modeling approach. In: Proceedings of WSDM, pp. 97–106 (2015)
20. Reyes, A., Rosso, P., Veale, T.: A multidimensional approach for detecting irony in Twitter. Lang. Resour. Eval. **47**(1), 239–268 (2013)
21. Ritchie, G.: Can computers create humor? AI Mag. **30**(3), 71–81 (2009)
22. Shahaf, D., Horvitz, E., Mankoff, R.: Inside jokes: identifying humorous cartoon captions. In: Proceedings of KDD, pp. 1065–1074 (2015)
23. Stock, O., Strapparava, C.: Getting serious about the development of computational humor. In: Proceedings of IJCAI, pp. 59–64 (2003)
24. Taylor, J.M., Mazlack, L.J.: Computationally recognizing wordplay in jokes. In: Proceedings of CogSci, pp. 1315–1320 (2004)
25. Tsur, O., Davidov, D., Rappoport, A.: ICWSM-A great catchy name: semi-supervised recognition of sarcastic sentences in online product reviews. In: Proceedings of ICWSM, pp. 162–169 (2010)
26. Valitutti, A., Toivonen, H., Doucet, A., Toivanen, J.M.: "Let everything turn well in your wife": generation of adult humor using lexical constraints. In: Proceedings of ACL, vol. 2, pp. 243–248 (2013)
27. Wen, M., Baym, N., Tamuz, O., Teevan, J., Dumais, S., Kalai, A.: OMG UR funny! Computer-aided humor with an application to chat. In: Proceedings of ICCC, pp. 86–93 (2015)
28. Yan, R., Song, Y., Wu, H.: Learning to respond with deep neural networks for retrieval-based human-computer conversation system. In: Proceedings of SIGIR, pp. 55–64 (2016)

29. Yan, Z., Duan, N., Bao, J., Chen, P., Zhou, M., Li, Z., Zhou, J.: DocChat: an information retrieval approach for chatbot engines using unstructured documents. In: Proceedings of ACL, pp. 516–525 (2016)
30. Yang, D., Lavie, A., Dyer, C., Hovy, E.: Humor recognition and humor anchor extraction. In: Proceedings of EMNLP, pp. 2367–2376 (2015)
31. Zhang, R., Liu, N.: Recognizing humor on Twitter. In: Proceedings of CIKM, pp. 889–898 (2014)

A Component-Level Analysis of an Academic Search Test Collection.

Part I: System and Collection Configurations

Florian Dietz and Vivien Petras[✉]

Berlin School of Library and Information Science, Humboldt-Universität zu Berlin,
Dorotheenstr. 26, 10117 Berlin, Germany
florian.dietz@alumni.hu-berlin.de, vivien.petras@ibi.hu-berlin.de

Abstract. This study analyzes search performance in an academic search test collection. In a component-level evaluation setting, 3,276 configurations over 100 topics were tested involving variations in queries, documents and system components resulting in 327,600 data points. Additional analyses of the recall base and the semantic heterogeneity of queries and documents are presented in a parallel paper. The study finds that the structure of the documents and topics as well as IR components significantly impact the general performance, while more content in either documents or topics does not necessarily improve a search. While achieving overall performance improvements, the component-level analysis did not find a component that would identify or improve badly performing queries.

Keywords: Academic search · Component-level evaluation · GIRT04

1 Introduction

The basis of success for every IR system is the match between a searcher's information need and the system's content. Factors that contribute to the success of such a match have been studied at length: the underlying information need, the searcher's context, the type of query, the query vocabulary and its ambiguity, the type, volume and structure of the searched documents, the content of the documents, the kind of expected relevance of the documents and finally - the primary focus of IR - the IR system components, i.e. the preprocessing steps, ranking algorithm and result presentation. While all these aspects have been shown to impact search performance, it is also common knowledge that a successful configuration of these aspects is highly contextual. There is no one-size-fits-all solution.

Earlier initiatives such as the Reliable Information Workshop [12] and the TREC Robust Track [23] used TREC test collections to study what causes differences in search performance. They showed that search performance depends on the individual searcher, the search task, the search system and the searched documents. This did not only motivate new research on processing difficult queries, but it also spawned a new research field in query performance prediction [2].

© Springer International Publishing AG 2017
G.J.F. Jones et al. (Eds.): CLEF 2017, LNCS 10456, pp. 16–28, 2017.
DOI: 10.1007/978-3-319-65813-1_2

Grid- or component-level evaluation initiatives [7,11] moved into another direction, focusing on the evaluation of system component configurations to identify optimal retrieval settings.

This paper presents a component-level study in academic search. Academic search presents different challenges from the collections and information needs previously studied: the queries and their vocabulary can be highly technical and domain-specific, and, often, the searched documents just contain the bibliographic metadata of scientific publications. The study utilizes component-level evaluation aspects to find the causes for differences in search performance using the whole pipeline of the search process including the query, the documents and the system components. In a parallel paper [6], we present an analysis of indicators used in query performance prediction to delve even deeper into the causes for successful or unsuccessful search performance based on the queries, particularly trying to identify badly performing queries.

The goal of the research was not to find the best configuration (some state-of-the-art ranking algorithms were not even considered) but to find the most predictive factors for performance differences for this test collection. Future work should then be able to use this approach to extrapolate from the analysis of one collection to compare it with other collections in this domain.

The paper is structured as follows: Sect. 2 describes the area of academic search and discusses relevant research on component-level evaluation. Section 3 describes the test collection GIRT4 (used in the CLEF domain-specific track from 2004–2008) and the experimental set-up including the test configurations used. Section 4 describes the components that were analyzed for their predictive power in determining search performance. Section 5 concludes with an outlook on future work.

2   Component-Level Evaluation in Academic Search

Academic search is defined as the domain of IR, which concerns itself with searching scientific data, mostly research output in the form of publications [17]. It is one of the oldest search applications in IR. Not only were bibliographic information systems one of the first automated information retrieval systems (e.g. Medline [19]), but the first systematic IR evaluations, the Cranfield retrieval experiments, were also performed with an academic search collection [4]. The late 1990s and 2000s saw a renewed interest in these collections when digital libraries became a prominent research topic [3]. Academic search differs from previously tested search environments - mostly newspaper or web documents with general information needs [2] - in significant aspects. Academic search output is still comparatively small: between 1.5 [25] and 2.5 million [24] new articles were reported for 2015 globally. Most academic search collections are focused on one or a small number of disciplines and are therefore significantly smaller.

Documents in an academic search collection have a particular organization - either just the bibliographic metadata or the structure of a scientific publication with further references. Bibliographic metadata could be enriched with technical vocabulary (such as the MeSH keywords in PubMed), which support searching in the technical language of the documents [21]. When searching the full-text of publications, the references can be a major source for search success [18].

Information needs and their query representations academic search are different as well. While queries were found to be the same in length or longer than in standard web search engines [10], the content differs more dramatically. Particularly, queries contain technical terms or search for particular document components (such as author, title, or keywords) that are specific to the type of documents searched [13]. It appears logical that with highly specific information needs and small document collections, the number of relevant documents for any information need in this domain is also low.

Finally, these different documents and queries also demand different processing [22] and different ranking algorithms [1]. The CLEF domain-specific track, first established as a track in TREC8, provided a test collection to evaluate IR systems in this domain [14]. This study uses the CLEF domain-specific test collection GIRT4, which was released in 2004.

As a possible solution to the problem of finding the root causes for IR challenges of this type of test collection, component-based evaluation might be a suitable approach. For analyzing their impact on search performance, it is important to understand the impact of individual IR system components and parameters which may make a difference in retrieval. To measure the effect of those factors independent from others, a sufficiently dimensioned amount of data is necessary. Component-based evaluation takes a parameter and averages the measured effects of all other parameters while keeping the respective factor in focus [11].

The amount of generated ground truth data is important. Scholer and Garcia [20] report that a diversity of factors is needed to make evaluation effective. They criticize the evaluation methods for query performance prediction, because usually only one retrieval system is used. The concentration on just one system distorts the results, as the effect of a different system can make significant changes in terms of prediction quality. This is also true when searching for root causes for query failure. However, testing in a large dimensional space of IR system components requires a large-scale effort [7,8]. Kürsten [16] used GIRT4 as a test collection for a component-level analysis.

Component-level evaluation has not been studied extensively in academic search. De Loupy and Bellot [5] analyzed query vocabulary and its impact on search performance in the Amaryllis test collection, a French collection of bibliographic documents, which was also used in CLEF. Kürsten's component-level evaluation of the GIRT4 collection [16] found that utilizing the document structure (different document fields) did not impact retrieval performance. Other aspects were not considered in detail. In this paper, query and document structure, IR system preprocessing filters and ranking algorithms will be analyzed to determine which factors contribute most to search success.

3 Study Configurations

3.1 Test Collection and Test IR System

Following the Cranfield approach [4], the GIRT4 test collection consists of documents (metadata about social science publications), matching topics representing information needs and relevance assessments [15]. The GIRT4 English collection contains 151,319 English-language documents, consisting of structured bibliographic data: author, title, year, language code, country code, controlled keyword, controlled method term, classification text and abstract. Most fields have very little searchable text. Only very few documents contain an abstract (less than 13% of the collection).

GIRT4 queries are TREC-style topics, prepared by expert users of the documents. The topics contain three fields, which can be utilized for search: title, description and narrative. The binary relevance assessments are based on pooled retrieval runs from the CLEF domain-specific track. As topic creators and relevance assessors were not the same, certain information about the searcher (e.g. their context) remains unknown and can therefore not be measured or evaluated. Altogether, 100 topics and their relevance assessments from the years 2005–2008 were used.

For the experiments, the open source retrieval toolkit Lemur[1] was used. The software offers different retrieval ranking algorithms, which can be adjusted and further specified by several parameters: the Vector Space Model (VSM), Okapi BM25 (BM25) and Language Modeling (LM) with either Dirichlet-Prior-Smoothing, Jelinek-Mercer-Smoothing, or absolute discount smoothing. Lemur also provides the Porter and Krovetz stemmers and stopword list integration for preprocessing of documents and queries. The Lemur toolkit allows easy configuration of system components, which allows a component-level evaluation.

For evaluation, the trec_eval program[2] was used. As most analyses were done on a topic-by-topic basis, average precision (AP) per query was chosen as a metric. All experiments were performed on an Ubuntu operating system. The result sets were automatically structured and evaluated with Python scripts.

3.2 Component-Level Configurations

Following the component-level evaluation approach, different configurations of document fields, topic fields and IR system components were compared.

For the document collection, different combinations of the title (DT), abstract (AB) and an aggregated keyword field (CV), which consisted of the controlled keywords, method terms and classification terms were compared. All other fields were discarded due to a lack of relevant content when comparing them to the information needs represented in the topics. All possible document field combinations make a total of seven document configurations. However, the AB-only

---

[1] https://www.lemurproject.org/lemur.php, last accessed: 04-30-2017.
[2] http://trec.nist.gov/trec_eval, last accessed: 04-30-2017.

variant was not analyzed, because too few documents would remain in the collection for retrieval.

The three topic fields title (T), description (D) and narrative (N) were also used in every possible configuration, totaling seven topic configurations. Although the narrative was originally not intended for retrieval but to help the relevance assessors determine the most relevant documents per query, it was still used as a query field.

Every preprocessing option provided by the Lemur toolkit was included as well: Porter stemmer, Krovetz stemmer, no stemming, use of a stopword list, and no stopword list. A general stopword list for the English language was used, adjusted with a small number of non-relevant topic words (such as: find, documents) to improve the performance of all topic fields.

To create a reliable amount of results, the inclusion of multiple retrieval models is a critical requirement [20]. All Lemur ranking models were used for the experiments. To analyze the impact of model parameters, the VSM term weights in documents and queries were parameterized (different TF/IDF variants). Overall, 13 different ranking approaches were tested.

Table 1 summarizes the possible configurations that were used for experiments. For each topic, 3,276 configurations were tested, totaling in 327,600 data points for the 100 topics.

Table 1. Component-level configurations

Component	Configuration variables	Configurations
Document fields	DT, AB, CV	6
Topic fields	T, D, N	7
Stemming	Krovetz, Porter, none	3
Stopwords	List, none	2
Ranking models	VSM, BM25, LM (different smoothing)	13

3.3   Analysis Steps

For every topic, the AP over every configuration was calculated to reach an impression of the topic's performance. The AP differs widely across the test set. Some topics perform very well overall, while others seem to fail in every tested configuration. As a starting point, we separated between good and bad topics similarly to [9]: the median of the APs (per topic over all configurations) represents the threshold between well and badly performing topics.

The analysis was divided in several parts. One part consisted of an overall evaluation of all components of the retrieval process reported in this paper. Another part looked more specifically at various query and collection factors, where a relation to retrieval performance was assumed [6].

Every single aspect (document and topics fields, preprocessing components, ranking algorithms) was looked at while keeping every other component in the tested configuration stable. To measure the impact of a component on the retrieval success, all results with a specific component in the configuration were compared against all results without the tested component. The significance of the impact of a specific configuration compared to others was measured using the Wilcoxon signed-rank test. When a specific configuration component significantly increased the average AP per topic, we concluded that this component had an impact on the search performance.

4 Analyzing Component Performance

This section reports the results for specific IR process aspects under consideration. Section 4.1 studies the impact of specific document or topic field configurations. Section 4.2 compares the impact of IR system preprocessing components and ranking algorithms.

4.1 Document and Topic Structure

In academic search, where the amount of textual content is limited, the appropriate use of document and topic structures is important. By including different fields in the retrieval process, the searchable content is influenced. This section determines the impact of different field configurations while keeping every other component stable.

Documents. For the document fields, there are six different configurations to compare. Intuitively, the more text is available, the better the performance should be. Table 2 compares the term counts for the respective document fields. The title field contains a higher number of unique terms than the controlled term field and should thus yield more available terms for retrieval. The abstract field contains the highest number of unique terms. However, because only 13% of all collection documents contain abstracts, its impact may not be as high. A retrieval run just on the abstracts was not attempted as too few documents would have been available to search.

Table 2. Number of terms per document field

	Title	Controlled term fields	Abstract
Terms	2, 157, 680	4, 841, 399	3, 776, 509
Terms w/o stopwords	1, 445, 065	4, 380, 116	1, 871, 727
Unique terms	38, 919	4, 326	69, 526
Unique terms w/o stopwords	38, 359	4, 196	68, 923

Table 3 shows the MAP over all topics for the different document field configurations. The MAP is averaged over every possible retrieval component configuration with the respective document field. The combination of all fields contains the most searchable text, but does not achieve the best search performance.

Table 3. Document field configurations and MAP

DT	CV	DT + CV + AB	DT + AB	CV + AB	DT + CV
0.1205	0.1242	0.1776	0.1153	0.1175	0.1961

While the title and controlled term fields perform similarly (according to a Wilcoxon signed-rank test, the difference is not significant, DT vs. CV: $Z = -1.2103$, $p = 0.226$), adding the abstract text tends to slightly deteriorate the performance although the difference is not significant for either combination (Wilcoxon signed-rank test for DT vs. DT + ABS: $Z = -0.2785$, $p = 0.779$; CV vs. CV + AB: $Z = -1.0177$, $p = 0.308$). The best configuration for search performance seems to be a combination of the title and controlled term fields. It performs significantly better than the combination of all fields (DT + AB vs. DT + CV + AB: $Z = -3.8612$, $p = 0.000$) showing a negative impact of the abstract field after all. The better performance of the combined title and keyword fields shows that the controlled terms are not only different from the title terms, but add relevant content to the documents.

About a third of the relevant documents contain an abstract (compared to only 13% of the documents in the whole collection), so the finding that the abstract field deteriorates the search performance is even more puzzling. Abstracts may have a negative effect by containing misleading terms for many queries, but a high number of abstracts in the relevant documents suggests that for a small number of queries, abstract terms provide relevant matching input. It is important to note that the number of abstracts in relevant documents could concentrate on the relevant documents for a small number of queries - the differences between the number of relevant documents per query are surprisingly high. One possible explanation is also the fact that the academic search collection and topics seem to rely on a combination of highly specific words and more general method terms.

Topics. The topic fields suffer from similar problems. The three fields - title, description and narrative - are different from each other. A first point to observe is their different lengths (Table 4). As expected from TREC-style topics, the title field is shorter than the description, which is shorter than the narrative. Observing the average length after stopword removal shows that the title field consists of mostly content-bearing terms, while the description and narrative fields are reduced by half.

Table 4. Average number of terms per topic field

	Title	Description	Narrative
Terms	3.79	12.67	32.28
Terms w/o stopwords	2.83	6.04	14.64

Table 5. Topic field configurations and MAP

T	D	N	TD	TN	DN	TDN
0.1704	0.1256	0.0849	0.1786	0.1380	0.1308	0.1657

Table 5 lists the MAP of all topic field configurations while keeping the other factors stable. A similar image to the document field analysis emerges. The shortest field (title) achieves the best results when compared on an individual field basis, while the longest field (narrative) performs significantly worse. The combination of title and description appears to achieve even better results than the title field alone, but the difference is not significant (T vs. TD: $Z = -1.1828$, $p = 0.238$). The addition of the longer narrative field to the title and description field configuration seems to deteriorate the performance, but the difference is also not significant (TD vs. TDN: $Z = -1.8326$, $p = 0.067$).

A possible explanation for these results could be that the description of many topics just repeats the title terms, which may improve the term weights, but does not add content-bearing terms. Although the narrative contains the highest number of terms, it has a mostly negative effect on retrieval performance, probably because the field also contains instructions for the relevance assessors and may add terms that divert from its topical intent.

The analyses of the impacts of document and topic structures show that the content of either components can have a decisive impact on the search performance. More terms do not automatically lead to a better performance - this is true for both documents and topics, although the specific structure of the test collection needs to be taken into account. For all retrieval scenarios, the impact of topic and document terms needs to be looked at in combination, because both factors are connected. Another paper [6] focuses on such combinatorial analyses.

4.2   IR System Components

After analyzing the document and topic structure, this section takes a closer look at the preprocessing steps and the influence of the ranking algorithms.

Stopwords. A simple factor to analyze is the influence of the stopword list as there are only two datasets to compare - all experiments with or without applying the stopword list. For this test collection, the removal of stopwords improved the AP by 30% on average, a significant difference.

The stopword list helps retrieval in two ways. One, it helps to reduce the amount of terms that need to be searched. After stopword removal, ideally, only content-bearing terms should remain. It also helps in optimizing term weights, because the important keywords are more exposed. Especially longer queries benefit from this effect. The positive impact seems to affect both documents and topics.

An example for the positive effect of stopword removal is topic 128-DS[3], which receives a boost of 70% in AP (before stopword removal, averaged AP = 0.1883, after, AP = 0.3196). The number of topic terms is reduced from 52 to 23, removing terms such as "their" and also explanatory phrases like "relevant documents". Left over are stronger verbs like "discussed", which can help to identify the scientific methods applied in the academic publication.

The narrative field length changes most with stopword removal (Table 4). Table 6 shows that it also benefits most from it in retrieval performance when compared to the other fields. The lowest effect is observable for the title field, because stopword removal does not change the query as much.

Table 6. MAP for topic fields with and without stopwords

	T	D	N
All terms	0.1682	0.1088	0.0669
w/o stopwords	0.1725	0.1424	0.1028

Removing stopwords can also have negative effects. A small number of topics suffered from the removal of supposedly unimportant terms. The problem are terms, which might be unimportant in a different context but are content-bearing in these topics. There a two very figurative examples, which show this particular problem. The title field of topic 177-DS has the following terms: "unemployed youths without vocational training". After stopping, the word "without" is removed, reversing the information intent. Consequently, the AP suffers a drop of over 10%. Another example is topic 151-DS, searching for "right-wing parties". After applying the stopword list, the term "right" is removed.

These examples might occur more often in academic search with queries in a specific or highly technical language, which may be adversely effected by a conventional stopword list. The usage of the stopword list is dependent on the relationship between information intent, query terms and the document collection. For topic 151-DS, the AP stays relatively stable although the actual information need is not represented anymore. This is because the collection does not contain a lot of documents distinguishing right-wing and left-wing parties, which means the same documents are retrieved. In larger document collections, these distinctions may have a much bigger impact.

Stemming. Stemmers have been shown to increase the search performance, because different word forms are reduced (plural and conjugated forms are

---

[3] T: Life Satisfaction; D: Find documents which analyze people's level of satisfaction with their lives.; N: Relevant documents report on people's living conditions with respect to their subjective feeling of satisfaction with their personal life. Documents are also relevant in which only single areas of everyday life are discussed with respect to satisfaction.

stemmed to their respective stems), unifying the vocabulary and thus making it easier to match queries and documents. In the test collection, stemming has positive effects. While the Porter stemmer leads to an average improvement of 52% in AP over all configurations, the improvement of the Krovetz stemmer is around 30%.

Also stemmers can have negative effects on the search performance, deriving from the same cause as the positive effects. In unifying the vocabulary to stems, errors occur when the stemming is too strict or too soft. Looking at the topic terms and comparing the original forms to the Porter and Krovetz stems, one can see a variety of over- and understemming occurring. The examples are relatively rare and do not have the biggest influence on performance, but in some cases are measurable. A figurative example is topic 210-DS (Table 7).

Table 7. Topic 210-DS: Stemming of Narrative Terms. Original Narrative: The activities of establishing business in the new German federal states are relevant. What characterized the boom in start-up businesses following reunification?

	Original	Porter	Krovetz
Narr. terms	Activities establishing business german federal states characterized boom start businesses reunification	Activ establish busi german feder state character boom start busi reunif	Active establish busy german federal state characterize boom start businesse reunification
AP	0.0184	0.0619	0.0159

While the AP is low overall, the search performance is greatly improved by using the Porter stemmer (AP +230%), but suffers when stemmed with the Krovetz stemmer (AP –13%). The Porter stemmer reduces the number of unique terms, while Krovetz does not. The Porter stemmer changes term weights, because the important terms appear more often, Krovetz just changes their form. Even worse is the transformation of the variations for "business", one of the key terms for this topic. While the plural form is stemmed to "businesse", the singular is changed to "busy". This might be the cause for the negative effect of Krovetz: the new stem distorts the informational intent as well as the term weights.

Analyzing the effect of the Krovetz and Porter stemmers over all documents in the collection, this observation remains stable. Porter drastically reduces the amount of unique terms while Krovetz stems much more cautiously. This means that Porter has a more significant effect on retrieval because of its dual impact: reducing the word forms also changes the term weights.

Ranking Algorithms. The influence of the ranking algorithms was analyzed as well. While the study did not aim at finding the best ranking algorithm for the test collection, it tested whether the optimization of parameters significantly impacted the search performance for the test collection.

For one ranking algorithm, the Vector Space Model, all available variations in term weighting parameters were tested. Altogether, Lemur offers three different term weighting options for documents and queries: raw frequency, logarithmic frequency or the weighting function of the Okapi retrieval model. Nine different configurations were evaluated for the study. While different document term weighting parameters show an impact on the search performance, different query term weighting schemes did not significantly change the results, probably because query term frequencies are small. According to the averages over all configurations per ranking model, the best weighting option for the Vector Space Model is the Okapi term weighting for document terms (averaged AP = 0.1717). It significantly outperforms both the logarithmic weighting schemes ($Z = -8.5064$, $p = 0.000$) and the raw frequency weighting for document terms ($Z = -8.4858$, $p = 0.000$).

When comparing the best Vector Space Model configuration to the other ranking models, the differences are small. Table 8 shows the average AP over all experiments performed with the respective ranking model. There is no significant difference between the different models, although some differences may have been observed if the other ranking algorithms were optimized for the test collection. All of the analyzed ranking algorithms use term weights to determine the relevant documents and thus resemble each other if no other ranking signals are used.

Table 8. MAP for different Retrieval Models (LM-ads = Language modeling with absolute discounting, LM-jms = Language modeling with Jelinek-Mercer smoothing, LM-ds = Language modeling with Dirichlet smoothing)

LM-ads	LM-jms	LM-ds	BM25	VSM
0.1700	0.1666	0.1452	0.1654	0.1717

Looking at the variations in search performance on a topic-by-topic basis, the ranking algorithms do not impact good or bad queries. While some models manage to improve the performance of the good queries, queries that were categorized as performing badly over all configurations also performed badly over all ranking models. The RIA workshop [12] also reported that badly performing queries do not improve when ranked by a different retrieval model, so different components may be more important here.

5 Conclusion

The study has shown that for this particular academic search collection in a component-level evaluation:

– Document collection fields have an impact on success (but more text does not necessarily lead to better performance);

- Topic fields also have an impact (but longer queries tend to decrease performance);
- Applying a stopword list has a significant positive impact on search success;
- Stemming can have a significant impact depending on the chosen stemmer (Porter better than Krovetz);
- Different ranking algorithms based on term weights did not show a significant impact.

The study confirmed previous research that a variety of factors - and particularly their combined (interfering or compounding) influence - impact the search performance. While changing components did change the retrieval performance overall, it did not improve the performance of bad queries. As a matter of fact, the study could not identify a single aspect or component where bad queries would significantly be improved when changing the component.

At first view, these results are frustrating - no matter what component was looked at, a strong correlation between good or bad queries could not be found.

More work could be invested into analyzing different IR ranking models who could deal with documents with sparse and ambiguous text such as LSI or query enrichment strategies such as blind relevance feedback. However, first we will focus on query performance indicators, which delve deeper into the terminology of queries and documents to see whether we can identify badly performing queries this way [6].

Since this component-level evaluation was performed on one academic search collection, comparison with other test collections is necessary to extrapolate from the results achieved here to the domain in general. This will also be designated as future work.

References

1. Behnert, C., Lewandowski, D.: Ranking search results in library information systems - considering ranking approaches adapted from web search engines. J. Acad. Librariansh. **41**(6), 725–735 (2015)
2. Carmel, D., Yom-Tov, E.: Estimating the query difficulty for information retrieval. Synth. Lect. Inf. Concepts Retr. Serv. **2**(1), 1–89 (2010)
3. Chowdhury, G.: Introduction to Modern Information Retrieval. Facet, London (2010)
4. Cleverdon, C.: The Cranfield tests on index language devices. In: Aslib Proceedings, vol. 19, pp. 173–194. MCB UP Ltd. (1967)
5. De Loupy, C., Bellot, P.: Evaluation of document retrieval systems and query difficulty. In: LREC 2000, Athens, pp. 32–39 (2000)
6. Dietz, F., Petras, V.: A component-level analysis of an academic search test collection. Part II: query analysis. In: CLEF 2017 (2017). doi:10.1007/978-3-319-65813-1_3
7. Ferro, N., Harman, D.: CLEF 2009: Grid@CLEF pilot track overview. In: Peters, C., Nunzio, G.M., Kurimo, M., Mandl, T., Mostefa, D., Peñas, A., Roda, G. (eds.) CLEF 2009. LNCS, vol. 6241, pp. 552–565. Springer, Heidelberg (2010). doi:10.1007/978-3-642-15754-7_68

8. Ferro, N., Silvello, G.: A general linear mixed models approach to study system component effects. In: SIGIR 2016, pp. 25–34. ACM (2016)

9. Grivolla, J., Jourlin, P., de Mori, R.: Automatic classification of queries by expected retrieval performance. In: Predicting Query Difficulty Workshop. SIGIR 2005 (2005)

10. Han, H., Jeong, W., Wolfram, D.: Log analysis of an academic digital library: user query patterns. In: iConference 2014. iSchools (2014)

11. Hanbury, A., Müller, H.: Automated component–level evaluation: present and future. In: Agosti, M., Ferro, N., Peters, C., de Rijke, M., Smeaton, A. (eds.) CLEF 2010. LNCS, vol. 6360, pp. 124–135. Springer, Heidelberg (2010). doi:10.1007/978-3-642-15998-5_14

12. Harman, D., Buckley, C.: Overview of the reliable information access workshop. Inf. Retr. **12**(6), 615–641 (2009)

13. Khabsa, M., Wu, Z., Giles, C.L.: Towards better understanding of academic search. In: JCDL 2016, pp. 111–114. ACM (2016)

14. Kluck, M., Gey, F.C.: The domain-specific task of CLEF - specific evaluation strategies in cross-language information retrieval. In: Peters, C. (ed.) CLEF 2000. LNCS, vol. 2069, pp. 48–56. Springer, Heidelberg (2001). doi:10.1007/3-540-44645-1_5

15. Kluck, M., Stempfhuber, M.: Domain-specific track CLEF 2005: overview of results and approaches, remarks on the assessment analysis. In: Peters, C., Gey, F.C., Gonzalo, J., Müller, H., Jones, G.J.F., Kluck, M., Magnini, B., Rijke, M. (eds.) CLEF 2005. LNCS, vol. 4022, pp. 212–221. Springer, Heidelberg (2006). doi:10.1007/11878773_25

16. Kürsten, J.: A generic approach to component-level evaluation in information retrieval. Ph.D. thesis, Technical University Chemnitz, Germany (2012)

17. Li, X., Schijvenaars, B.J., de Rijke, M.: Investigating queries and search failures in academic search. Inf. Process. Manag. **53**(3), 666–683 (2017)

18. Mayr, P., Scharnhorst, A., Larsen, B., Schaer, P., Mutschke, P.: Bibliometric-enhanced information retrieval. In: de Rijke, M., Kenter, T., Vries, A.P., Zhai, C.X., Jong, F., Radinsky, K., Hofmann, K. (eds.) ECIR 2014. LNCS, vol. 8416, pp. 798–801. Springer, Cham (2014). doi:10.1007/978-3-319-06028-6_99

19. McCarn, D.B., Leiter, J.: On-line services in medicine and beyond. Science **181**(4097), 318–324 (1973)

20. Scholer, F., Garcia, S.: A case for improved evaluation of query difficulty prediction. In: SIGIR 2009, pp. 640–641. ACM (2009)

21. Vanopstal, K., Buysschaert, J., Laureys, G., Stichele, R.V.: Lost in PubMed. Factors influencing the success of medical information retrieval. Expert Systems with Applications **40**(10), 4106–4114 (2013)

22. Verberne, S., Sappelli, M., Kraaij, W.: Query term suggestion in academic search. In: de Rijke, M., Kenter, T., Vries, A.P., Zhai, C.X., Jong, F., Radinsky, K., Hofmann, K. (eds.) ECIR 2014. LNCS, vol. 8416, pp. 560–566. Springer, Cham (2014). doi:10.1007/978-3-319-06028-6_57

23. Voorhees, E.M.: The TREC robust retrieval track. ACM SIGIR Forum **39**, 11–20 (2005)

24. Ware, M., Mabe, M.: The STM report: an overview of scientific and scholarly journal publishing (2015). http://www.stm-assoc.org/2015_02_20_STM_Report_2015.pdf

25. Web of Science: Journal Citation Report. Thomson Reuters (2015)

A Component-Level Analysis of an Academic Search Test Collection.
Part II: Query Analysis

Florian Dietz and Vivien Petras[(⊠)]

Berlin School of Library and Information Science,
Humboldt-Universität zu Berlin, Dorotheenstr. 26, 10117 Berlin, Germany
florian.dietz@alumni.hu-berlin.de, vivien.petras@ibi.hu-berlin.de

Abstract. This study analyzes causes for successful and unsuccessful search performance in an academic search test collection. Based on a component-level evaluation setting presented in a parallel paper, analyses of the recall base and the semantic heterogeneity of queries and documents were used for performance prediction. The study finds that neither the recall base, query specificity nor ambiguity can predict the overall search performance or identify badly performing queries. A detailed query analysis finds patterns for negative effects (e.g. non-content-bearing terms in topics), but none are overly significant.

Keywords: Academic search · Query analysis · Query prediction · GIRT04

1 Introduction

Several IR research areas study the aspects that contribute to a successful search performance of an IR system. Grid- or component-level evaluation initiatives [5,6] focus on the large-scale and fine-grained evaluation of system component configurations for optimal retrieval. Query performance prediction studies [2] zoom in on individual pre- and post-retrieval indicators to predict the level of difficulty for queries in a particular system configuration to perform well [8,9].

This paper reports on a search performance study combining both component-level evaluation approaches with query performance prediction on a particular academic search test collection. The study's objective is to find the causes for differences in search performance using the whole pipeline of the search process including the query, the documents and the system components. The research question is stated as: "Which aspects in the search process best predict performance differences in an academic search collection?"

Academic search is defined as the domain of IR, which concerns itself with searching scientific data, mostly research output in the form of publications [12]. Both information needs and document collections present unique challenges in academic search: document collections are smaller than web search or newspaper collections; document and query terminology can be highly technical and

© Springer International Publishing AG 2017
G.J.F. Jones et al. (Eds.): CLEF 2017, LNCS 10456, pp. 29–42, 2017.
DOI: 10.1007/978-3-319-65813-1_3

domain-specific; documents may contain only sparse text (just metadata); and information needs can be very specific with subsequent fewer relevant results in smaller collections [4].

In a parallel paper [4], we present the system configurations that were tested in a component-level setting, while in this paper, query performance indicators for an academic search collection are studied.

The paper is structured as follows: Sect. 2 discusses relevant previous research on query performance indicators. The next section briefly describes the test collection GIRT4, a test collection representative for academic search, and the experimental set-up for the query performance analysis. Section 4 describes the factors that were analyzed for their predictive power in determining search performance. The paper concludes with some speculations on the nature of academic search and the performance success factors of IR systems in this context.

2   Query Performance Prediction

Query performance is an important topic in IR, as it influences the usefulness of an IR system - no matter in which domain. The field of query performance or query performance prediction addresses query behavior in IR. It is a known fact that some queries nearly always perform well while other examples fail in every circumstance [19]. Query performance prediction tries to distinguish between those good and bad queries without necessarily going through the actual retrieval process. The goal is to help make retrieval more robust so that a user always gets at least somewhat relevant results. In Carmel and Yom-Tov [2], the authors provide a broad overview of state of the art query performance predictors and an evaluation of published work.

The TREC Robust track was established to study systems that could deal robustly even with difficult queries [18,20]. Unlike this paper, most of the research coming out of the Robust track attempted to predict the performance, but the causes for the performance variations are not focused on.

Query performance research can be divided into two different areas: pre-retrieval prediction and post-retrieval prediction. While pre-retrieval aspects are analyzed before the actual retrieval process and are much easier to compute, post-retrieval indicators are often found to be more precise, but are much more time consuming to calculate.

The inverse document frequency (IDF) of query terms is used quite often for pre-retrieval performance prediction. The average IDF (avgIDF) of a query and its relation to performance [3] takes the relative term weights of all query terms and averages them. The maximum IDF (maxIDF) per query is very similar as it takes the highest IDF per term of a query and matches it with the average precision [17]. Several IDF indicators will be tested in this study.

An example for an effective post-retrieval indicator is the Clarity Score [3], which measures the coherence of the vocabulary of the result list of a query and the vocabulary of the whole collection with regards to the query terms. The assumption is that if the language of the retrieved documents differs from the general language of the collection, the degree of difference is an indicator for performance. A pre-retrieval version of the Clarity Score is the Simplified Clarity Score (SCS) [9], which compares term statistics of the query and the document collection, showing a strong correlation to performance for TREC collections. The SCS will also be calculated for this study.

Other approaches zoom in on the specific linguistic features of queries. Mothe and Tanguy [14] analyzed 16 different linguistic features of a query to predict the search performance. Among those were the number of words per query (i.e. query length), the average word length per query, the syntactic structure of a query or the average polysemy value. The latter averages the number of meanings all of the query terms can adopt. This indicator is based on the synsets from the linguistic database WordNet, which was used to get the number of synsets for every query term. The same procedure was also tested in this study.

This paper utilizes query prediction indicators not for performance prediction, but to identify which factors cause a good or bad performance. As Buckley [1] motivates, it is more important to try to identify root causes for query failure first than to come up with new approaches to improve the retrieval techniques for such queries - especially in domains where the failure causes are still relatively unsolved as is the case for academic search environments studied here.

While component-level evaluation as presented in another paper [4] focuses on the IR system components, but neglects to delve more deeply into query or document analyses, this paper delves deeper into the query and document analysis.

3   Study Configurations

3.1   Test Collection and Test IR System

For the analysis, the GIRT4 test collection, containing bibliographic metadata records on social science publications such as journal or conference articles, was used. The GIRT4 test collection consists of 151,319 documents with author, title, different types of keywords and abstract information and a 100 matching topics and relevance assessments, which were used in the CLEF domain-specific tracks from 2005–2008 [11]. The GIRT4 queries are TREC-style topics, consisting of three fields with different amounts of content (title, description and narrative). The relevance assessments were created with pooled retrieval runs from the CLEF domain-specific track. All three query fields were used for retrieval in all possible combinations. Three document fields were used: document title, controlled vocabulary (consisting of controlled keywords, controlled method terms and classification text) and the abstracts. Most of the document fields contain very little searchable content. Some documents contain more content than others, making them more likely to be retrieved. The test collection and component-level experimental setup is described in more detail in [4].

For the experiments, the open source retrieval kit Lemur[1] was used. For the component-level evaluation [4], nearly every possible option of Lemur in

---

[1] https://www.lemurproject.org/lemur.php, last accessed: 04-30-2017.

combination with the document and topic fields was used for retrieval, resulting in a total of around 327,600 data points. The retrieval experiments conducted with Lemur were evaluated with the trec_eval program[2]. All experiments were performed on an Ubuntu operating system and automatically structured with Python scripts.

3.2   Analysis Steps

For evaluating the query performance on a topic-by-topic basis, the component-level approach from [4] proved to be the optimal basis. Analog to the first part, the average AP over every configuration was calculated to classify the topics by their respective performance. While some topics perform very well, others seem to have vast problems independent of the configuration that was used.

One aspect of the analysis addresses the problem of an insufficient recall base, i.e. the number of relevant documents per topic. The retrieval performance was compared to the number of relevant documents per topic to verify if the number of relevant documents impacts the search performance.

The linguistic specificity of the collection documents and queries and the query ambiguity were also examined. Specificity determines the uniqueness of a term for a collection [7]. Different values for query term specificity were calculated. The query length was also associated with its query's respective retrieval performance. As the narrative was not meant for retrieval, containing a lot of descriptive terms for the relevance assessors, only the lengths of title (short queries) and the combination of title and description (longer queries) were used for evaluation.

The next part delves deeper, using IDF measures. The IDF has been shown to have a relation to search performance, thus the measure was used in different variations to see if there are any connections to performance. Another measure already tested in other articles, the Simplified Clarity Score (SCS) was calculated for every query and compared to its AP [9].

The last part of the query analysis addressed the linguistic ambiguity of the query terms. For this, the WordNet database's synsets were utilized, as was suggested by Mothe and Tanguy [14]. Synsets are not synonyms, but the number of senses a word can have when used in natural language. As an example, a word like "put" has a lot of synsets and is thus considered more ambiguous, while a word like "soccer" has a small number of synsets, making it more specific. The ambiguity of a query is then measured by extracting the number of synsets for each query term. The results of these analysis parts were checked for significance by computing the Spearman rank correlation coefficient. The Simplified Clarity Score was also tested with the Pearson correlation efficient to compare the results to previous research. When one of the factors showed a significant correlation with the averaged AP per topic (over all configurations), we concluded that this aspect impacts the search performance.

---

[2] http://trec.nist.gov/trec_eval, last accessed: 04-30-2017.

4   Analyzing Query Performance

This section reports the results for specific query and collection aspects under consideration. Section 4.1 studies the impact of the recall base per topic. Sections 4.2 and 4.3 focus on the linguistic features of queries and the collection by analyzing the impact of query specificity and ambiguity.

4.1   Recall Base

Buckley [1] states that the most important factor for search performance is probably the collection, i.e. the documents that are available. The recall base is the number of relevant documents in a collection available for a particular query. It is the basis for every successful search - without relevant documents, an information need cannot be fulfilled.

As a first step in the study, the available recall base for each topic was checked to see whether it had an influence on the performance. In the GIRT4 collection, some of the topics have very few relevant documents. Topic 218-DS[3], for example, has only three relevant documents within the GIRT4 collection and performs badly in any configuration.

To verify that a small recall base is not the reason for bad search performance, the average AP per topic over all configurations was correlated with the number of relevant documents. Figure 1 shows the distribution of averaged AP for the number of relevant documents per topic. The figure reveals no discernible correlation. The statistical verification also shows this trend, but is inconclusive ($r_s = 0.17$, $p < 0.1$).

The explanation for search failure may not be the small number of relevant documents, but the content and structure of those relevant documents. Looking at the stopped topic 218-DS (which has only three relevant documents in the collection), one can find the following terms: "generational differences internet significance internet communication differences people ages utilize significance function internet communication areas society private life work life politics generation gap means communication". The gist of this query concentrates on generational differences in using the internet. The narrative emphasizes internet use in everyday life. Looking at the content of the three relevant documents, they do not contain the term "internet", but use a different wording - "media", for example. The query cannot be matched with the relevant documents because those documents do not contain the terms used in the query. Vice versa, the collection contains about 30 other documents containing both "internet" and "generation", none of which was assessed as relevant. Vocabulary problems (synonymous, but unmatched term use in queries and documents) as well as insufficient relevance

---

[3] T: Generational differences on the Internet; D: Find documents describing the significance of the Internet for communication and the differences in how people of different ages utilize it.; N: Relevant documents describe the significance and function of the Internet for communication in different areas of society (private life, work life, politics, etc.) and the generation gap in using various means of communication.

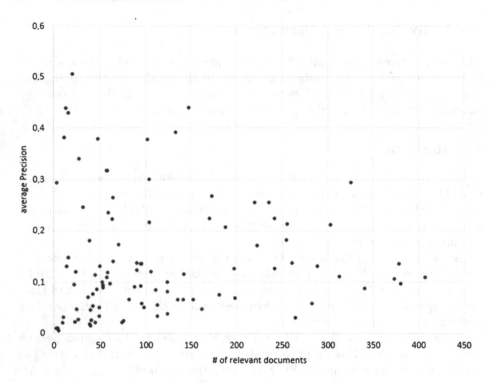

Fig. 1. Recall base per topic compared to AP

assessments might be the cause for the performance problems with this test collection.

4.2 Topic Specificity

Topic specificity refers to the uniqueness of the query terms in relationship to the document collection. It is two-sided: it considers the vocabulary that topics and documents share. Specificity has been found to influence the search performance [7] as it considers the number of matchable terms as well as the discriminatory nature of those terms in the retrieval process. Topic specificity can be measured in many different ways.

Query Length. The easiest way to calculate specificity is the query length. Intuitively, query length makes sense as an indicator for a good query. The more terms one issues to an IR system, the more accurate the results could be. However, this theoretical assumption does not hold true in every environment. For example, synonym expansion may add more noise to a query because of the inherent ambiguity of terms or because relevant documents only use one of the synonymous words.

In this study, query length was defined as the number of terms - other definitions calculate the number of characters a query consists of [14]. Query length based on the number of non-stop query terms was already tested on a TREC collection, showing no significant correlation with AP [9].

The average query length varies widely between the different topic fields [4]. Our analysis showed that query length correlates very weakly with AP over all configurations when all query terms are considered ($r_s = -0.20$, $p < 0.005$) and still weakly, but slightly better, when stopwords are removed ($r_s = -0.28$, $p < 0.005$). Stopword removal improves the predictive power of query length as only content-bearing words are considered. The negative correlation seems to support this - the longer the query is, the more general terms can be found. The amount of non-relevant words affects the performance.

Table 1 shows the impact of shorter and longer queries on AP by comparing queries using only the title (short query) or the title and description together (longer query) as was also calculated by [9]. The correlation is stronger once stopwords are removed from the queries, but very similar. The correlations for stopped queries are somewhat stronger than what [9] found. It is possible that for academic search, query length may be a weak indicator for search performance.

Table 1. Spearman rank correlation for query length and AP for shorter and longer queries (T = title, TD = title + description)

	All terms		w/o stopwords	
	T	TD	T	TD
r_s	−0.31	−0.19	−0.35	−0.37
p	0.001	0.046	0.000	0.000

Inverse Document Frequency. In contrast to query length, IDF has been shown to predict the search performance [7,9]. An IDF measure shows how rare query terms might be in the collection. A somewhat rarer query term is more helpful for discriminating between relevant and non-relevant documents. IDF was calculated in its original form [16] with Python scripts, although it should not make any difference when computed in a modified form [15]. These indicators were calculated:

- maxIDF (maximum IDF of all query terms)
- minIDF (minimum IDF of all query terms)
- avgIDF (average of query term IDFs)
- devIDF (standard deviation per query term IDF)
- sumIDF (sum of all query term IDFs)

Table 2 lists the results for the correlation of the IDF indicators with AP after applying a stopword list while searching document title and the controlled vocabulary fields. There are no significant correlations between the various IDF indicators and AP, except avgIDF, which shows a very weak correlation. Although [7]

Table 2. Spearman rank correlation for IDF indicators and AP (using DT + CV documents fields, the Porter stemmer & a stopword list)

	maxIDF	minIDF	avgIDF	devIDF	sumIDF
r_s	0.14	0.08	0.24	0.15	−0.17
p	0.160	0.405	0.015	0.143	0.098

reports maxIDF as a relatively stable indicator, the results here could not really be used to determine a query's search performance. The term weights alone do not provide enough information about the possible performance of a query.

Simplified Clarity Score. As another measure of specificity, the Simplified Clarity Score (SCS), introduced by [9], was also tested. This pre-retrieval indicator is based on the Clarity Score [3], but focuses on the whole collection instead of a retrieved result list. It is calculated by the following formula:

$$\text{SCS} = \sum_Q P_{ml}(wjQ) \cdot \log_2(\frac{P_{ml}(w|Q)}{P_{coll}(w)})$$

$P_{ml}(w|Q)$ is given by Qtf/QL, where Qtf is the frequency of a term w in query Q and QL is the query length. It is the maximum likelihood model of a query term w in query Q. $P_{coll}(w)$ defines the collection model and is given by $tf_{coll}/token_{coll}$, which is the term frequency of the term in the whole collection divided by the overall number of terms in the collection.

The SCS was calculated for eight different configurations. Four topic field combinations were tested against the document title and controlled vocabulary (DT + CV) fields, because this is the best document field combination for optimal retrieval performance [4]. Forms with and without stopword removal for each configuration were evaluated to test the impact of query length[4]. The correlation results for the SCS with the respective AP of all configurations with the tested topic fields are listed in Table 3.

No matter which topic field combination is tested, the SCS of a query does not correlate with its performance using the Spearman Rank Correlation. We verified the results with the Pearson Correlation (which was used in [9]), although not all of the involved variables may have a linear distribution. These calculations show a weak correlation for stopped topics, but are smaller than those reported by [9] (except for the TDN topic field combination, which is not significant). This could be due to numerous reasons. One cause might be differences in collection size. Also, the average lengths of the topic fields in this study are longer than in [9], possibly weakening the effect of the SCS. Most likely, it is again the nature of the

---

[4] Note that the number of data points becomes smaller with every fixed factor such as DV + CV documents. Thus, the amount of data per query for testing SCS is smaller than in other calculations in this article. Still, the data is comparable to the amount of data used by [9].

Table 3. Simplified clarity score per topic field and AP (correlation with Spearman and Pearson)

	All terms				w/o stopwords			
	T	D	TD	TDN	T	D	TD	TDN
r_s	0.1	0.12	0.12	0.13	0.16	0.17	0.17	0.04
p	0.346	0.215	0.226	0.195	0.107	0.927	0.096	0.681
r_p	0.15	0.23	0.22	0.16	0.19	0.26	0.27	0.08
p	0.126	0.017	0.022	0.106	0.055	0.009	0.007	0.404

test collection, resulting in variations because of specific terms in this academic search context, which distort the calculations of term discriminativeness.

4.3   Semantic Heterogeneity of Topic Terms

Query ambiguity is another possible indicator for search performance. Both WordNet [14] and Wikipedia [10] were suggested as good resources to study the semantic heterogeneity of topic terms. This study used WordNet for analyzing topic term ambiguity. The count of synsets for every topic term and its respective part of speech (noun, verb or adjective) were downloaded. For each topic, the average synsets (avgSyn) and the sum of synsets (sumSyn) were calculated and further grouped by the three word classes to get an indicator for query ambiguity (each also with and without stopword removal).

WordNet appeared to be the best option for academic queries, as it offers a good mixture of technical terms and normal language. Hauff [7] warns that Wordnet might not be reliable enough to use when predicting performance, because the database has problems with proper names and year dates. In this study, missing terms such as these were assigned the equivalent of one synset uniformly as dates and proper names (in this case mostly authors) would take on only one meaning.

The results for the average number of synsets per topic after stopword removal are shown in Fig. 2. The Spearman Rank Correlation test showed no significant correlation between AP and the average number of synsets ($r_s = -0.09$, $p < 0.38$ w/o stopword removal; $r_s = -0.13$, $p < 0.19$ with stopword removal).

When comparing the absolute number of synsets over all topic terms, a slightly different image emerges as shown in Fig. 3. The correlation test revealed a weak correlation, which was slightly stronger after stopword removal ($r_s = -0.25$, $p < 0.011$ w/o stopword removal; $r_s = -0.28$, $p < 0.005$ with stopword removal). Similar observations appear when correlating the indicators of the individual word classes with the AP.

Query length does not significantly change the correlation. It was tested by comparing the results for shorter (T) and longer queries (TD) as in Sect. 4.2. As before, the average synsets indicator does not have a significant correlation with performance. Regardless of query field or query length, the correlation is

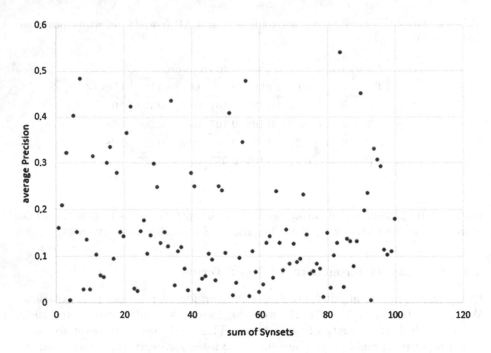

Fig. 2. Average number of synsets per topic after stopword removal compared to AP

not significant. For the sum of all synsets per query term, the correlation test reveals similar results compared to the sum of the synsets of all query terms - with shorter queries showing just the slightest increase in correlation strength $(T = r_s = -0.29, p < 0.002; D = r_s = -0.24, p < 0.015; TD = r_s = -0.27, p < 0.006)$.

Overall, there is a very weak correlation between the total number of synsets of a query and their respective AP over all configurations. This could be due to numerous causes. A reason might be that there is indeed no strong connection between the synsets of a query and the possible performance. A logical next step would be to test this with other sources or on different document fields. An assumption is that the controlled term fields contain more technical language and are therefore less ambiguous than the title or abstract fields.

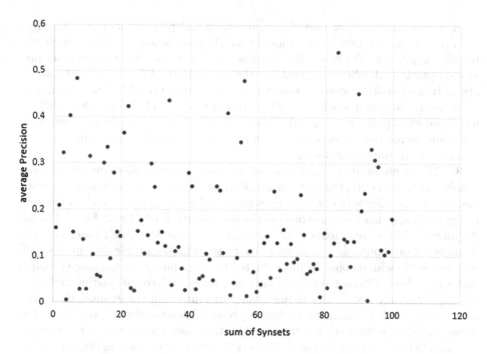

Fig. 3. Sum of synsets per topic after stopword removal compared to AP

5 Conclusion

The query analysis has shown that for this particular academic test collection:

- The recall base (number of relevant documents per topic) does not impact the search performance or distinguishes between good or bad topics;
- Query length could be a weak indicator for retrieval success as opposed to other studies using non-academic search test collections;
- Inverse document frequency in contrast to other studies does not predict search performance in any of its variations;
- The Simplified Clarity Score in contrast to other studies also does not predict search performance;
- The semantic heterogeneity (as expressed by the WordNet synsets) also does not predict the performance of the query.

For GIRT4 at least, the conventional query performance predictors could not distinguish between good and bad topics. Summarizing the results of the component-level analysis [4] and the query performance analysis presented here, we could not identify a query aspect, test collection or IR system component which would help in distinguishing successful from unsuccessful queries, which would also explain search performance.

The analyses did show that the performance is dependent more on the query - collection combination than on any other IR process aspect, for example the ranking algorithm. The test collection has some aspects that differed from previously researched collections, showing the importance of domain-specific evaluation. In particular, document structure and the mixture of technical language with general language appeared an interesting research challenge when individual topic examples were studied. However, to validate this observation, more academic search test collections need to be analyzed with the same indicators to extrapolate to the whole domain.

GIRT4 is an example for a typical academic search collection with bibliographic metadata only, but even for academic search, it contains few documents, making it a particularly challenging example. Most academic search collections today contain at least the abstracts of the documents, GIRT4 only for 13% of the documents. More documents would certainly increase the recall base, but also introduce more language problems. In contrast to bibliographic documents, full-text academic search could be more related to other query performance research and could show different results. The recall base analyses of queries and their requisite relevant documents showed that without synonym expansion, queries might not match relevant documents especially because text is so sparse. The same analysis also showed that the test collection may also contain other relevant documents that were not assessed. A larger test collection (such as iSearch, which also contains full-text documents [13]) could alleviate some of these problems.

So what is the logical next step? For this type of search context - sparsity of text and technical language - the term matching aspect in the IR process is crucial. There are only few relevant documents available for any information need and they may contain other terminology than the queries. Future research could concentrate on the complex language problem of considering the collection and topics in tandem. For the IR process, ranking algorithms such as LSI and query feedback approaches could alleviate the synonym problem somewhat. For query performance indicators, future research might concentrate on semantic parsing and understanding of queries and documents that goes beyond the discrimination value of a term (as represented by IDF or SCS) or its language ambiguity (as represented by WordNet synsets) in the collection. Another idea would be to combine factors and look at the possible effects on performance, but those (average) effects shown here would probably still appear. A formal aspect that could also be changed was the distinction of good and bad queries: just taking the median of the averaged AP for all topics to distinguish between good and bad performing topics might be too coarse as a threshold.

The research presented here and in [4] may not yet be able to make representative statements about the nature of the academic search problem or list particular requirements for an IR system to improve performance, but it designed a comprehensive process for analyzing a domain-specific search problem, combining both component-level evaluation of system and collection configurations and query performance analyses. This is a step towards a structured and holistic search process analysis for a specific domain, taking the searchers and their information needs, the collection and the IR system into account.

References

1. Buckley, C.: Why current IR engines fail. Inf. Retr. **12**(6), 652–665 (2009)
2. Carmel, D., Yom-Tov, E.: Estimating the query difficulty for information retrieval. Synth. Lect. Inf. Concepts Retr. Serv. **2**(1), 1–89 (2010)
3. Cronen-Townsend, S., Zhou, Y., Croft, W.B.: Predicting query performance. In: SIGIR 2002, pp. 299–306. ACM (2002)
4. Dietz, F., Petras, V.: A component-level analysis of an academic search test collection. Part I: system and collection configurations. In: Jones, G.J.F., et al. (eds.) CLEF 2017. LNCS, vol. 10456, pp. 16–28. Springer, Cham (2017). doi:10.1007/978-3-319-65813-1_2
5. Ferro, N., Harman, D.: CLEF 2009: Grid@CLEF pilot track overview. In: Peters, C., di Nunzio, G.M., Kurimo, M., Mandl, T., Mostefa, D., Peñas, A., Roda, G. (eds.) CLEF 2009. LNCS, vol. 6241, pp. 552–565. Springer, Heidelberg (2010). doi:10.1007/978-3-642-15754-7_68
6. Hanbury, A., Müller, H.: Automated component–level evaluation: present and future. In: Agosti, M., Ferro, N., Peters, C., de Rijke, M., Smeaton, A. (eds.) CLEF 2010. LNCS, vol. 6360, pp. 124–135. Springer, Heidelberg (2010). doi:10.1007/978-3-642-15998-5_14
7. Hauff, C.: Predicting the Effectiveness of Queries and Retrieval Systems. Ph.D. thesis, University of Twente, Netherlands (2010)
8. Hauff, C., Azzopardi, L., Hiemstra, D.: The combination and evaluation of query performance prediction methods. In: Boughanem, M., Berrut, C., Mothe, J., Soule-Dupuy, C. (eds.) ECIR 2009. LNCS, vol. 5478, pp. 301–312. Springer, Heidelberg (2009). doi:10.1007/978-3-642-00958-7_28
9. He, B., Ounis, I.: Query performance prediction. Inf. Syst. **31**(7), 585–594 (2006)
10. Katz, G., Shtock, A., Kurland, O., Shapira, B., Rokach, L.: Wikipedia-based query performance prediction. In: SIGIR 2014, pp. 1235–1238. ACM (2014)
11. Kluck, M., Stempfhuber, M.: Domain-specific track CLEF 2005: overview of results and approaches, remarks on the assessment analysis. In: Peters, C., et al. (eds.) CLEF 2005. LNCS, vol. 4022, pp. 212–221. Springer, Heidelberg (2006). doi:10.1007/11878773_25
12. Li, X., Schijvenaars, B.J., de Rijke, M.: Investigating queries and search failures in academic search. Inf. Process. Manag. **53**(3), 666–683 (2017). http://doi.org/10.1016/j.ipm.2017.01.005
13. Lykke, M., Larsen, B., Lund, H., Ingwersen, P.: Developing a test collection for the evaluation of integrated search. In: Gurrin, C., He, Y., Kazai, G., Kruschwitz, U., Little, S., Roelleke, T., Rüger, S., van Rijsbergen, K. (eds.) ECIR 2010. LNCS, vol. 5993, pp. 627–630. Springer, Heidelberg (2010). doi:10.1007/978-3-642-12275-0_63
14. Mothe, J., Tanguy, L.: Linguistic features to predict query difficulty. In: Predicting Query Difficulty Workshop. SIGIR 2005, pp. 7–10 (2005)
15. Robertson, S.E.: Understanding inverse document frequency: on theoretical arguments for IDF. J. Doc. **60**, 503–520 (2004)
16. Robertson, S.E., Sparck Jones, K.: Relevance weighting of search terms. In: Willett, P. (ed.) Document Retrieval Systems, pp. 143–160. Taylor Graham (1988)

17. Scholer, F., Williams, H.E., Turpin, A.: Query association surrogates for web search: research articles. J. Am. Soc. Inf. Sci. Technol. **55**(7), 637–650 (2004)
18. Voorhees, E.M.: Overview of the TREC 2003 robust retrieval track. In: TREC, pp. 69–77 (2003)
19. Voorhees, E.M.: The TREC robust retrieval track. ACM SIGIR Forum **39**, 11–20 (2005)
20. Yom-Tov, E., Fine, S., Carmel, D., Darlow, A.: Learning to estimate query difficulty: including applications to missing content detection and distributed information retrieval. In: SIGIR 2005, pp. 512–519. ACM (2005)

Improving the Reliability of Query Expansion for User-Generated Speech Retrieval Using Query Performance Prediction

Ahmad Khwileh[(✉)], Andy Way, and Gareth J.F. Jones

School of Computing, ADAPT Centre, Dublin City University, Dublin 9, Ireland
ahmad.khwileh2@mail.dcu.ie, {away,gjones}@computing.dcu.ie

Abstract. The high-variability in content and structure combined with transcription errors makes effective information retrieval (IR) from archives of spoken user generated content (UGC) very challenging. Previous research has shown that using passage-level evidence for query expansion (QE) in IR can be beneficial for improving search effectiveness. Our investigation of passage-level QE for a large Internet collection of UGC demonstrates that while it is effective for this task, the informal and variable nature of UGC means that different queries respond better to alternative types of passages or in some cases use of whole documents rather than extracted passages. We investigate the use of Query Performance Prediction (QPP) to select the appropriate passage type for each query, including the introduction of a novel Weighted Expansion Gain (WEG) as a QPP new method. Our experimental investigation using an extended adhoc search task based on the MediaEval 2012 Search task shows the superiority of using our proposed adaptive QE approach for retrieval. The effectiveness of this method is shown in a per-query evaluation of utilising passage and full document evidence for QE within the inconsistent, uncertain settings of UGC retrieval.

1 Introduction

An ever increasing amount of user-generated content (UGC) is being uploaded to online repositories. UGC can take various forms, the particular focus of this paper is on audio-visual UGC where the information is mainly in the spoken media stream. For example, a recorded lecture or discussion, which we refer to as user generated speech (UGS). UGS content often has highly variable characteristics of length, topic and quality depending on its purpose, source and recording conditions. A significant challenge for effective use of UGS is the development of information retrieval (IR) methods to enable users to locate content relevant to their needs efficiently. The diversity of UGS means that the general IR problem of query-document vocabulary mismatch is made worse. A standard approach to addressing this problem is query expansion (QE); in which the user's query is modified to include additional search terms taken from the top ranking documents from an initial search run [3].

© Springer International Publishing AG 2017
G.J.F. Jones et al. (Eds.): CLEF 2017, LNCS 10456, pp. 43–56, 2017.
DOI: 10.1007/978-3-319-65813-1_4

While QE techniques have been shown to be highly effective in improving IR performance for professionally created spoken content [7], there has been little work reported on its application to UGS retrieval. The variable nature of UGS means it is likely to be susceptible to query drift in the expansion process where the focus of the query moves away from that intended by the searcher [16]. This means that careful selection of feedback terms is required to maintain the query focus in QE for UGS. Previous work suggests that the use of sub-topical passages taken from feedback documents can be more effective for QE than whole documents [1,9,15]. However, it is not obvious what the most suitable passages for QE are, this will be particularly so in the case of highly variable UGS, and whether suitable passages can be extracted reliably from feedback documents. Also, for some queries, where relevant feedback documents are short or even long but have a single-topic, using full document evidence can actually be more effective for expansion, and use of passages thus undesirable [9].

Query Performance Prediction (QPP) provides effective post-retrieval methods to estimate query performance by analyzing the retrieval scores of the most highly ranked documents [13,20,24]. While well researched for IR tasks, there is no reported work in implementing QPP methods for the prediction of QE performance. In this paper, we investigate the use of QPP methods for QE with UGS, in particular we examine its potential to select the most effective passage extraction scheme to maximise QE effectiveness for individual queries. Our study uses two well established QPP schemes and a novel QPP method designed for QE tasks. Our experimental investigation with an extended IR task based on the blip10000 internet UGC collection [19], shows our method to be more effective for QPP in QE than existing QPP methods developed for IR. We further demonstrate how our QPP method can be integrated into QE to adaptively select the most suitable passage evidence for QE for individual queries. QPP methods can be divided into pre- and post-retrieval methods, the former seeking to predict performance without actually carrying out an IR pass and the latter seeking to predict performance following an IR pass. Since QE is carried out as a post-retrieval process, in this study we restrict ourselves to the investigation of post-retrieval QPP methods. The remainder of this paper is structured as follows. Section 2 outlines some relevant previous work on speech retrieval. Section 3 describes the UGS test collection used for our experiments. Section 4 describes our experimental settings and initial passage-Based QE investigation. Section 5 describes our proposed QE QPP framework, Sect. 6 reports the results of using adaptive QE for UGS retrieval and Sect. 7 concludes.

2 Previous Work on QE for Spoken Content

Research in spoken content retrieval has a long history which has explored a range of techniques to improve search effectiveness [17]. However, within this work relatively little research has focused on QE. It was established in the TREC Spoken Document Retrieval (SDR) that QE can be highly effective for professionally prepared news stories which can be manually segmented easily from new broadcasts [7]. Later work at CLEF explored QE in the context of personal testimonies in an oral history collection [18]. However, while the content was more informal in the latter case, it was manually segmented into semantically focused sections, and most of the value in retrieval actually cames from manually created expert summaries for each passage. UGS content of the type considered in this paper is unlikely to have such rich descriptive annotations, and we focus on the use of automated content segmentation methods for speech transcripts such as those studied in [6].

Table 1. Word-level length statistics for blip10000 documents based on ASR transcripts.

Measure	Value
Standard deviation	2399.5
Avg. Length	703.0
Median	1674.8
Max	20451.0
Min	10.0

3 UGS Test Collection

In our study we first establish baseline IR effectiveness figures for QE using alternative passage methods, before moving on to explore QPP methods and their use in QE for speech retrieval. In this section we introduce the test collection used in our study.

3.1 UGS Document Set

Our experimental investigation is based on the blip10000 collection of UGC crawled from the Internet [19]. It contains 14,838 videos which were uploaded to the social video sharing site blip.tv by 2,237 different uploaders and covers 25 different topics. Much of this content is in the form of documentaries and other material whose main informational content is within the spoken data stream. To complement the data itself a transcript of the spoken content was created as part of the published collection by LIMSI[1]. The blip1000 collection exemplifies the challenges of searching UGS highlighted earlier. As shown in Table 1, the documents range in length from 10 words to over 20 K words, with an average of 703 words and a standard deviation of 2399 words.

3.2 Query and Relevance Sets

Our test collection is derived from the query set created for the MediaEval 2012 Search and Hyperlinking task[2]. This task was a known-item search task

---

[1] https://archives.limsi.fr/tlp/topic6.html.
[2] http://www.multimediaeval.org/mediaeval2012/.

involving search for a single previously seen relevant video (the *known-item*). This task provided 60 English queries collected using the Amazon Mechanical Turk[3] (MTurk) crowdsourcing platform. For our experiments, we created an adhoc version of this task developed by khwileh et al. [12]. The creation of these adhoc queries is explained in detail in [12].

4 Initial Investigation of QE for UGS Retrieval

Our first investigation examines the effectiveness of traditional QE approaches [3,14,22,23] using full-document evidence and previously proposed passage-based QE methods [9,15]. We begin by describing the methods used for retrieval, document-based QE, and passage-based QE techniques, and then give the results obtained for our experiments.

4.1 Retrieval Model

Our retrieval runs use the PL2 model [8] which is a probabilistic retrieval model from the *the Divergence From Randomness (DFR)* framework. We selected this model since it achieved the best results for our task in preliminary experiments comparing alternative models. This finding is consistent with previous studies, such as [2], which showed that PL2 has less sensitivity to wide variation in document length compared to other retrieval models. The PL2 model used in our investigation is explained in detail in [11,12].

4.2 Query Expansion

For QE, we used the DFR QE mechanism [8] to weight the top extracted terms from the top documents. DFR QE generalizes Rocchio's method to implement several term weighting models that measure the informativeness of each term in a relevant or pseudo relevant set of documents. DFR first applies a term-weighting model to measure the informativeness of the top ranking terms in the top ranking documents retrieved in response to a particular query. The DFR term-weighting model infers the informativeness of a term by the divergence of its distribution in the top documents from a random distribution. We use the DFR weighting model called Bo1, which is a parameter-free DFR model which uses BoseEinstein statistics to weight each term based on its informativeness. This parameter-free model has been widely used and proven to be the most effective [2,8]. The weight $w(t)$ of a term t in the top-ranked documents using the DFR Bo1 model is shown in Eq. 1, where tf_x is the frequency of the term in the pseudo-relevant list (top-n ranked documents). P_n is given by F/N; F is the term frequency of the query term in the whole collection and N is the number of documents in the whole collection.

$$w(t) = tf_x . \log_2(\frac{1 + P_n}{P_n}) + \log_2(1 + P_n) \qquad (1)$$

---

[3] http://www.mturk.com/.

The query term weight qt_w obtained from the initial retrieval is further adjusted based on $w(t)$ for both the newly extracted terms and the original ones, where $w_{max}(t)$ is indicated by the maximum value of $w(t)$ of the expanded query terms obtained as follows

$$qt_w = qt_w + \frac{w(t)}{w_{max}(t)}$$

4.3   Passage-Based QE

To implement passage-based QE, we follow a similar approach to that proposed in [15], where the *best-scoring passage* from each document for the query is utilised for QE instead of the full document. After initial retrieval, retrieved documents are re-ranked based on the score of their highest scoring passage. This best scoring passage for each of the top-ranked documents is then used for expansion. We explore alternative types of passage creation including those investigated in [6,15] as follows.

Semantic C99 segmentation (C99) based on the lexical cohesion within the ASR transcript using a well-established topic segmetation algorithm is C99 described in details in [4]. C99 uses a matrix-based ranking and a clustering approach, and assumes the most similar words belong to the same topic. We evaluate the effectiveness of this technique to achieve effective topic passages for the speech transcripts in QE. We used the C99 implementation in the UIMA Text segmentor[4].

Discourse Speech passage (SP) where consecutive silence bounded utterances from the same speaker are extracted as passages [6]. We hypothesise that silence points can be useful for detecting topic boundaries where each point is considered a passage and can be used as QE evidence. We used the speech passages detected by the ASR system whenever a speaker silence or a change of speaker is detected.

Window-based segmentation uses fixed-length sequences (windows) of words. We tested no-overlapping fixed length predefined passages where we segmented the transcripts into 20, 50, 100 and 500 words. We also evaluate two other window-based approaches studied in [15]:

- *Half-overlapped fixed-length passages* where the first window starts at the first word in a document, and the subsequent windows start at the middle of the previous one.
- *Arbitrary passages (TexTile)* which has also been shown to be the most effective for passage retrieval tasks [5,15]. Arbitrary passages are also of fixed-length and very similar to the overlapping passages with one key difference, that passages can start at any word in the document. The passage size is set before query time for the arbitrary passages, we tested different sizes of 20, 50, 100, and 500 words and three different starting points namely the 25th word, the 10th word, and the beginning of every sentence as detected by the ASR system.

---

[4] https://code.google.com/p/uima-text-passageer/.

4.4 Experimental Results

For all the window-based passages, we only report results for the best runs from the explored passages sizes 20, 50, 100, 500 due to space limitations. For our task, we found that the best performing passage size for fixed length windows is 500 words (fix500), while for half-overlapping windows (over) it is 100 words. For the arbitrary passages (TexTile), the best was to take 50 words passages starting at every sentence boundary. For QE parameters, we examined *top-term* expansion using 3, 5, 10 and 15 terms, and for the *top-doc* we examined use of the top 3 ranked documents and the 5 to 55 ranked documents in increments of 5 documents.

Table 2 shows results for full-document QE (ASR) and passage based QE runs (fix500, TexTile, over, SP and C99), in terms of MAP, Recall and P@10 for each QE run with optimal top-terms and top-doc settings. The results indicate that most passages achieve better performance than the baseline (ASR) QE runs on average. However none of these improvements was actually statistically significant[5]. In order to better understand the results of these different QE runs, we show the changes in MAP value (ΔMAP) for each passage-QE run relative to the baseline run for each query in Fig. 1. These results indicate that each QE run has different performance for each query; some have increased performance over the baseline where passage-based QE was more effective, while others show decreased performance over the baseline where document-based QE was better.

Table 2. Results for passage-based and document-based QE runs

	MAP	Recall	MAP@10	Terms	Docs
No QE	0.5887	0.9000	0.5450	–	–
ASR	0.6109	0.9167	0.5517	3	3
C99	0.6151	0.9000	0.5717	3	50
SP	0.6014	0.9167	0.5550	5	15
fix500	0.6156	0.9000	0.5667	5	35
TexTile	0.6166	0.9167	0.5617	10	50
over	0.6158	0.9000	0.5700	5	45

The ineffectiveness of the passage-based QE can be attributed to the fact that the segmentation process is sometimes ineffective where poor detection of topic boundaries can also harm the rank of good documents and lead to reduced scores for possibly good terms for QE. It can also be the case that segmentation is not actually needed for some queries, as discussed in [9], since some of the relevant documents only contain a single-topic and thus using full document evidence is often better for expansion.

---

[5] Confirmed by running query-level paired t-test comparison at the 0.05 confidence level [21].

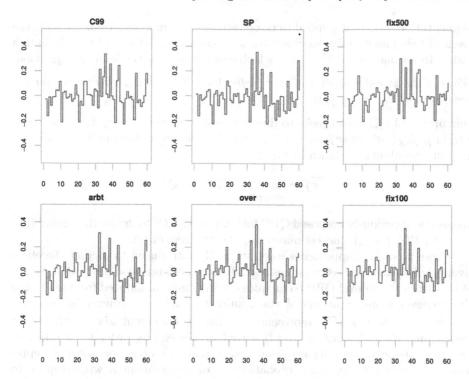

Fig. 1. ΔMAP between full document and passage-based QE for every query

From this analysis, we hypothesize that if we could predict which QE scheme to apply for a query, then we could automatically select the one which will give the best performance. In the next section we introduce QPP techniques, and consider their application to our QE task.

5 Query Performance Prediction for QE

This section begins by introducing the standard QPP methods used in our investigation and then motivates and introduces our new method for QPP, this is followed by the results of our experimental investigation.

5.1 Query Performance Prediction Methods

QPP metrics are designed to estimate the performance of a retrieval list based on heuristics in the absence of relevance judgement information [13,20,24]. We propose to utilise QPP techniques to predict the performance of QE. In this section, we review two existing state-of-the-art QPP metrics, and then introduce a novel QPP metric designed for QE.

For a query q, let the list of retrieved pseudo relevant documents be $D_q^{[prf]}$. Let $S_q^{[prf]}$ be a corresponding list of feedback passages S produced by any segmentation technique applied to these retrieved documents. Assuming $\mathcal{Q}$ is the event of being an effective feedback list for expansion, the goal of the prediction task is to estimate $\mathcal{P}(D_q^{[prf]}|q, qr)$ which seeks to answer the following question:

What is the probability $\mathcal{P}(.)$ of this feedback list (prf) being effective $\mathcal{Q}$ for the expansion of this query (q)?

Our proposed QE model seeks to find the best feedback list $\overline{prf}$ for each query q from prf_{lists} where $prf_{lists} \in \{D_{prf}, S_{prf}\}$ by selecting the one that gives the highest probability as shown in Eq. (2):

$$\overline{prf} = \underset{prf \in prf_{lists}}{\mathrm{argmax}} \; \mathcal{P}(prf|q, \mathcal{Q}) \tag{2}$$

We use two previously proposed QPP techniques [20, 24] to devise the probability function $\mathcal{P}(prf|q, \mathcal{Q})$ that estimates the effectiveness of QE.

These prediction approaches were originally introduced to predict the effectiveness of a particular retrieval method M using post-retrieval prediction. State-of-the-art post-retrieval QPP techniques use information induced from analysing the retrieval scores $Score(d)$ of the results set $D_q^{[res]}$ produced by retrieval method M, where $D_q^{[res]}$ represents the list of document ids together with their ranks $\mathcal{R}i$ and scores $Score(d)$ sorted in descending order for a query q. In probabilistic terms, the resultant score $Score(d)$ of a document d represents the estimated relevance probability r of a document d with respect to q $Score(d) \equiv \mathcal{P}(d|q, r)$. These methods [20, 24] are based on analysing the performance of the top k ranked documents which includes all documents that have rank $\mathcal{R}i$ that is less than k ($\forall d \epsilon D_q^{[res]} d_{\mathcal{R}i}$ where $0 \leqslant ri \leqslant k$). Two existing QPPs measures which have been found to be effective for IR tasks are used in our study: Weighted Information Gain (WIG) [24] and Normalised Query Commitment (NQC) [20].

5.1.1   WIG

WIG [24] is based on the weighted entropy of the top k ranked documents. WIG works by comparing the scores of the top-k documents $\forall d \epsilon D_q^{[k]} Score(d)$ to that obtained by the corpus $Score(D)$. Essentially the top-k documents here are assumed to be relevant and $Score(D)$ is assumed to be characteristic of a non-relevant document. WIG is often combined with a query length normalisation $\frac{1}{\sqrt{|q|}}$ to make the scores comparable over different queries, defined in Eq. (3).

$$WIG(q, M) = \frac{1}{k} \sum_{d \epsilon D_k} \frac{1}{\sqrt{|q|}} (Score(d) - Score(D)) \tag{3}$$

5.1.2   NQC

NQC [20] is based on estimating the potential amount of query drift in the list of top k documents by measuring the standard deviation of their retrieval scores.

A high standard deviation indicates reduced topic drift and hence improved retrieval defined in Eq. (4) where $\bar{\mu} = \frac{1}{k}\sum_{d\epsilon D_q^{[k]}} Score(d)$.

$$NQC(q, M) = \frac{1}{Score(D)}\sqrt{\sum_{d\epsilon D_q^{[k]}} \frac{1}{k}(Score(d) - Score(\bar{\mu}))} \qquad (4)$$

Both WIG and NQC are tuned to have a strong linear relationship with the MAP of the query in which the only variable that needs to be set is the top-k documents that are required for the prediction.

5.1.3   Weighted Expansion Gain (WEG)

We introduce a modified version of the WIG predictor which we refer to as Weighted Expansion Gain (WEG) for predicting QE performance. The $Score(D)$ in WIG is a very rough model of non-relevant content for comparison with pseudo-relevant passages, for WEG we modify the definition of the top-k documents $D_q^{[k]}$ for each query to be composed of two subsets $D_q^{[prf]}$ and $D_q^{[nprf]}$ defined as follows: $D_q^{[prf]}$ is the set of prf feedback documents that are assumed relevant and effective for expansion $\forall d\epsilon D_q^{[res]}$ ($d_{\mathcal{R}i}$ where $0 \leqslant \mathcal{R}i \leqslant prf < k$), and $D_q^{[nprf]}$ is the set of documents that are assumed non-relevant and ineffective for expansion. These documents are ranked among the top-k documents and right after the prf documents ($prf < nprf < k$) as in $\forall d\epsilon D_q^{[res]}$ ($d_{\S\mathcal{R}i}$ where $prf < \mathcal{R}i \leqslant nprf$).

Although the goal is different, our predictor relies on a hypothesis that is somewhat similar to that proposed in [20] by assuming that topic drift is caused by *misleading* documents that are high ranked in retrieval because they are *similar but not relevant* to the query. The WEG predictor aims to analyse the quality of the prf documents by measuring the likelihood that they will have topic drift. This is estimated by measuring the weighted entropy of the prf documents against the top-ranked yet non-pseudo $nprf$ set of documents. Unlike WIG, which uses the *centroid* of general non-relevant documents $Score(D)$, WEG uses the *centroid* of the $nprf$ document scores:

$$Cnprf \equiv Cent(Score(D_q^{[nprf]})) \equiv \frac{1}{nprf} \sum_{d\epsilon D_q^{[nprf]}} Score(d)$$

as a reference point for estimating of the effectiveness of the prf documents in QE, as shown in Eq. (5).

$$WEG(q, D_{prf}) = \frac{1}{prf} \sum_{d\epsilon D_{prf}} \frac{1}{\sqrt{|q|}}(Score(d) - Cnprf) \qquad (5)$$

WEG requires three parameters: the number of prf documents, the number of top-k documents and the number of $nprf$ documents to perform the actual estimation. In fact, the number of $nprf$ documents does not need to be explicitly selected since calculated as $nprf = k - prf$. We use this predictor as a function to estimate the performance for each list of feedback documents and find the one that maximises effectiveness for list of prf as shown in Eq. 2. The final formal probabilistic representation of our WEG predictor is explained in Eq. (6).

$$\mathcal{P}(D_q^{[prf]}|q, \mathcal{Q}) \equiv WEG(q, D_{prf}) = \frac{1}{prf} \sum_{d \in D_{prf}} \mathcal{P}(d|q, r) - Cent(\mathcal{P}(D_q^{[nprf]})|q, r)) \quad (6)$$

5.2   Evaluation

We evaluate the three predictors in terms of their effectiveness for predicting QE performance. To do this we measure the Pearson correlation coefficient ρ between the actual Average Precision (AP) after expansion QE(AP) values for queries in a given test set, and the values assigned to these queries by each prediction measure (WIG, NQC, WEG). A higher correlation value indicates increased prediction performance. For retrieval and QE, we again used the PL2 and BO1 models. We explore 5 different combinations of {top-terms,top-docs} for QE: {(10, 30), (3, 10), (3–3), (5–5), (10–3)} to assess the relationship between the QE(AP) and the predictors. Similar to the work studied in [13,20] we assume that all scores follow uniform distributions and calculate the $Cnprf$ accordingly.

To set the k parameter for these predictors we use the *cross-validation paradigm*, also proposed in [20]. Each of the query set was randomly split into training and testing sets (30 queries each). During the training, k for WEG, NQC and WIG for each corpus is tuned using manual data sweeping through the range of [5, 100] with an interval of 5, and through the range of [100, 500] with an interval of 20. We performed 20 different splits to switch the roles between the training and testing sets and report the best performing k parameter that yield optimal prediction performance (as measured by Pearsons correlation) by taking the average between the 20 different runs.

The WEG predictor also requires the prf parameter to be set, which we set automatically based on the number of feedback documents used for each QE. $nprf$ is set as the difference between both prf and k values for each run. As explained in the previous section, for the WEG predictor the k parameter *must* be always higher than the prf value.

Table 3. Correlation Coefficients between QE(AP) for each QE run vs each QPP. Percentages in () indicates the difference between WEG and WIG correlation values. All reported correlation are significant at the 0.05 confidence level **except** those marked with *.

QE(terms-docs)	10t-30d	3t-10d	3t-3d	5t-5t	10t-3d
WEG	**0.403** (+74%)	**0.550** (+22%)	**0.533** (+23%)	**0.533** (+28%)	**0.528** (+18%)
WIG	0.232*	0.449	0.435	0.418	0.449
NQC	0.170*	0.379	0.378	0.377	0.379

Table 3 shows prediction results measured using the optimal parameters obtained using the cross-validation evaluation paradigm. The results reveal that all predictors have strong ability to estimate the QE performance for most runs. However, WEG is shown to significantly outperform the other two predictors across all QE runs and on the three data collections. This can be attributed to the fact that WEG is tuned to focus on the actual prf and $nprf$ documents, which have the highest influence on the QE performance. This, on one hand, helps WEG to identify the effectiveness of QE that can in fact distinguish the relevant prf documents from the non relevant ones, but on the other hand, raises efficiency concerns about WEG, since it takes almost twice the time required for WIG/NQC tuning.

Furthermore, although they are designed to predict the original query performance (AP), WIG and NQC provided good estimation of QE(AP). This in fact confirms the relationship between the original query performance AP and QE(AP). In other words, we found that if the AP is too low, the QE process does not have a good prf list to extract useful expansion terms. The same finding about this relationship was previously discussed in detail by He et al. [10]. This relationship is also confirmed in terms of the optimal k parameters that maximise the correlation for both predictors with QE(AP); where the WIG k parameters is optimal in [5–25] documents while NQC k parameters were between (70–150) which similar to that reported in [20] for predicting the AP, i.e. the optimal k are always higher for NQC than WIG, while the optimal k values for WEG vary across different QE runs and depend on the number of prf documents considered for expansion. In general, our experiments reveal that for lower prf values (3, 5), the best prediction is achieved by setting k in [150–250], while for higher values ($prf \geqslant 10$) the best k is in [50–100]. We found that for better prediction, the size of the $nprf$ set must be at least 3 times the size of the prf set.

6   Adaptive Passage-Based QE for UGS Retrieval

To further examine our proposed QPP technique, we use it as the basis of an adaptive QE algorithm to select between the alternative methods explored in Sect. 4. This simple adaptive QE approach applies Eq. (2) by calculating the WEG in Eq. (6) for every candidate prf list for each query. It then selects the $\overline{prf}$ that achieves the highest value for expansion.

We use the optimal QE parameter value which was shown previously in Table 2 to generate the candidate prf sets for each query. The WEG predictor is used to perform the actual prediction. To make them comparable, we also normalize the scores of each feedback list using: $\frac{Score(d)-\mu}{\sigma}$ before calculating the actual prediction value of each feedback list, where μ is the average score of the retrieved results list and σ is the overall standard deviation. We experiment with two combinations between full ASR document evidence (ASR) and each passage type (TexTile, over, SP, C99, fix500) to evaluate the robustness of the adaptive approach (e.g. ASRTexTile seeks to predict between both ASR and TexTile evidences, ASRover predicts between both ASR and over passage

Table 4. Performance for adaptive QE runs. * indicates statistically significant improvement at (0.05) over the best non-adaptive QE run from Table 4.

QE-run	MAP	Recall	Acc	O-MAP
ASRTexTile	0.6328	0.9197	0.6333	0.6533
ASRover	0.6312	0.92	0.6000	0.6565
ASRfix500	0.6338	0.9187	0.8146	0.6440
ASRC99	0.6382	0.92	0.7833	0.6511
ASRSP	0.6265	0.9187	0.6167	0.6469
ALL	**0.6690***	**0.9733***	0.7367	0.7424
ALL-SP	**0.7140***	**0.9851***	0.8134	0.7424

evidence, ASRC99 predicts between both C99 and ASR). In addition, we test combining all evidence (all passages including the ASR) together using the QE run (ALL) and picking the best to evaluate the effectiveness of the adaptive QE when there are multiple available options. We also experiment with having all passages excluding the SP passages (ALL-SP); since it is the least effective passage evidence for QE (as shown in Table 2), yet the most difficult to normalise/predict due to its varying quality and length which results in varying score distributions.

The results of these different adaptive QE runs are shown in Table 4. As well as MAP, we also report the O-MAP which is obtained using the optimal selection of the best feedback list based on relevance judgement information for each query. The Acc column represents the prediction accuracy of the adaptive selection for each QE. It can be seen that all adaptive QE runs improve over the regular QE runs reported in Table 2 which use single feedback evidence for expansion. The best performance is obtained by using all options for QE (ALL) and (ALL-SP) obtaining a statistically significant (tested at the 0.05 confidence level) improvement of 10% and 15% (respectively) over all the regular traditional QE runs reported in Table 2, and a 14%, 21% increase over the original retrieval performance (no QE) run. The prediction accuracy (Acc) for each run was between 60% and 82%, showing that there is still room for improvement compared to the O-MAP score. It is important to note that the aim of this adaptive approach is to evaluate the prediction in a search task; previously proposed prediction approaches were only tested and evaluated using correlation measures similar to those in Table 3.

7   Conclusions and Further Studies

This paper has examined the effectiveness of document-based and passage-based QE methods for search of a UGS archive. Our investigation reveals that none of the approaches evaluated is always better for all queries. We found that in the UGC settings, the variation in the length, quality and structure of the relevant documents can harm the effectiveness of both techniques across different queries. We then examined QPP techniques for selecting the best QE retrieval results and presented WEG, a new QPP technique specifically designed for prediction of the effectiveness of QE. An empirical evaluation demonstrates the effectiveness of our predictor WEG over the state-of-art QPP techniques of WIG and NQC. We then showed how this technique can be utilised to implement an adaptive QE technique in UGS retrieval by predicting the best evidence (passage/document) for expanding each query.

Our prediction technique relies on the obtained retrieval scores which follow different distributions (i.e. gamma distribution). The assumed uniform normal distribution may not fit these scores correctly for all queries. An alternative would be to assume different score distributions such as log-normal, gamma and others to improve the prediction quality. We leave investigation of this possibility for future work. Also, for a further performance optimisation could be to include more query/document related features for the prediction, such as TF distribution in passages, the IDF of top terms in feedback documents and other signals.

Acknowledgments. This research was partially supported by Science Foundation Ireland in the ADAPT Centre (Grant 13/RC/2106) (www.adaptcentre.ie) at Dublin City University.

References

1. Allan, J.: Relevance feedback with too much data. In: Proceedings of the 18th Annual International ACM SIGIR Conference on Research and Development in Information Retrieval, pp. 337–343. ACM (1995)
2. Amati, G., Van Rijsbergen, C.J.: Probabilistic models of information retrieval based on measuring the divergence from randomness (2002)
3. Carpineto, C., Romano, G.: A survey of automatic query expansion in information retrieval (2012)
4. Choi, F.Y.Y.: Advances in domain independent linear text segmentation. In: Proceedings of NAACL 2000 (2000)
5. Eskevich, M.: Towards effective retrieval of spontaneous conversational spoken content. Ph.D. thesis, Dublin City University (2014)
6. Eskevich, M., Jones, G.J.F., Wartena, C., Larson, M., Aly, R., Verschoor, T., Ordelman, R.: Comparing retrieval effectiveness of alternative content segmentation methods for internet video search. In: 2012 10th International Workshop on Content-Based Multimedia Indexing (CBMI), pp. 1–6. IEEE (2012)
7. Garofolo, J., Auzanne, G., Voorhees, E.: The TREC spoken document retrieval track: a success story. In: Proceedings of RIAO 2000, pp. 1–8 (2000)
8. Gianni, A.: Probabilistic models for information retrieval based on divergence from randomness. Ph.D. thesis, Department of Computing Science, University of Glasgow (2003)
9. Gu, Z., Luo, M.: Comparison of using passages and documents for blind relevance feedback in information retrieval. In: Proceedings of the 27th Annual International ACM SIGIR Conference on Research and Development in Information Retrieval, pp. 482–483. ACM (2004)

10. He, B., Ounis, I.: Studying query expansion effectiveness. In: Boughanem, M., Berrut, C., Mothe, J., Soule-Dupuy, C. (eds.) ECIR 2009. LNCS, vol. 5478, pp. 611–619. Springer, Heidelberg (2009). doi:10.1007/978-3-642-00958-7_57

11. Khwileh, A., Ganguly, D., Jones, G.J.: Utilisation of metadata fields and query expansion in cross-lingual search of user-generated internet video (2016)

12. Khwileh, A., Jones, G.J.: Investigating segment-based query expansion for user-generated spoken content retrieval. In: 2016 14th International Workshop on Content-Based Multimedia Indexing (CBMI), pp. 1–6. IEEE (2016)

13. Kurland, O., Shtok, A., Hummel, S., Raiber, F., Carmel, D., Rom, O.: Back to the roots: a probabilistic framework for query-performance prediction. In: Proceedings of the 21st ACM International Conference on Information and Knowledge Management, pp. 823–832. ACM (2012)

14. Lam-Adesina, A.M., Jones, G.J.F.: Dublin City University at CLEF 2005: cross-language speech retrieval (CL-SR) experiments. In: Peters, C., et al. (eds.) CLEF 2005. LNCS, vol. 4022, pp. 792–799. Springer, Heidelberg (2006). doi:10.1007/11878773_87

15. Liu, X., Croft, W.B.: Passage retrieval based on language models. In: Proceedings of the Eleventh International Conference on Information and Knowledge Management, pp. 375–382. ACM (2002)

16. Mitra, M., Singhal, A., Buckley, C.: Improving automatic query expansion. In: Proceedings of the 21st Annual International ACM SIGIR Conference on Research and Development in Information Retrieval, pp. 206–214. ACM (1998)

17. Larson, M., Jones, G.J.F.: Spoken content retrieval: a survey of techniques and technologies (2011)

18. Pecina, P., Hoffmannová, P., Jones, G.J.F., Zhang, Y., Oard, D.W.: Overview of the CLEF-2007 cross-language speech retrieval track. In: Peters, C., et al. (eds.) CLEF 2007. LNCS, vol. 5152, pp. 674–686. Springer, Heidelberg (2008). doi:10.1007/978-3-540-85760-0_86

19. Schmiedeke, S., Xu, P., Ferrané, I., Eskevich, M., Kofler, C., Larson, M.A., Estève, Y., Lamel, L., Jones, G.J.F., Sikora, T.: Blip10000: a social video dataset containing SPUG content for tagging and retrieval. In: Proceedings of the 4th ACM Multimedia Systems Conference, pp. 96–101. ACM (2013)

20. Shtok, A., Kurland, O., Carmel, D.: Predicting query performance by query-drift estimation. In: Azzopardi, L., Kazai, G., Robertson, S., Rüger, S., Shokouhi, M., Song, D., Yilmaz, E. (eds.) ICTIR 2009. LNCS, vol. 5766, pp. 305–312. Springer, Heidelberg (2009). doi:10.1007/978-3-642-04417-5_30

21. Smucker, M.D., Allan, J., Carterette, B.: A comparison of statistical significance tests for information retrieval evaluation. In: Proceedings of the Sixteenth ACM Conference on Information and Knowledge Management, pp. 623–632. ACM (2007)

22. Terol, R.M., Palomar, M., Martinez-Barco, P., Llopis, F., Muñoz, R., Noguera, E.: The University of Alicante at CL-SR track. In: Peters, C., et al. (eds.) CLEF 2005. LNCS, vol. 4022, pp. 769–772. Springer, Heidelberg (2006). doi:10.1007/11878773_84

23. Wang, J., Oard, D.W.: CLEF-2005 CL-SR at Maryland: document and query expansion using side collections and thesauri. In: Peters, C., et al. (eds.) CLEF 2005. LNCS, vol. 4022, pp. 800–809. Springer, Heidelberg (2006). doi:10.1007/11878773_88

24. Zhou, Y., Croft, W.B.: Query performance prediction in web search environments. In: Proceedings of the 30th Annual International ACM SIGIR Conference on Research and Development in Information Retrieval, pp. 543–550. ACM (2007)

Optimized Convolutional Neural Network Ensembles for Medical Subfigure Classification

Sven Koitka[1,2]([✉]) and Christoph M. Friedrich[1]([✉])

[1] Department of Computer Science,
University of Applied Sciences and Arts Dortmund (FHDO),
Emil-Figge-Strasse 42, 44227 Dortmund, Germany
{sven.koitka,christoph.friedrich}@fh-dortmund.de
[2] Department of Computer Science, TU Dortmund University,
Otto-Hahn-Str. 14, 44227 Dortmund, Germany
http://www.inf.fh-dortmund.de

Abstract. Automatic classification systems are required to support medical literature databases like *PubMedCentral*, which allow an easy access to millions of articles. *FHDO Biomedical Computer Science Group (BCSG)* participated at the *ImageCLEF 2016 Subfigure Classification Task* to improve existing approaches for classifying figures from medical literature. In this work, a data analysis is conducted in order to improve image preprocessing for deep learning approaches. Evaluations on the dataset show better ensemble classification accuracies using only visual information with an optimized training, in comparison to the mixed feature approaches of BCSG at ImageCLEF 2016. Additionally, a self-training approach is investigated to generate more labeled data in the medical domain.

Keywords: Convolutional neural networks · Deep learning · Ensembles · ImageCLEF · Modality classification · Resizing strategies · Self-training · Subfigure classification

1 Introduction

The amount of publicly available medical literature is constantly growing. Websites, like *PubMedCentral*[1], provide an easy access to medical publications. As of April 2017 PubMedCentral offers over 4.3 million articles. Those are queried on text basis and are available as a PDF download or HTML version.

However, a term based search query may result in too many unrelated articles, which is why image based search engines are becoming increasingly popular. One example is *Open-i*[2] [3], which allows the search of all images from five different collections including PubMedCentral to enable case related search queries. All images are furthermore categorized into modalities and contain annotations.

---

[1] https://www.ncbi.nlm.nih.gov/pmc/ (last access: 19.04.2017).
[2] https://openi.nlm.nih.gov/ (last access: 19.04.2017).

© Springer International Publishing AG 2017
G.J.F. Jones et al. (Eds.): CLEF 2017, LNCS 10456, pp. 57–68, 2017.
DOI: 10.1007/978-3-319-65813-1_5

Technologies used by Open-i have been developed and evaluated in form of tasks at *ImageCLEF*, as stated on the project website[3]. The latest *ImageCLEF 2016 Medical Task* [8] evaluated among others the modality classification of medical subfigures. Since results of this subtask were satisfying, no renewal has been announced. Besides ImageCLEF 2015 and 2016 this task also existed at ImageCLEF 2012-2013 with a sightly different class hierarchy.

FHDO Biomedical Computer Science Group (BCSG) participated in the subfigure classification task at ImageCLEF 2015 and 2016 with top placements in both years [10,13]. Approaches based on *Deep Convolutional Neural Networks* lacked proper tuning of hyperparameters and preprocessing. Therefore part of this work is to optimize the adopted deep learning training process to achieve higher accuracies in comparison to ImageCLEF 2016.

Since deep learning requires a large amount of data in order to train generalizable models, a *self-training* [2] approach is part of this evaluation. In comparison to state-of-the-art datasets like *ImageNet* [17], the ImageCLEF datasets are very small. Therefore using a semi-supervised approach to generate more but noisy labeled data could help to train useful convolutional filters for medical literature figures.

The participation of BCSG in ImageCLEF 2016 will be discussed in Sect. 2 with distinction to results presented in this work. In Sect. 3, some problems of the dataset are shown and potential improvements for future datasets are mentioned. Furthermore in Sect. 4, preprocessing strategies for images of the ImageCLEF 2016 subfigure classification dataset are discussed and evaluated in Sect. 5. Finally, conclusions are drawn in Sect. 6 with potential future research topics.

2   Revisiting Participation at ImageCLEF 2016

BCSG participated in the subfigure classification subtask of the ImageCLEF 2016 Medical Task, which was the second participation after ImageCLEF 2015 [13] in this subtask. The aim of this subtask was to provide an automatic system that predicts one out of the 30 modalities for a given medical literature subfigure [8]. Further analysis of the ImageCLEF 2015 dataset was conducted in [14]. In 2016, traditional feature engineering and deep learning approaches had been evaluated. However, the latter haven't been tuned properly due to time and resource limitations.

The official evaluation results of this subtask are visualized in Fig. 1. It shows top placements among all three categories, visual, textual, and mixed features for BCSG. The run which achieved the lowest accuracy was an Inception-V1 [19], also known as GoogLeNet, trained from scratch with the caffe framework [9] and DIGITS[4] front-end. Surprisingly, the best visual run was an unmodified *ResNet-152* [5], which was used as a feature extractor. The activation vectors

---

[3] https://ceb.nlm.nih.gov/ridem/iti.html (last access: 24.04.2017).
[4] https://github.com/NVIDIA/DIGITS (last access: 20.04.2017).

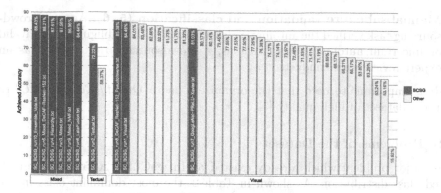

Fig. 1. Official evaluation results of the ImageCLEF 2016 subfigure classification sub-task (image from [10]).

from the last fully connected layer were fed into a linear neuron layer trained with the *Projection Learning Rule* [15].

Since deep learning based approaches are becoming increasingly popular, this work will focus on optimizing the general approach to train convolutional neural networks on the ImageCLEF 2016 subfigure classification dataset. According to [10,13,14], textual features can improve the overall accuracy consistently as these are independent of the visual content of an image. However, this work optimizes the image preprocessing and therefore will not be evaluated in conjunction with textual information from image captions.

3 Dataset Analysis

In this section the subfigure classification dataset is analyzed for possible problems which occur during training a convolutional neural network. Additionally, the data generation process will be reviewed and potential improvements for future datasets are mentioned.

3.1 Data Generation Process

The overall data generation process was explained by [7] and consisted of eight steps. The most relevant parts for this analysis are:

- **Automatic compound figure detection and separation (1, 2):** Every image was checked by a system if it is a compound or non-compound figure. In the latter case an automatic separation of subfigures was computed.
- **Manual compound figure separation (3):** All compound figures were checked using a crowdsourcing task. An additional manual separation step was done in case an automatic separation wasn't successful.
- **Automatic subfigure classification (4):** A classification of all separated subfigures was performed using various visual features and a *k-Nearest-Neighbor (k-NN)* approach.

– **Manual subfigure validation and classification (5, 6):** Again a crowd-sourcing task verified the automatic classification of the subfigures. For invalid or uncertain automatic classifications, a crowdsourcing task as well as an expert review were conducted.

It is important to note that the organizers had a *Quality Control (QC)* to ensure a dataset of high quality [7].

3.2 Problems of the Dataset

One major problem which already existed at ImageCLEF 2015, is the unbalanced class distribution. As shown in Table 2, the most dominant class in the 2016 dataset is *GFIG* with 2954 images, which is 43.60% of all training images. On the other hand, classes like *GPLI* (1 image), *DSEM* (5 images), or *DSEE* (8 images) are underrepresented. Due to this imbalance, BCSG and others [20] decided to extend the training set with images from the ImageCLEF 2013 modality classification task [6]. It is a smaller but overall a better-balanced dataset. Since the dataset from ImageCLEF 2013 contained 31 classes, images from the compound figure class *COMP* were removed.

A lot of images are of a lower resolution than originally published, as visualized in Fig. 2. Both Fig. 2(a) and (b) suffer from low resolution and JPEG compression artifacts. As the text is nearly unreadable it is difficult or impossible to use *Optical Character Recognition (OCR)* to extract textual features from images. This was one attempted approach of BCSG to generate textual features from images during the model creation phase for ImageCLEF 2016. Since both images are available as vector graphics in their respective original literature, this is a disadvantage in the data collection process. For reference, both images were collected in a higher resolution for the visualization purpose in Fig. 2(c) and (d).

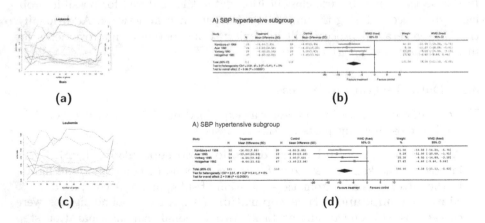

(a)   **(b)**

(c)   **(d)**

Fig. 2. Many images from the ImageCLEF 2016 dataset [8] are of low resolution and have JPEG compression artefacts like 2(a), (b), (c) and (d) show higher resolution versions taken from literature [12,16]. These articles where published under Creative Commons 2.0 (https://creativecommons.org/licenses/by/2.0/).

(a) 600×51 (b) 587×98

Fig. 3. Two example images with unnecessary padding from the ImageCLEF 2016 dataset [8] (frames were added for better visualization). More than 50% of the overall image content is uninformative background colour.

As previously described all compound figures had been separated automatically and verified by crowdsourcing. Figure 2(a) is one example where this verification process failed. On the right side is the y-axis of the next subfigure, which is missing on the next image in the dataset. Although both plot axes are not connected and it is hard for the automatic process to separate this correctly, this should have been detected by the multi-user verification.

Another common problem is unnecessary and uninformative background. In Fig. 3 are two examples of such occurrences. While this may be less problematic for feature engineered approaches, it is a problem for neural networks. All images have to be rescaled to a common image size. In other words, for these examples, the rescaled images contain over 50% of whitespace where convolutional filters cannot detect any edges.

3.3 Suggestions for Future Datasets

Although the overall classification accuracies at ImageCLEF 2016 were relative high, there is still room for improvements. Some of them could be:

- Higher/Original image resolution for bitmap images. Vector images could be rendered with a uniform DPI setting.
- Bitmap images should be exported as PNG images. JPEG compression artefacts occur often with high contrast edges, which are found in plots.
- Distribution of vector based figures in a vector image format.
- Non-linear separation boundary for compound figure separation. For example some images cannot be separated linearly and thus axis labels are truncated.
- Better class distribution with a minimum amount of samples per class. Classes like *GPLI* cannot be learned from just one single training sample.

4   Methodology

Image sizes differ in both dimensions, since the subfigure images are automatically/manually separated from a compound figure. As shown in Fig. 4 the aspect ratio of all images is broadly spread. Interestingly, there is a border at around approximately 600 pixel width, which might be caused by an user interface during the crowdsourcing process. Additionally, as already mentioned in Sect. 3.2, a lot of images contain unnecessary background. Most of the images can be cropped and even a few to approximately 30% of the overall image.

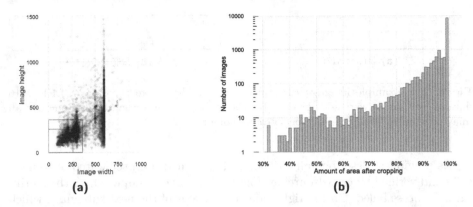

Fig. 4. Visualization of (a) image aspect ratios compared to 256 × 256 respectively 360 × 360 images and (b) remaining image areas after auto cropping.

Therefore an aspect-ratio aware resizing is proposed in conjunction with automatic background colour cropping. For automatic cropping, a connected component analysis based on the edge colours can be used. When using the aspect ratio for resizing, empty space occurs in the target image which has to be filled with an appropriate colour. For instance, this can be chosen from the cropped areas, for example, the most dominant colour. If the image cannot be cropped, then white should be a good background colour, since literature usually has a white background. Compared with squash resizing, the outcome is clearer and less disturbed, as shown in Fig. 5.

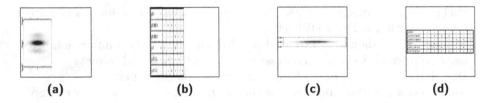

Fig. 5. (a, b) Both images from Fig. 3 with *squash* resizing as used in DIGITS by default and (c, d) with the proposed preprocessing and resizing method.

5   Evaluation

All adopted convolutional neural networks were trained using *Tensorflow Slim*[5], which is a wrapper framework for *Tensorflow*[6] [1]. Besides network architectures, several pre-trained models were published, which can be used to fine-tune on other datasets. The preconfigured optimizer settings were used (RMSprop with

---

[5] https://github.com/tensorflow/models/blob/master/slim (last access: 09.03.2017).
[6] https://www.tensorflow.org (last access: 09.03.2017).

rms_momentum $= 0.9$ and rms_decay $= 0.9$), except for the learning rate decay, which was changed to polynomial decay with power 2, and the L_2 regularization weight, which was set to 0.0001.

To fine-tune a pre-trained model, the training process was split into two separate optimization steps. First, only the top layer (Logits) was trained for 10 epochs with lr $= 0.0001$ and batch_size $= 64$ to adapt the randomly initialized layer to both existing information flow and dataset domain. Finally, all layers were trained for 20 epochs with lr $= 0.01 \rightarrow 0.0001$ using the polynomial decay as described above. The pre-trained models were published as part of Tensorflow Slim and were trained on the ImageNet dataset [17]. For the evaluation process, the Inception-V4 and Inception-ResNet-V2 [18] network architectures were used. Both recently achieved state of the art results on the ImageNet dataset.

According to [10], the training set was extended to add images for under-represented classes and to increase the overall number of images for feeding a deep neural network. For training, all images from the *ImageCLEF 2013 Modality Classification* [6] training and test set, and the *ImageCLEF 2016 Subfigure Classification* [8] training set were used. Classification accuracies are single-crop evaluations on the *ImageCLEF 2016 Subfigure Classification* test set.

All results using the proposed resizing method are presented in Table 1. In total $n = 5$ trained models were evaluated per configuration using one single center crop, as well as 10 crops (center plus four corner crops and their horizontal flips). Unfortunately, the results do not show a significant boost in terms of achieved accuracy between squash and the proposed method. There is only a slight improvement for the Inception-ResNet-V2.

Combining all $n = 5$ trained models to one ensemble per configuration leads to overall higher accuracies, as stated in Table 1. Surprisingly, in ensembles, the proposed method is consistently better than using *squash* for resizing.

With 88.48% the best ensemble result outperforms the highest accuracy from BCSG at ImageCLEF 2016 (88.43%) using visual features instead of mixed features (see Fig. 1 for reference). Compared to run 8 at ImageCLEF 2016, the best visual run (85.38%), there is a difference of +3.05% accuracy. This might be caused by the optimized preprocessing and resizing, but also by newer architectures and optimized fine-tuning of pre-trained networks. When comparing this result to the baseline of squash resizing, the relative gain of the proposed

Table 1. Classification accuracies (%) of the proposed resizing method ($n = 5$). Results are stated as mean ± standard deviation as well as the ensemble accuracy in parenthesis. Squash refers to the default resizing method in DIGITS and is used as a baseline.

	Inception-V4		Inception-ResNet-V2	
	Center-crop	10-crops	Center crop	10 crops
Squash	86.87 ± 0.32(87.93)	**87.11 ± 0.24(88.05)**	86.25 ± 0.36(86.99)	86.82 ± 0.42(87.37)
Aspect-ratio aware	**86.88 ± 0.57(88.12)**	**87.11 ± 0.51(88.29)**	**86.67 ± 0.50(87.45)**	86.87 ± 0.38(87.61)
+Auto-cropping	86.64 ± 0.45(88.09)	86.98 ± 0.56(**88.48**)	86.66 ± 0.71(87.35)	**87.02 ± 0.74(87.73)**
+No upscaling	86.36 ± 0.32(87.28)	86.44 ± 0.31(87.35)	84.34 ± 2.06(86.97)	85.05 ± 1.22(86.82)

preprocessing is +0.43%. Source code to replicate the experiments is provided on Github[7].

5.1   Results Without Additional Images from ImageCLEF 2013

Results stated above, as well as the results in [10], were achieved using an extended training dataset. In all experiments, all 30 classes were extended with images from the ImageCLEF 2013 Medical Task [6] training and test set. Since other authors [11] noted the lack of comparability, additional evaluations using only the ImageCLEF 2016 [8] dataset for training will be provided.

For better comparability, models were trained on only images from the ImageCLEF 2016 dataset using the exact same optimizer configuration as described in Sect. 5 and aspect-ratio aware resizing combined with automatic cropping. These models achieve a top-1 accuracy of 83.96 $\pm$ 1.52% (84.18 $\pm$ 1.28%) for the Inception-ResNet-V2 and 83.14 $\pm$ 0.10% (83.40 $\pm$ 0.09%) for the Inception-V4 network (single-crop and 10-crop evaluation results).

Compared to the ensemble results in [11], with a top-1 accuracy of 82.48%, the presented method is better in terms of achieved accuracy. Combining the single-crop models by taking the mean probabilities produces an ensemble top-1 accuracy of 85.55% for the Inception-ResNet-V2 and 85.33% for the Inception-V4 network. Interestingly, the 10-crop ensembles differ only in <0.05%, which might be caused by the reduced number of random crop samples during training for underrepresented classes.

Both purely CNN ensembles outperform the mixed classifier approach from [11], as summarized in Table 2. The Inception-V4 ensemble trained only on images from the 2016 dataset has a better or equal F1-Score in 24 of 30 cases. However, using a two-sample one-tailed t-test could not prove a significantly better result. Contrary, the ensemble trained on images from the ImageCLEF 2013 and 2016 dataset, outperform in 29 of 30 cases regarding the F1-Score. A statistical hypothesis testing with $\alpha = 0.05$ yielded a significantly better approach over [11] with $p = 0.014$. It also shows a strong trend with $p = 0.054$ to be superior to the ensemble without additional training data.

The multi-classifier approach, with features from CNNs, described in [11] is nevertheless interesting and should be further investigated with newer state of the art network architectures. Unfortunately, the authors chose a network architecture which lacks all the improvements of several years of research in the field of deep learning. Using a CNN as a feature extractor has been described by many authors [4, 21] and is worth being part of further research.

5.2   Self-training Using a PubMedCentral Database Snapshot

In May 2016, a snapshot of the complete public available *PubMedCentral Open Access Subset*[8] was gathered using the FTP service within approximately two

---

Table 2. Comparison of two ensembles, generated from fine-tuned Inception-V4 models, against the ensemble approach described in [11]. Bold values indicate the best value between all three ensembles. Precision, recall and F1-score are macro averaged.

Class	# Samples			Inception-V4 (2013+2016)			Inception-V4 (only 2016)			External ensemble [11]		
	Training set	Extended training set	Test set	Precision	Recall	F1-score	Precision	Recall	F1-score	Precision	Recall	F1-score
D3DR	201	271	96	**96.05**	**76.04**	**84.88**	91.67	68.75	78.57	69.31	72.92	71.01
DMEL	208	279	88	**62.96**	**38.64**	**47.89**	56.52	29.55	38.81	39.62	23.86	29.79
DMFL	905	971	284	82.21	94.37	87.87	74.59	**95.07**	83.59	73.45	91.55	81.50
DMLI	696	908	405	**91.63**	**91.85**	**91.74**	91.13	91.36	91.25	87.94	**91.85**	89.86
DMTR	300	366	96	**73.17**	62.50	67.42	68.69	**70.83**	**69.74**	48.39	62.50	54.44
DRAN	17	89	76	96.05	**96.05**	**96.05**	**96.23**	67.11	79.07	92.31	31.58	47.06
DRCO	33	56	17	**100.00**	**64.71**	**78.57**	40.00	11.77	18.18	41.67	29.41	34.48
DRCT	61	360	71	**91.89**	**95.78**	**93.79**	77.91	94.37	85.35	80.26	85.92	82.99
DRMR	139	326	144	93.92	**96.53**	**95.21**	**93.95**	86.81	90.25	75.29	90.97	82.39
DRPE	14	33	15	75.00	**40.00**	52.17	**100.00**	**40.00**	**57.14**	100.00	13.33	23.53
DRUS	26	171	129	96.06	**94.57**	**95.31**	98.89	68.99	81.28	98.59	54.26	70.00
DRXR	51	465	18	**62.50**	55.56	**58.82**	50.00	38.89	43.75	34.38	**61.11**	44.00
DSEC	10	135	8	0.00	0.00	0.00	0.00	0.00	0.00	0.00	0.00	0.00
DSEE	8	38	3	100.00	66.67	80.00	0.00	0.00	0.00	100.00	**100.00**	**100.00**
DSEM	5	24	6	0.00	0.00	0.00	0.00	0.00	0.00	0.00	0.00	0.00
DVDM	29	136	9	100.00	44.44	**61.54**	100.0	33.33	50.00	62.50	**55.56**	58.82
DVEN	16	100	8	72.73	**100.00**	**84.21**	100.0	62.50	76.92	100.00	12.50	22.22
DVOR	55	217	21	73.33	52.38	**61.11**	52.94	42.85	47.37	52.00	**61.90**	56.52
GCHE	61	142	14	100.00	**100.00**	**100.00**	100.0	**100.00**	**100.00**	92.86	92.86	92.86
GFIG	2954	3158	2085	**90.34**	99.14	**94.54**	87.70	99.52	93.24	88.80	**99.23**	93.73
GFLO	20	134	31	100.00	**32.26**	**48.78**	90.00	29.03	43.90	71.43	16.13	26.32
GGEL	344	429	224	**95.48**	**84.82**	**89.84**	94.27	80.80	87.02	95.03	76.79	84.94
GGEN	179	268	150	**86.91**	**48.67**	**62.39**	84.21	42.67	56.64	71.74	22.00	33.67
GHDR	136	236	49	**47.73**	**42.83**	**45.16**	35.71	40.82	38.10	29.09	32.65	30.77
GMAT	15	40	3	0.00	0.00	0.00	0.00	0.00	0.00	0.00	0.00	0.00
GNCP	88	221	20	56.25	**45.00**	50.00	**64.29**	**45.00**	**52.94**	36.84	35.00	35.90
GPLI	1	51	2	100.00	**100.00**	**100.00**	0.00	0.00	0.00	0.00	0.00	0.00
GSCR	33	144	6	100.00	16.67	**28.57**	100.00	16.67	**28.57**	50.00	16.67	25.00
GSYS	91	196	75	**55.10**	**36.00**	**43.55**	50.00	13.33	21.05	33.33	6.67	11.11
GTAB	79	173	13	**57.90**	**84.62**	**68.75**	50.00	76.92	60.61	50.00	46.15	48.00
=	6775	10137	4166	**80.62**	**62.00**	**72.90**	77.95	48.23	62.93	59.16	46.11	51.83

weeks, to perform medical classifications at a larger scale. The articles of the dataset are published under the "Creative Commons or similar license". Extracting all images from the archives, which were referenced in the .nxml file as a figure, resulted in 4,187,425 images in total.

An Inception-V1 network was trained using the *Compound Figure Detection* dataset from ImageCLEF 2016 [8]. In total 1,591,702 images where classified as non-compound images and therefore used for further training. An automatic process to separate compound figures into subfigures could be used. However, the workload to verify the correctness of this separation process would have been unmanageable.

All non-compound figures had been labeled in the following step by an ensemble of three neural networks trained on images from the subfigure classification dataset from ImageCLEF 2013 and 2016. Part of the ensemble was an unmodified AlexNet, an unmodified Inception-V1, as well as a modified version of the

Inception-V1 including Xavier initialization and PReLU as activation function, which also has been used as one individual run in [10].

Self-training refers to a semi-supervised domain adaption technique, which labels data for further training [2] purposes. In this example, the pre-trained models from Tensorflow Slim were fine-tuned from the ImageNet domain into the medical domain. The focus is to adjust the convolutional filters to the target domain, rather than the overall achieved classification accuracy. For accuracy optimizations a second training step was conducted using the domain-adapted network weights. The training set was reduced to the labeled images from Image-CLEF 2013 and 2016.

Due to computational limitations, the chosen split for self-training was limited to 321,491 images. No labeled images from ImageCLEF were used at this stage. The chosen optimization hyperparameters were identical to Sect. 5. In contrast to previous training runs, label-smoothing of 0.2 was applied as the self-labeled training data is highly noisy and therefore the network shall not train overconfident. For the Inception-ResNet-V2, $n = 5$ models were trained and achieved $87.87 \pm 0.02\%$ using center-crop and $87.87 \pm 0.01\%$ using 10-crops. Compared with the results in Table 1, the models have a lower standard deviation and seem to be more stable. Combining all five models results in a ensemble classification accuracy of 88.38% for center-crop and 88.45% for 10-crop, which is an improvement of 0.72/0.90% over the ensemble results in Table 1 for the Inception-ResNet-V2.

6   Conclusion

In this work, problems of the ImageCLEF 2016 subfigure classification dataset were investigated. Based on these findings an adjusted image preprocessing and resizing method was formulated for training convolutional neural networks. With an ensemble of five fine-tuned Inception-V4 models, an overall accuracy of 88.48% could be achieved. This is an overall improvement of +3.05% compared to the best visual run during ImageCLEF 2016. Furthermore, this purely visual approach is also slightly better compared to the best mixed feature run of BCSG at ImageCLEF 2016.

Investigations on using additional images from ImageCLEF 2013 for training set extension showed a statistically significant boost in terms of achieved per-class F1-Scores. Despite the fact that the approach from [11] has lower F1-Scores compared to the presented ensembles, the general idea to use features from neural networks should still be further investigated.

Finally, self-trained neural networks seemed to be more stable since the convolutional filters are better adapted to the target domain. Additional evaluations using more images from the PubMedCentral database could be conducted.

References

1. Abadi, M., Agarwal, A., Barham, P., Brevdo, E., Chen, Z., Citro, C., Corrado, G.S., Davis, A., Dean, J., Devin, M., Ghemawat, S., Goodfellow, I.J., Harp, A., Irving, G., Isard, M., Jia, Y., Józefowicz, R., Kaiser, L., Kudlur, M., Levenberg, J., Mané, D., Monga, R., Moore, S., Murray, D.G., Olah, C., Schuster, M., Shlens, J., Steiner, B., Sutskever, I., Talwar, K., Tucker, P.A., Vanhoucke, V., Vasudevan, V., Viégas, F.B., Vinyals, O., Warden, P., Wattenberg, M., Wicke, M., Yu, Y., Zheng, X.: TensorFlow: large-scale machine learning on heterogeneous distributed systems. CoRR abs/1603.04467 (2016). http://arxiv.org/abs/1603.04467
2. Csurka, G. (ed.): Domain Adaption in Computer Vision Applications. Advances in Computer Vision and Pattern Recognition, 1 edn. Springer International Publishing (2017). http://www.springer.com/de/book/9783319583464
3. Demner-Fushman, D., Antani, S., Simpson, M., Thoma, G.R.: Design and development of a multimodal biomedical information retrieval system. J. Comput. Sci. Eng. 6(2), 168–177 (2012)
4. Donahue, J., Jia, Y., Vinyals, O., Hoffman, J., Zhang, N., Tzeng, E., Darrell, T.: DeCAF: a deep convolutional activation feature for generic visual recognition. In: Jebara, T., Xing, E.P. (eds.) Proceedings of the 31st International Conference on Machine Learning (ICML 2014), pp. 647–655, JMLR Workshop and Conference Proceedings (2014)
5. He, K., Zhang, X., Ren, S., Sun, J.: Deep residual learning for image recognition. In: The IEEE Conference on Computer Vision and Pattern Recognition (CVPR), June 2016
6. García Seco de Herrera, A., Kalpathy-Cramer, J., Demner Fushman, D., Antani, S., Müller, H.: Overview of the ImageCLEF 2013 medical tasks. In: Working Notes of CLEF 2013 (Cross Language Evaluation Forum), CEUR Workshop Proceedings, vol. 1179, September 2013
7. García Seco de Herrera, A., Schaer, R., Antani, S., Müller, H.: Using crowdsourcing for multi-label biomedical compound figure annotation. In: Carneiro, G., et al. (eds.) LABELS/DLMIA -2016. LNCS, vol. 10008, pp. 228–237. Springer, Cham (2016). doi:10.1007/978-3-319-46976-8_24
8. García Seco de Herrera, A., Schaer, R., Bromuri, S., Müller, H.: Overview of the ImageCLEF 2016 medical task. In: CLEF2016 Working Notes, CEUR Workshop Proceedings, CEUR-WS.org, Évora, Portugal (2016). http://ceur-ws.org
9. Jia, Y., Shelhamer, E., Donahue, J., Karayev, S., Long, J., Girshick, R., Guadarrama, S., Darrell, T.: Caffe: convolutional architecture for fast feature embedding. In: Proceedings of the 22nd ACM International Conference on Multimedia (MM 2014), NY, USA, pp. 675–678. ACM, New York (2014)
10. Koitka, S., Friedrich, C.M.: Traditional feature engineering and deep learning approaches at medical classification task of ImageCLEF 2016. In: Working Notes of CLEF 2016 - Conference and Labs of the Evaluation forum, Évora, Portugal, 5–8 September, 2016. CEUR-WS Proceedings Notes, vol. 1609, pp. 304–317, July 2016
11. Kumar, A., Kim, J., Lyndon, D., Fulham, M., Feng, D.: An ensemble of fine-tuned convolutional neural networks for medical image classification. IEEE J. Biomed. Health Inform. 21(1), 31–40 (2017)
12. Lê Cao, K.A., Boitard, S., Besse, P.: Sparse PLS discriminant analysis: biologically relevant feature selection and graphical displays for multiclass problems. BMC Bioinform. 12(1), 253 (2011)

13. Pelka, O., Friedrich, C.M.: FHDO biomedical computer science group at medical classification task of imageclef 2015. In: Working Notes of CLEF 2015 - Conference and Labs of the Evaluation forum, Toulouse, France, 8–11 September 2015. CEUR-WS Proceedings Notes, vol. 1391 (2015)

14. Pelka, O., Friedrich, C.M.: Modality prediction of biomedical literature images using multimodal feature representation. GMS Med. Inform. Biometry Epidemiol. (MIBE) **12**(1), Doc4 (2016)

15. Personnaz, L., Guyon, I., Dreyfus, G.: Collective computational properties of neural networks: new learning mechanisms. Phys. Rev. A (General Physics) **34**(5), 4217–4228 (1986)

16. Ried, K., Frank, O.R., Stocks, N.P., Fakler, P., Sullivan, T.: Effect of garlic on blood pressure: a systematic review and meta-analysis. BMC Cardiovasc. Disorders **8**(1), 13 (2008)

17. Russakovsky, O., Deng, J., Su, H., Krause, J., Satheesh, S., Ma, S., Huang, Z., Karpathy, A., Khosla, A., Bernstein, M., Berg, A.C., Fei-Fei, L.: ImageNet large scale visual recognition challenge. Int. J. Comput. Vis. **115**(3), 211–252 (2015)

18. Szegedy, C., Ioffe, S., Vanhoucke, V., Alemi, A.A.: Inception-v4, Inception-ResNet and the impact of residual connections on learning. In: ICLR 2016 Workshop (2016)

19. Szegedy, C., Liu, W., Jia, Y., Sermanet, P., Reed, S., Anguelov, D., Erhan, D., Vanhoucke, V., Rabinovich, A.: Going deeper with convolutions. In: The IEEE Conference on Computer Vision and Pattern Recognition (CVPR), June 2015

20. Valavanis, L., Stathopoulos, S., Kalamboukis, T.: IPL at CLEF 2016 medical task. In: Working Notes of CLEF 2016 - Conference and Labs of the Evaluation forum, Èvora, Portugal, 5–8 September 2016. CEUR-WS Proceedings Notes, vol. 1609, pp. 413–420 (2016). http://ceur-ws.org/Vol-1609/

21. Yosinski, J., Clune, J., Bengio, Y., Lipson, H.: How transferable are features in deep neural networks? In: Ghahramani, Z., Welling, M., Cortes, C., Lawrence, N.D., Weinberger, K.Q. (eds.) Advances in Neural Information Processing Systems 27. pp. 3320–3328. Curran Associates, Inc. (2014)

IRIT-QFR: IRIT Query Feature Resource

Serge Molina[1], Josiane Mothe[1(✉)], Dorian Roques[1], Ludovic Tanguy[2],
and Md Zia Ullah[1]

[1] IRIT, UMR 5505, CNRS & Univ. Toulouse, Toulouse, France
`Josiane.Mothe@irit.fr`
[2] CLLE-ERSS, UMR 5263, CNRS & Univ. Toulouse, Toulouse, France

Abstract. In this paper, we present a resource that consists of query features associated with TREC adhoc collections. We developed two types of query features: linguistics features that can be calculated from the query itself, prior to any search although some are collection-dependent and post-retrieval features that imply the query has been evaluated over the target collection. This paper presents the two types of features that we have estimated as well as their variants, and the resource produced. The total number of features with their variants that we have estimated is 258 where the number of pre-retrieval and post-retrieval features are 81 and 171, respectively. We also present the first analysis of this data that shows that some features are more relevant than others in IR applications. Finally, we present a few applications in which these resources could be used although the idea of making them available is to foster new usages for IR.

Keywords: Information systems · Information Retrieval · Query features · IR resource · Query feature analysis

1 Introduction

Query features are features that can be associated with any query. They have been used in information retrieval (IR) literature for (1) query difficulty prediction and (2) selective query expansion; however, they can be useful for other applications.

In this paper, we present a resource that we have developed, which associates features to queries considering several TREC collections and which considers many different approaches: linguistic versus statistic-based, pre- and post-retrieval, and collection-dependent and -independent. This resource is to be made available to the IR community.

In the literature of query difficulty prediction, query features are categorized into two groups, according to the fact that the feature can be calculated prior any search (pre-retrieval feature) or not (post-retrieval feature) [2]. An example of a pre-retrieval feature is IDF_Max which is calculated as the maximum of the IDF term weight (as computed when indexing the document collection) over the query terms. High IDF means the term is not very frequent, thus high IDF_Max

© Springer International Publishing AG 2017
G.J.F. Jones et al. (Eds.): CLEF 2017, LNCS 10456, pp. 69–81, 2017.
DOI: 10.1007/978-3-319-65813-1_6

for a query means that this query contains at least one non-frequent term. On the other hand, an example of a post-retrieval feature is NQC (Normalized query commitment), which is based on the standard deviation of the retrieved document scores [13]. A high standard deviation means that the retrieved documents obtained very different scores meaning the retrieved document set is not homogeneous.

As for pre-retrieval features, we can also make a distinction between features that can be calculated independently to any document collection and the ones that need the document collection in some way (obviously, post-retrieval features are collection dependent since they are calculated over a retrieved document set). Going back to IDF_Max, it is obviously dependent on the document collection. On the other hand, SynSet (the average number of senses per query term as extracted from WordNet) [10] is collection-independent, since it only requires access to the query terms in order to be calculated.

In our work, we extract both pre- and post-retrieval features. We also distinguish between collection dependent and collection independent features. The details of the feature definitions can be found in Sect. 2 as well as the collections on which the features are already available.

In Sect. 3, we provide the first analysis of the data. In Sect. 4, we introduce a few applications that make use of such features. Finally, Sect. 5 concludes this paper.

2   Query Features

In this section, we describe the query features that we have estimated as well as their variants, including pre- and post-retrieval features.

2.1   Document Collection Independent Pre-retrieval Features: WordNet-Based and Other Linguistic Features

WordNet-Based Features (Pre-retrieval). WordNet-based features are pre-retrieval and document collection independent.

WordNet is a linguistic resource that interlinks senses of words (represented as sets of synonyms, or synsets) and labels the semantic relations between word senses [9]. The original (Princeton) version contains more than $117,000$ synsets and more than $150,000$ unique entries (source: https://wordnet.princeton.edu/ on the 5th of March 2017).

Figure 1 presents an extract of WordNet for the term "tiger." WordNet distinguishes different relationships between terms as follows:

1. Synonyms: words that denote the same concept and are interchangeable in many contexts. Synonyms are terms that belong to the same Synset, such as "tiger" and "panthera tigris."
2. Hyponyms/Hypernyms: these relationships link more generic synsets to specific ones. While "Panthera tigris" is a *hypernym* of "Bengal tiger"; the latter is a *hyponym* of the former.

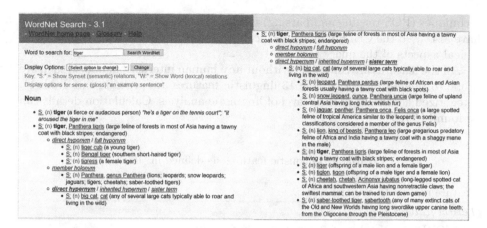

Fig. 1. Extract of WordNet for the term **tiger**. The right-side part of the figure provides the sister terms from the term tiger while the left-side part provides the two senses of the word and details for the second sense as well as direct hyponyms, holonym, and direct hypernym.

3. Meronym/Holonyms: correspond to the part-whole relation. Y is a *meronym* of X means Y is a part of X; in that case, X is a *holonym* of Y.
4. Sister-terms: *Sister-terms* are terms that share the same hypernym.

   We use this resource as follows: for each query term, we count the associated terms of each type (e.g. number of sister terms for each query term) and then aggregate the obtained values for a given relationship over query terms for a given query. Since a query may contain several query terms, given a query and a relationship type (e.g. synonym or sister-term), we calculate the following aggregations to get a single value of the feature variant for each query:

- minimum, maximum, mean, and total: the minimum (maximum, average, and total) number of the terms associated with the query terms when using the relationship, over the query terms;
- Q1, median, Q3: for each query term, we calculate the number of associated terms using the relationship. These numbers are first sorted in increasing order; then the set is divided into quartiles. Q1 (median, Q3) is the value that makes at least one quarter (2 quarters, 3 quarters) of the numbers having a score lower than Q1 (median, Q3).
- Standard deviation (std) and variance (or std^2): standard deviation refers to the square root of the mean of the squared deviations of the number of terms associated with the query terms from their mean and variance is the squared value of the standard deviation.

Other Linguistic Features (Pre-retrieval). We also consider the linguistic features as defined by Mothe and Tanguy [10]. These are also pre-retrieval features and collection independent. The queries have been analysed using generic techniques (POS tagging and parsing) from the Stanford CoreNLP suite [8]. We have also used the CELEX morphological database [1] for assessing the morphological aspects of the query terms.

All the features are computed without any human interaction, and as such are prone to processing errors. The 13 linguistic features are presented in Table 1[1], categorized according to their level of linguistic analysis. Calculation details can be found in [10].

Table 1. Linguistic features as defined in [10].

Feature name	Description
Lexical features	
NBWORDS	Number of words (terms) in the query
LENGTH	Word length in number of characters
MORPH	Average number of morphemes per word (according to CELEX)
SUFFIX	Number of suffixed words (based on suffixes extracted from CELEX)
PN	Number of proper nouns (according to CoreNLP's POS tagger)
ACRO	Number of acronyms
NUM	Number of numeral values (dates, quantities, etc.)
UNKNOWN	Number of unknown tokens (based on WordNet)
Syntactical features	
CONJ	Number of conjunctions (according to CoreNLP's POS tagger)
PREP	Number of prepositions (idem)
PP	Number of personal pronouns (idem)
SYNTDEPTH	Syntactic depth (maximum depth of the syntactic tree, according to CoreNLP parser)
SYNTDIST	Syntactic links span (average distance between words linked by a dependency syntactic relation)

2.2   Document Collection Dependent Pre-retrieval Features

Finally, as pre-retrieval features, we also consider IDF (Inverse document frequency), which is extracted from the document indexing file (we use LEMUR index for that since it gives a direct access to it). Moreover, IDF statistics across IR tools are consistently used in previous research [5]. As opposed to the other pre-retrieval features presented upper, IDF is collection dependent. We calculate the same 9 variants as previously: minimum, maximum, mean, total (sum), Q1, median, Q3, standard deviation, and variance of term-IDF score over the query terms.

---

[1] [10] paper presents the SynSet feature which is one of the features based on WordNet and thus included in Sect. 2.1.

2.3   Post-retrieval Features

Post-retrieval features are by definition collection dependent. These features are extracted either from the query-document pairs or retrieved documents.

Letor-Based Post-retrieval Features. Letor features have been used in learning to rank applications [12]. In Letor, these features are associated with query-document pairs. For example, *BM25.0* corresponds to the score as obtained using BM25 model for a given query; it is thus attached to a query-document pair. We use Terrier platform[2] to calculate the Letor features. The Terrier platform has implemented the Fat component, which allows to compute many features in a single run [7]. More details on the Letor features can be found on Letor collection description[3]. One of them is PageRank which can be calculated for linked documents only (for this reason this feature cannot be calculated for the TREC Robust collection).

In the feature names (Table 2), SFM stands for SingleFieldModel and means that the value corresponds to score, which a document obtained using the mentioned search model (LM stands for Language Model, DIR for Dirichlet smoothing and JM for Jelinek-Mercer smoothing). A .0 means that the calculations have been made on the document's title only; while .1 means they have used the entire document content. The features that do not contain SFM are measures calculated from the occurrences of query terms in the retrieved documents (e.g. mean_tf is the mean of TF (term frequency) of query terms in the considered document).

To make the Letor features usable as query features, we have aggregated them over the retrieved documents for a given query. For example, we calculate the mean of the BM25 scores over the retrieved document list for the considered query. We have used the same 9 aggregation functions as presented in Sect. 2.1 (minimum, maximum, mean, total (*Nbdoc*), Q1, median, Q3, standard deviation, and variance). Nbdoc is not an aggregation value since it corresponds to the number of documents retrieved for the given query given the retrieval model used.

PageRank Features. We also calculate two PageRank features: PageRank_prior and PageRank_rank, when dealing with linked documents, that means, for WT10G and GOV2 collections. We use Lemur implementation of the PageRank feature. To generate the variants of the PageRank features, we have used the same 9 aggregation functions as previously over the retrieved documents for a query.

---

[2] http://terrier.org/docs/v4.0/learning.html.
[3] https://www.microsoft.com/en-us/research/project/mslr/.

Table 2. Post-retrieval features as defined for Letor in [7,12]

Feature name	Description
Calculated using Terrier module	
WMODEL.SFM.Tf.0 and .1	The value of the TF score for the query and the document title/body
WMODEL.SFM.TF_IDF.0 and .1	The value of the TF*IDF score for the query and the document title/body
WMODEL.SFM.BM25.0 and .1	The value of BM25 score for the query and the document title/body
MODEL.SFM.DirLM.0 and .1	The score value for the language model with Dirichlet smoothing for the query and the document title/body
QI.SFM.Dl.0 and .1	Number of terms in the document title/body
Dirichlet.Mu1000	The score value for the language model with Dirichlet smoothing with the smoothing parameter = 1000, for the whole document
JM.col.λ.0.4doc.λ0.0	The score value for the language model with Jelinek-Mercer smoothing, with a collection lambda of 0.4
Calculated using Lemur	
sum_tf_idf_full	The sum of TF*IDF values for the query terms
mean_tf_idf_full	The mean of TF*IDF values for the query terms
sum_tf_full	The sum of TF values for the query terms
mean_tf_full	The mean of TF values for the query terms
pagerank_rank	The rank of document based on PageRank scores
pagerank_prior	The log probability of PageRank scores

2.4 Collections for Which the Features Have Been Estimated

In total, we calculated 258 individual features. So far, these features have been calculated on three TREC data collections from the adhoc task: Robust, WT10G, and GOV2.

For Robust collection[4], TREC competition provided approximately 2 GB of newspaper articles including the Financial Times, the Federal Register, the Foreign Broadcast Information Service, and the LA Times [14]. The TREC WT10G collection is composed of approximately 10 GB of Web/Blog page documents [6]. The GOV2 collection includes 25 million web pages, which is a crawl of .gov domain [3].

The three test collections consist also of topics that comprise a topic title which we use as the query. There are 250 topics in the Robust collection, 100 topics in the WT10G collection, and 150 topics in the GOV2 collection.

---

[4] http://trec.nist.gov/data/robust.html.

Now that we have either implemented new code or gathered codes to estimate the query feature values, there is no limit to calculate the features for other collections; that we plan to do in the next months. ClueWeb09B and Clueweb12B are the short term targets. Moreover, we will continue to gather new query features.

We make available the feature resource to foster new usages for IR, the resource is available to download at http:/doi.org/10.5281/zenodo.815319 (proper user agreements). If you use this resource in your research, it is required to cite the following paper:

S. Molina, J. Mothe, D. Roques, L. Tanguy, and M. Z. Ullah. IRIT QFR: IRIT Query Feature Resource. In Experimental IR Meets Multilinguality, Multimodality, and Interaction 8th International Conference of the CLEF Association, CLEF2017, Dublin, Ireland, September 11–14, 2017, Proceedings, volume 10439, 2017.

3 Analysis of the Resource

In this section, we provide some elements of the descriptive analysis of the resources we have built and presented in the previous sections.

3.1 Descriptive Analysis

In order to have an idea of the trends of feature variants, in Fig. 2, we show the boxplots associated to 4 query features and their variants. In a given boxplot, each query makes a contribution. For the 2 linguistic features, we did not plot the variance since the high values would have flattened the others. For the 2 post-retrieval features, we did not plot the number of retrieved documents for the same reason.

Since we can assume that the queries are diverse in many senses in the TREC collections (e.g. in terms of difficulty, in terms of specificity, ...), one interesting insight can be to know how much the different features and variants vary according to the queries.

On Fig. 2, we can see that the Synonyms and Sister terms features (which are calculated on the query only) have the same trends when considering their different variants.

When considering the BM25.0 (calculated on document titles only), we can see on Fig. 2 that most of the queries got a null value for the min, Q1, median, and Q3 variants. The feature variant that varies the most is the variance and in a little smaller extend the max variant. The null value for the min, Q1, and median variants holds for all the features calculated on the title but one that got negative value which is the Jelinek.Mercer.collectionLambda0.4.documentLambda0.0 feature (see Fig. 3).

When considering the BM25.1 (calculated on the entire documents), we can see on Fig. 2 that the values are higher than when calculated on the title only (which is indeed an expected result), and that the null phenomenon does not

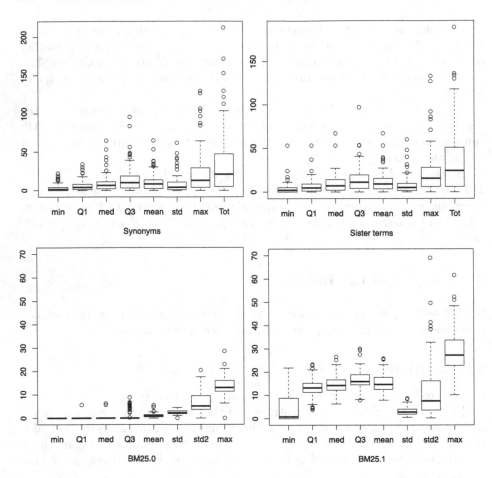

Fig. 2. Boxplots of the Synonyms, Sister_Termes, BM25.0, and BM25.1 features and their variants - WT10G Collection.

hold on. Still, the variance is the variant that varies the most, however, max and min values are also on a quite large scale.

Figure 3 presents the median variant for several Letor features. On the left side part of the figure, which represents the values when the title of the documents is considered, we can see that the values are not at the same scale and that the Jelinek-Mercer smoothing is (i) the one that varies the most and (ii) the only one that is negative. On the right side of the figure, we removed the Jelinek-Mercer values since they would have hidden the other values variation. We can see as for BM25 in Fig. 3 that the values vary more when considering the entire document than when considering the title only.

The NbDoc (Number of documents retrieved) variant is also somehow interesting: while it is often equal to its maximum value 1,000 (this value comes from

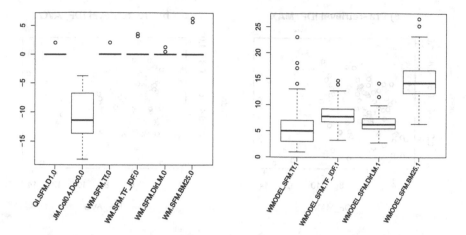

Fig. 3. Boxplots of the median variant for various features when calculated on title only (0) and on the entire document (1) - WT10G Collection. For the variants calculated on the title only, let us mention that the null values hold for min and Q1 as well.

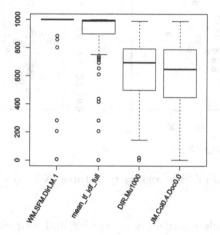

Fig. 4. Boxplots of the NbDoc (number of retrieved documents) variant for a few features from Letor features - WT10G Collection.

the way we configured Terrier when calculating the features), for a few queries, its value is lower. Figure 4 displays the values for a few models.

In the various previous figures, we display the results for WT10G; but the same type of conclusions can be made using Robust and GOV2.

4 Applications

One possible application as mentioned previously in this paper is query difficulty prediction. Figure 5 displays the plots of NDCG (Y-axis) as calculated from a

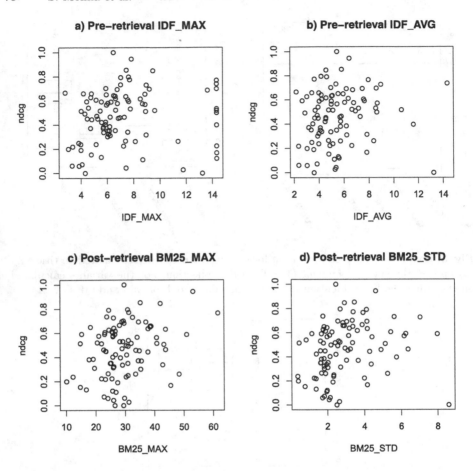

Fig. 5. Plots of NDCG (Y-axis) and query features (X-axis) - WT10G Collection.

BM25 run using default parameters in Terrier and four of the query features (X-axis) using WT10G collection.

Alternatively, Pearson correlation can be calculated in order to measure the link between a single feature and actual system effectiveness. More concretely, we investigate the combination of query predictors in order to enhance prediction. As for query predictors, we consider the features presented in this paper.

We developed another application in [4], where query features are used in a machine learning model based on learning to rank principle in order to learn which system configuration among a variety of configurations should be used to best treat a given query. The candidate space is formed of tens of thousands of possible system configurations, each of which sets a specific value for each of the system parameters. The learning to rank model is trained to rank them with respect to an IR performance measure (such as nDCG@1), thus emphasizing the importance of ranking "good" system configurations higher in the ranked list.

Moreover, the approach makes a query-dependent choice of system configuration, i.e. different search strategies could be selected for different types of a query; based on query features. In that study, a subset of the features we present in this paper have been used (linguistic features from [10] and IDF variants). The paper shows that this approach is feasible and significantly outperforms the grid search way to configure a system, as well as the top performing systems of the TREC tracks.

In current work, we are developing a new method that aims like in [4], at optimizing the system configuration on a per-query basis [11]. Our method learns the configuration models in a training phase and then explores the system feature space and decides what should be the system configuration for any new query. The experiments on TREC 7 & 8 topics from adhoc task show that the method is very reliable with good accuracy to predict a system configuration for an unseen query. We considered about 80,000 different system configurations.

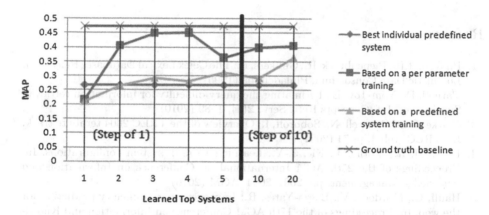

Fig. 6. MAP - A comparison between the best predefined system, the PPL method, the predefined system classifier, and the ground truth method. The predictive functions are trained on top 1, 2, 3, 4, 5, 10, and 20 systems per query (X-axis). (Color figure online)

Figure 6 shows a comparison between the method we proposed in [11] (red square line) and the weak baseline (the configuration that provides the best results in average over the queries) in one hand and the ground truth on the other hand (when the best system is used for each query), represented by the blue and purple straight lines respectively. We also compare the results with a fair baseline that corresponds to a method that learns on a limited predefined set of systems only (green triangles). Figure 6 reports the MAP over the set of unseen test queries (averaged over the 10 draws resulting from 10-folds cross-validation) using the predicted configuration.

5 Conclusion

This paper presents a new resource that associates many features to queries from TREC collections. We distinguish between pre- and post-retrieval features as well as between collection-dependent and collection-independent features. Some features have linguistic basis while others are based on statistics only. We use features from the literature but also new features that are generated from both WordNet linguistic resource and Letor learning to rank document-query pairs features.

We have already used some of these features in applications related to system configuration selection, query difficulty prediction, and selective query expansion, but we think these resources could also be used for other applications.

In our future work, we aim at developing this resource for other collections. We are targeting ClueWeb09B and ClueWeb12B, but also other collections such as TREC Microblog-based collections.

References

1. Baayen, R.H., Piepenbrock, R., Gulikers, L.: The Celex Lexical Database (Release 2). Linguistic Data Consortium, Philadelphia (1995)
2. Carmel, D., Yom-Tov, E.: Estimating the query difficulty for information retrieval. Synth. Lect. Inf. Concepts Retr. Serv. **2**(1), 1–89 (2010)
3. Clarke, C.L., Craswell, N., Soboroff, I.: Overview of the TREC 2004 terabyte track. In: TREC, vol. 4, p. 74 (2004)
4. Deveaud, R., Mothe, J., Nie, J.-Y.: Learning to rank system configurations. In: Proceedings of the 25th ACM International on Conference on Information and Knowledge Management, pp. 2001–2004. ACM (2016)
5. Hauff, C., Murdock, V., Baeza-Yates, R.: Improved query difficulty prediction for the web. In: Proceedings of the 17th ACM Conference on Information and Knowledge Management, pp. 439–448. ACM (2008)
6. Hawking, D.: Overview of the TREC-9 web track. In: TREC (2000)
7. Macdonald, C., Santos, R.L., Ounis, I., He, B.: About learning models with multiple query-dependent features. ACM Trans. Inf. Syst. (TOIS) **31**(3), 11 (2013)
8. Manning, C.D., Surdeanu, M., Bauer, J., Finkel, J., Bethard, S.J., McClosky, D.: The stanford CoreNLP natural language processing toolkit. In: Association for Computational Linguistics (ACL) System Demonstrations, pp. 55–60 (2014)
9. Miller, G.A.: Wordnet: a lexical database for English. Commun. ACM **38**(11), 39–41 (1995)
10. Mothe, J., Tanguy, L.: Linguistic features to predict query difficulty. In: ACM Conference on Research and Development in Information Retrieval, SIGIR, Predicting Query Difficulty-Methods and Applications Workshop, pp. 7–10 (2005)
11. Mothe, J., Washha, M.: Predicting the best system parameter configuration: the (per parameter learning) PPL method. In: 21st International Conference on Knowledge-Based and Intelligent Information & Engineering Systems (2017)
12. Qin, T., Liu, T.-Y., Xu, J., Li, H.: Letor: a benchmark collection for research on learning to rank for information retrieval. Inf. Retr. **13**(4), 346–374 (2010)

13. Shtok, A., Kurland, O., Carmel, D.: Predicting query performance by query-drift estimation. In: Azzopardi, L., Kazai, G., Robertson, S., Rüger, S., Shokouhi, M., Song, D., Yilmaz, E. (eds.) ICTIR 2009. LNCS, vol. 5766, pp. 305–312. Springer, Heidelberg (2009). doi:10.1007/978-3-642-04417-5_30

14. Voorhees, E.M.: The TREC robust retrieval track. In: ACM SIGIR Forum, vol. 39, pp. 11–20. ACM (2005)

Evaluating and Improving the Extraction of Mathematical Identifier Definitions

Moritz Schubotz[✉], Leonard Krämer, Norman Meuschke, Felix Hamborg, and Bela Gipp

University of Konstanz, Konstanz, Germany
{moritz.schubotz,leonard.kramer,norman.meuschke,
felix.hamborg,bela.gipp}@uni.kn

Abstract. Mathematical formulae in academic texts significantly contribute to the overall semantic content of such texts, especially in the fields of Science, Technology, Engineering and Mathematics. Knowing the definitions of the identifiers in mathematical formulae is essential to understand the semantics of the formulae. Similar to the sense-making process of human readers, mathematical information retrieval systems can analyze the text that surrounds formulae to extract the definitions of identifiers occurring in the formulae. Several approaches for extracting the definitions of mathematical identifiers from documents have been proposed in recent years. So far, these approaches have been evaluated using different collections and gold standard datasets, which prevented comparative performance assessments. To facilitate future research on the task of identifier definition extraction, we make three contributions. First, we provide an automated evaluation framework, which uses the dataset and gold standard of the NTCIR-11 Math Retrieval Wikipedia task. Second, we compare existing identifier extraction approaches using the developed evaluation framework. Third, we present a new identifier extraction approach that uses machine learning to combine the well-performing features of previous approaches. The new approach increases the precision of extracting identifier definitions from 17.85% to 48.60%, and increases the recall from 22.58% to 28.06%. The evaluation framework, the dataset and our source code are openly available at: https://ident.formulasearchengine.com.

1 Introduction

Mathematical formulae consist of *identifiers* (e.g. x, π or σ), *symbols* (e.g. $+$, $\leq$ or $\rightarrow$) and *other constituents*, such as numbers. Formally, any definition consists of three components:

1. the *definiendum*, which is the expression to be defined
2. the *definiens*, which is the phrase that defines the definiendum
3. the *definitor*, which is is the verb that links definiendum and definiens

The task in mathematical identifier definition extraction is to find definitions whose definiendum is a mathematical identifier and extract the definiens.

© Springer International Publishing AG 2017
G.J.F. Jones et al. (Eds.): CLEF 2017, LNCS 10456, pp. 82–94, 2017.
DOI: 10.1007/978-3-319-65813-1_7

As an example, we use an explanation of the Planck-Einstein relation (https://en.wikipedia.org/wiki/Planck-Einstein_relation) in Wikipedia (see Fig. 1). The Planck-Einstein relation $E = hf$ consists of three identifiers: E, h, f and two symbols: '$=$', '$\cdot$' (times). The text surrounding the formula contains the definiens for photon energy and wave frequency. In this case, the definiens photon energy is particularly specific, since it contains an intra-wiki link to a unique concept identified by the Wikidata item Q25303639. However, the explanatory text does not give a definition for the Planck constant h. A reader or an information system must infer this missing information from elsewhere, which poses a challenge to both a reader and a system.

Identifier-definiens pairs contain semantic information that can improve mathematical information retrieval (MIR) tasks, such as formula search and recommendation, document enrichment, and author support. To increase the accessibility of this valuable semantic information, we address the automated extraction of mathematical identifiers and their definiens from documents as follows. In Sect. 2, we review existing approaches for mathematical identifier definition extraction. In Sect. 3, we describe the development of an automated evaluation framework that allows for objective and comparable performance evaluations of extraction approaches. Furthermore, we describe how we used machine learning to create a new extraction approach by combining the well-performing features of existing approaches. In Sect. 4, we present the results of evaluating the extraction performance of existing and our newly developed extraction approach. In Sect. 5, we summarize the findings of our comparative performance evaluations and present suggestions for future research.

2 Related Work

This Section briefly reviews the following approaches to mathematical identifier definition extraction: (1) the statistical feature analysis of Schubotz et al. [13] (Sect. 2.1), (2) the pattern matching approach of Pagel et al. [9] (Sect. 2.2), (3) the machine learning approach of Kristianto et al. [7] (Sect. 2.3). See [10] for an extensive review of related work.

2.1 Statistical Feature Analysis (ST)

Schubotz et al. [13] proposed a *mathematical language processing (MLP)* pipeline and demonstrated the application of the pipeline for extracting the definiens of identifiers in formulae contained in Wikipedia. In summary, the MLP pipeline includes the following steps:

1. **Preprocessing:** Parse wikitext input format, perform tokenization, part-of-speech (POS) tagging and dependency parsing using an adapted version of the Stanford CoreNLP library. The modified library can handle mathematical identifiers and formulae by emitting special tokens for identifiers, definiens candidates and formulae.

> The Planck-Einstein relation connects the particular photonenergy E with its associated wave frequency f
>
> $$E = hf. \tag{1}$$
>
> | identifier | definiens |
> |---|---|
> | E | Q25303639 (Photon energy) |
> | h | NIL |
> | f | wave frequency |

Fig. 1. Excerpt of a sentence from Wikipedia explaining the Planck-Einstein relation (https://en.wikipedia.org/w/index.php?title=Planck_constant\&oldid=766777932) (top) and the extractable identifiers and corresponding definiens (bottom).

2. **Find identifiers** in the text.
3. **Find candidates for identifier-definiens pairs.**
4. **Score identifier-definiens pairs** using statistical methods.
5. **Identify and extract namespaces:** Cluster documents, map the clusters to document classification schemata and determine identifier definitions specific to each identified class in the schema, i.e. specific to a namespace (NS).

To score identifier-definiens pairs, Schubotz et al. used the scoring function shown in Eq. 2. The function considers the number of words Δ between the identifier and the definiens, the number of sentences n between the first occurrence of the identifier and the sentence that connects the identifier to the definiens, and the relative term frequency of the definiens t in document d.

$$R(\Delta, n, t, d) = \frac{\alpha R_{\sigma_\alpha}(\Delta) + \beta R_{\sigma_\beta}(n) + \gamma \mathrm{tf}(t, d)}{\alpha + \beta + \gamma} \mapsto [0, 1], \tag{2}$$

The parameters α, β and γ are used to weigh the influence of the three factors by making the following assumptions:

α The definiens and the identifier appear close to each other in a sentence.
β The definiens appears close to the first occurrence of the identifier in the text.
γ The definiens is used frequently in the document.

To derive α and β, Schubotz et al. used the zero-mean Gaussian normalization function $R_\sigma(\Delta) = \exp\left(-\frac{1}{2}\frac{\Delta^2 - 1}{\sigma^2}\right)$ to map the infinite interval of the distances Δ and n to $[0, 1]$. The parameters σ_α and σ_β control the width of the function.

Schubotz et al. report a precision of $p = .207$ and a recall of $r = .284$ for extracting identifier definitions [13]. The weighting parameters, the Gaussians and the threshold for the overall score must be manually adjusted to the specific use case, which can be a tedious process. The major advantages of the statistical approach are its language-independence and adjustability to different document collections.

2.2 Pattern Matching (PM)

Pagel et al. employed a pattern matching approach for POS tag patterns to extract identifier-definiens pairs from Wikipedia articles [9]. The lines 1–10 in Table 1 show the patterns, which were defined by domain experts. Patterns like <identifier> denote(s?) the <definiens>, which would match the definition 'h denotes the Planck constant' have a high probability of retrieving a true positive (tp) result. However, simpler patterns, such as <definiens> <identifier> have a high probability of producing false positive (fp). For instance, this pattern would match the apposition 'photon energy E', but also the phrase 'subsection a' in the description of a law.

Pagel et al. reported a precision of $p = .911$, and recall of $r = .733$ for their approach [9]. While the recall of such a pattern matching approach can easily be increased by adding additional patterns, the precision declines if the added patterns are too broad. Therefore, Pagel et al. concluded that using more than the patterns 1–10 in Table 1 does not significantly increase the performance [9].

The sentence patterns 3–10 in Table 1 achieved a high precision in the evaluation of Pagel et al. We consider these patterns promising candidates for inclusion in a hybrid approach that uses machine learning to combine the pattern matching approach of Pagel et al. and the statistical feature analysis of Schubotz et al. However, before applying the sentence patterns for extracting identifier definitions from a different corpus, the suitability of the patterns must be re-evaluated, since different text genres, e.g., encyclopedic article vs. scientific publication, may use different notational conventions. Furthermore, the pattern matching approach is language-dependent.

2.3 Machine Learning

Kristianto et al. proposed a machine learning approach to extract natural language descriptions for entire formulae from academic documents [7]. This extraction task is slightly different from extracting identifier-definiens pairs. Kristianto et al. associate each mathematical expression with a span of words that describes the expression. For example, for the sentence: "..the number of permutations of length n with exactly one occurrence of 2–31 is $\left(\frac{2n}{n-3}\right)$..", the gold standard of Kristianto et al. states that the correct description of "$\left(\frac{2n}{n-3}\right)$" is "the number of permutations of length n with exactly one occurrence of 2–31".

In contrast, the sentence contains only one identifier n, whose definiens is 'length'. Kristianto et al. used the native Standford CoreNLP library for their analysis, whereas Schubotz et al. modified the CoreNLP library to create their MLP pipeline (cf. Sect. 2.1). To identify formulae descriptions, Kristianto et al. defined a large set of features, which they classified into three groups:

1. *pattern matching*: features similar to those of Pagel et al. (see Sect. 2.2);
2. *basic*: features that consider the POS tags between pairs of identifier and definiens, as well as the POS tags in their immediate vicinity;
3. *dependency graph (DG)*: features related to the DG of a sentence.

Kristianto et al. used a support vector machine (SVM) [2] for a combined analysis of all features. Except for the features in the *pattern matching* group, their approach is applicable to documents in all languages supported by the Stanford CoreNLP library [8]. A significant drawback of the approach is the necessity to manually annotate a portion of the dataset to train the SVM classifier.

3   Methodology

The approaches we present in Sect. 2 perform different extraction tasks (extracting identifier definitions [9,13] vs. extracting formulae descriptions [7]) using different datasets (Wikipedia articles [9,13] vs. scientific publications [7]), which so far prevented a comparison of the reported precision an recall values.

To enable comparative performance evaluations for these and other extraction approaches, we created an open evaluation framework by extending the open source MLP and MIR framework Mathosphere introduced in [13]. Section 3.1 presents the evaluation framework and explains major improvements we made to Mathosphere's MLP pipeline. Section 3.2 describes how we used the developed framework to individually evaluate the three approaches we present in Sect. 2. Section 3.3 explains how we adapted and evaluated the approach of Kristianto et al. [7] for the task of extracting identifier definitions. Section 3.4 presents how we investigated the effect of considering Namespaces (NS) as part of identifier definition extraction [13].

3.1   Evaluation Framework

Our framework uses a subset of the dataset of the NTCIR-11 Math Retrieval Wikipedia task [12] and the gold standard created by Schubotz et al. for evaluating their statistical feature analysis approach (cf. Sect. 2.1) [13]. The dataset contains 100 formulae taken from 100 unique Wikipedia articles and contains 310 identifiers [12]. Every formula in the gold standard contains: (1) a unique query-id (qID); (2) the title of the document; (3) the id of the formula within the document (fid), which corresponds to the sequential position of the formula in the document; (4) the latex representation of the formula (math_inputtex).

The gold standard includes definiens for every identifier in a formula. In total, the gold standard includes 369 definiens for the 310 identifiers, or 575 definiens when counting wikidata links and link texts separately. However, distinguishing Wikidata links and link texts for the evaluation has a drawback. For example, the identifier c in the formula $f_c(z) = z^2 + c$ from the article on orbit portraits (https://en.wikipedia.org/w/index.php?title=Orbit_portrait\&oldid=729107245) is associated with two Wikidata concepts: parameter Q1413083 and coefficient Q50700. While this ambiguity can be interpreted as a shortcoming of insufficient concept specificity and definiteness of the Wikidata items as discussed by Corneli and Schubotz [3], we argue that current IR systems should be able to deal with such indefiniteness. In [13], each correctly extracted definition was regarded as a true positive. This can result in more

Table 1. All features used in the SVM to classify identifier-definiens pairs with rank, where possible. Feature groups 1–10: PM, 11–21: basic, 22–26: DG, 27–29: ST. The ranking was performed by comparing the merit of training with only one feature in isolation to training all features

#	Description	Merit	Rank
1	`<definiens> <identifier>` [9]	0.196	5
2	`<identifier> <definiens>` [9]	<.001	27
3	`<identifier> denote(s?) <definiens>` [9]	0.001	20
4	`<identifier> denote(s?) the <definiens>` [9]	0.001	19
5	`<identifier> (is|are) <definiens>` [9]	0.001	21
6	`<identifier> (is|are) the <definiens>` [9]	0.059	13
7	`<identifier> (is|are) denoted by <definiens>` [9]	<0.001	24
8	`<identifier> (is|are) denoted by the <definiens>` [9]	<0.001	25
9	`let <identifier> be denoted by <definiens>` [9]	<0.001	22
10	`let <identifier> be denoted by the <definiens>` [9]	<0.001	23
11	Colon between identifier and definiens [7]	0.037	15
12	Comma between identifier and definiens [7]	0.121	7
13	Other math expression or identifier between identifier and definiens [7]	0.122	6
14	Definiens is inside parentheses and identifier is outside parentheses [7]	0.016	16
15	Identifier is inside parentheses and definiens is outside parentheses [7]	0.060	12
16	Identifier appears before definiens [7]	0.015	17
17	Surface text and POS tag of two preceding and following tokens around the definiens candidate [7]	0.441	1
18	Unigram, bigram and trigram of feature 17 [7]	0.441	1
19	Surface text and POS tag of three preceding and following tokens around the identifier [7]	0.398	2
20	Unigram, bigram and trigram of feature 19 [7]	0.398	2
21	Surface text of the first verb that appears between the identifier and the definiens [7]	0.093	9
22	Distance between identifier and definiens in the shortest edge path between identifier and definiens of the dependency graph [7]	0.001	18
23	Surface text and POS tag of dependency with length 3 from definiens along the shortest path between identifier and definiens [7]	0.292	4
24	Surface text and POS tag of dependency with length 3 from identifier along the shortest path between identifier and definiens [7]	0.328	3
25	Direction of 24. Incoming to definiens or not [7]	0.064	11
26	Direction of 25. Incoming to identifier or not [7]	<0.001	26
27	Distance between the identifier and definiens in number of words [7,13]	0.064	10
28	Distance of the identifier-definiens candidate from the first appearance of the identifier in the document, in sentences [13]	0.101	8
29	Relative term frequency of the definiens [13]	0.044	14

than one correct definition for an identifier. Therefore, we evaluated using the following policy:

1. Use the number of identifiers (310) as truth.
2. True positive: at least one definition for the identifier was found.
3. Ignore: more than one correct definition was found.

4. False positive: a definition that is not in the set of possible definitions.
5. False negative: no definition was found for the identifier.

This policy assigns an optimal score $p = r = 1$ if (1) only one correct definiens is retrieved, and (2) if more than one correct definiens is retrieved. Using this policy, we could compare the precision, recall, and F_1 score of the statistical approach, the pattern matching approach, and the newly developed approach (cf. Sect. 2.3). We packaged the new evaluation method in a Java tool (https://github.com/leokraemer/mathosphere/tree/temp/evaluation) that evaluates .csv files of the form qId, title, identifier, definiens.

During the development of the evaluation framework, we discovered several weaknesses of the MLP pipeline. In step 3 (find candidates for identifier - definiens pairs, cf. Sect. 2.1), we discovered that certain operators, such as special cases of 'd' in integrals and the '∞' symbol were misclassified as identifiers. While addressing this issue, we also created unit tests with the data from the gold standard to prevent future regressions in the identifier extraction. In addition, we discovered that many false positives included the identifier 'a'. Thus, we improved the identifier detection for simple Latin charters using style information from the Wikitext markup.

3.2   Evaluating Existing Approaches

Using the evaluation framework, we could accurately judge the impact of changes in the MLP pipeline and develop a new approach for the identifier-definiens scoring. As a first experiment, we evaluated the statistical feature analysis (cf. Sect. 2.1) and the pattern matching approach (cf. Sect. 2.2) individually, with and without the improvements to the preprocessing steps of the MLP pipeline. Additionally, we evaluated the union of the identifier definiens tuples returned by both approaches.

3.3   New Machine Learning Approach (ML)

Following the idea of Kristianto et al. [7], we employed a support vector machine to combine the strengths of the statistical feature analysis of Schubotz et al. [13] (cf. Sect. 2.1) and the pattern matching approach of Pagel et al. [9] (cf. Sect. 2.2) as well as to implicitly tune the parameters of the approaches. The SVM accepts as input nominal features, e.g., whether an identifier appears before its definiens, and ordinal features, e.g., the relative term frequency of identifiers. Using a filter that converts strings to word vectors, we can also use the SVM to train on parts of the original sentences and POS-Tags. After the feature vector generation phase, we obtain 7902 feature vectors of which 244 are actual matches of the gold standard and 7658 are true negatives. We use a combination of oversampling of the minority class and undersampling the majority class approach to balance the data for training. We chose an radial basis function (RDF) kernel, due to the non-linear characteristics of some of the features. We found the best hyperparameters in cost $= 1$ and $\gamma \approx 0.0186$. We trained four different classifiers examine the performance of different feature classes: A classifier ML_ST_PM using only simple features (1–16 and 27–29 in Table 1), ML_no_DG without the features using the expensive dependency graph generation (1–21 and 27–29 in Table 1), ML_no_PM without the hand-crafted patterns (11–29 in Table 1) and ML_full with all features. The features are a combination of features used for the three approaches described in Sect. 2. Some features used by Kristianto et al. were too specific to the task of classifying formulae descriptions instead of identifier-definiens pairs and thus were ignored for our approach. The remaining features were adjusted to be compatible with the MLP pipeline (cf. Sect. 2.1).

For training, we used all extracted identifier-definiens candidates and annotated them with the information from the gold standard. We employed a 10-fold cross validation using the entire gold standard. We divided the test and the training sets on document level, i.e. we trained the model using the data from 90 documents and evaluated the model using the data from 10 other documents.

3.4 Evaluating the Influence of Namespaces

So far, we described approaches that operate on the level of individual documents to extract identifier definitions, i.e. approaches that implement the steps 1–4 of the MLP pipeline (cf. Sect. 2.1). We also executed and evaluated the computationally expensive step 5 of the MLP pipeline (namespace discovery), which requires to process the entire test collection. To enable an unbiased comparison, we build namespaces using each of the methods individually. In other words, we use each extraction approach to collect the identifier-definiens pairs from all documents. We then use the identifier-definiens pairs as features to cluster documents and label the obtained clusters with suitable categories from well-known topic categorizations, such as the Mathematics Subject Classification provided by the American Mathematical Society. For details please refer to [13].

4 Results

We re-evaluated the pattern matching approach (PM) of Pagel et al. (cf. Sect. 2.2) [9] and the statistical feature analysis (ST) of Schubotz et al. (cf. Sect. 2.1) [13] to obtain the '_before' results shown in Table 2. The '_before' suffix indicates the use of the MLP pipeline as presented in [13], i.e. before making the improvements described in Sect. 3.1.

While PM has not been evaluated using this gold standard before, we already evaluated ST using the same gold standard in the past [13]. However, in the previous evaluation of ST, we manually judged the relevance of the extracted identifier-definiens pairs and used a different policy to judge true positives than employed by the automated evaluation procedure in our framework. As described in Sect. 3.1, our framework ignores cases in which more than one correct definition is retrieved for calculating the performance metrics. Opposed to that, we counted each correctly extracted definition as a true positive in our previous

Table 2. Performance comparison of the pattern matching (PM), statistical feature analysis (ST), and machine learning (ML) methods. See Sect. 4 for details.

Baseline	tp	fp	Prec%	Rec%	$F_1\%$
ST_before	69	351	16.43	22.26	18.90
ST_after	70	322	17.85	22.58	19.94
PM_before	56	290	16.18	18.06	17.07
PM_after	56	199	22.00	18.06	19.80
Without namespaces					
PM_after ∪ ST_after	77	551	12.26	24.84	16.42
ML_ST_PM	60	181	24.90	19.35	21.78
ML_no_DG	79	171	31.60	25.48	28.21
ML_no_PM	86	111	43.65	27.74	33.92
ML_full	87	92	48.60	28.06	35.58
With namespaces					
ST_after + NS	75	340	18.07	24.19	20.69
ML_full + NS	93	118	43.66	30.00	35.56

work [13]. In our previous manual evaluation, we also counted several synonymous identifier-definiens relations as true positives. For example, we manually matched *'eigenfrequencies'* to the gold standard entry *'natural frequency'* and the wikidata item Q946764, which does not (yet) have an alias for *'eigenfrequencies'*. Realizing that both terms are synonyms is trivial for human assessors, but beyond the capabilities of our current automated evaluation framework. Since these limitations of the framework apply to all evaluated methods, the relative performance scores of the methods should be unaffected.

Due to the different evaluation policies, the measured performance of ST decreased from $p \approx .21$, $r \approx .28$, $F_1 \approx .24$ in our previous manual evaluation [13] to $p \approx .16$, $r \approx .22$, $F_1 \approx .19$ in the current automated evaluation. The absolute number of true positives (tp) declined by 18 from 88 to 70. Identifiers for which more than one definiens was found account for 9 fewer tp and the inability of the evaluation framework to resolve synonymous identifier-definiens pairs accounts for 8 fewer tp.

Our improvements to the MLP pipeline slightly increased the precision achieved by the pattern matching approach (PM) and the statistical feature analysis (ST). Refer to the results with the suffix '_after' in Table 2.

The PM and the ST approach extracted 48 identical and 29 different definientia, which means that combining both results will achieve a higher recall. The simple union (see Table 2) yields a higher recall, but the precision drops disproportionately. To create a combined classifier of both approaches that achieves better precision, we must rank nominal features, e.g., pattern matches, together with ordinal features, e.g., term frequency.

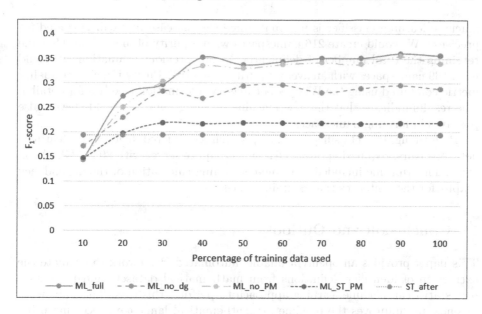

Fig. 2. Extraction performance (F_1-score) for different sizes of the training dataset.

Our new machine learning method (ML) extracted 87 definientia correctly with 92 false positives, resulting in $p = .4860$, $r = .2806$, $F_1 = .3558$. In addition, we trained the following classifiers using subsets of the features: ML_ST_PM indicates the combination of only statistical features (27–29 in Table 1) with the pattern matcher (1–10 in Table 1). This approach yielded $F_1 = .2178$, which is comparable to the performance of the approaches in the previous manual investigation. Adding the string features (11–21 in Table 1), but leaving out the computationally intensive dependency graph features increased the performance to $F_1 = .2821$ (see ML_no_DG in Table 2). Training with all features except for the patterns 1–10 in Table 1 yielded a good performance of $F_1 = .3276$ (see ML_no_PM in Table 2). When comparing this result to the full classifier (ML_full), which achieves $F_1 = .3558$, shows that the improvement achieved by including the pattern-based features is small. In other words, creating a well-performing classifier without the language-dependent pattern features is possible.

Figure 2 plots the F_1-scores for different sizes of the training dataset. The different models seem to converge to a fixed threshold with a decreasing gradient. This indicates that the classifier can be trained well despite the limited number of positive instances in the training data and the complexity of the features. Additionally, all ML classifiers outperform the best individual extraction method (ST or PM) if more than 20% of the training data is used.

Investigating the effects of considering namespaces for the identifier definition extraction, we could confirm the finding in [13] that namespaces improve both precision and recall for the statistical approach. However, the gain for the machine learning classifier is minimal, since the increase in recall is traded for precision. We could create 216 namespaces with a purity of more than 0.8 while retaining the total number of definientia. In our previous evaluation, we could form 169 namespaces with an average purity of 0.8 [13]. Purity is a cluster quality metric computed using the Wikipedia category information (see [13] for details). The results indicate that the new machine learning approach yields fewer false positives in forming namespaces.

Performing the classification task, which considered 5'400'702 identifier-definiens pairs, required approx. 3.5 h on a compute server with 80 2.6 Ghz Xeon cores. This runtime included the time-consuming calculation of the dependency graphs for the sentences that contain identifiers.

5 Conclusion and Outlook

This paper provides an openly available, automated framework to evaluate the extraction of identifier definitions from mathematical datasets and presents a new statistical machine learning approach to perform this task. The framework extends and improves the pipeline for mathematical language processing using Wikitext input that we presented in [13]. The evaluation framework uses parts of the dataset of the NTCIR-11 Math Retrieval Wikipedia task [12] and a manually created gold standard that contains definiens for all the 310 mathematical identifiers in the dataset.

Using the newly developed evaluation framework, we compared existing approaches for identifier definition extraction. The previously best-performing approach achieved a precision $p \approx .18$, recall $r \approx .23$, and $F_1 \approx .20$. Our newly developed machine learning approach significantly increased the extraction performance to $p \approx .49$, $r \approx .28$, $F_1 \approx .36$.

Despite the improvements of the preprocessing pipeline, we could see that the statistical feature analysis (ST) clearly outperforms the pattern matching (PM) approach. In addition, our machine learning (ML) approach significantly reduces the number of false positives, even without relying on language-dependent patterns. At its core, the developed machine learning approach relies heavily on features developed for the statistical feature analysis, which the new approach combines with a better method for tuning the extraction parameters.

Even the newly developed machine learning approach achieves a relatively low performance when compared to approaches for other information extraction tasks. The results indicate that a large potential for future improvements of identifier extraction approaches remains. While our newly proposed method significantly reduced the number false positives, new strategies are needed to further improve the number of true positives.

For future research, we advise against using our gold standard dataset for training purposes, since the gold standard contains identifier definitions that cannot be identified by any of the features we examined in this paper. This limitation lies rooted in the creation history of the gold standard, which involved tedious logical inference and the consultation of tertiary sources by the domain experts who created the gold standard. The experts deduced information, incorporated world knowledge and exhibited a higher fault tolerance than can be expected from automated systems. For instance, extracting the identifiers η, Q_1, Q_2 from the malformed input formula $\eta = \frac{workdone}{heatabsorbed} = \frac{Q1-Q2}{Q2}$, as the domains experts did when creating the gold standard, is likely a suboptimal training for future extraction approaches.

Future research should focus on increasing recall, because current methods exclusively find exact definitions for approx. 1/3 of all identifiers. New approaches may further improve the task at hand, e.g., by using logical deduction [11]. Likewise, the use of multilingual features of Wikipedia, e.g., by applying approaches like multilingual semantic role labeling [1], can prove beneficial. In the medium term, the semantic granularity of the corresponding Wikidata concepts should be considered [3]. Lastly, the proposed approach could also be applied in other domains. One use case would be to identify biased media coverage by analyzing the relations between words in image captions and texts of news articles (cf. [4,5]). Another idea would be to adapt the approach for resolution of abbreviations and synonyms [6].

In conclusion, by evaluating and combining existing approaches we achieved a significant performance improvement in extracting mathematical identifier definitions. We also identified several promising directions for future research to further improve the extraction performance.

References

1. Akbik, A., Guan, X., Li, Y.: Multilingual aliasing for auto-generating proposition banks. In: Calzolari, N., Matsumoto, Y., Prasad, R. (eds.) 6th International Conference on Computational Linguistics, Proceedings of the Conference: Technical Papers (COLING 2016), December 11–16, 2016, Osaka, Japan, pp. 3466–3474. ACL (2016)
2. Chang, C.-C., Lin, C.-J.: LIBSVM: a library for support vector machines. In: ACM Transactions on Intelligent Systems and Technology (TIST 2011) vol. 2, no. 3, p. 27 (2011)
3. Corneli, J., Schubotz, M.: math.wikipedia.org: a vision for a collaborative semiformal, language independent math(s) encyclopedia. In: Conference on Artificial Intelligence and Theorem Proving (AITP 2017) (2017)
4. Hamborg, F., Meuschke, N., Gipp, B.: Matrix-based news aggregation: exploring different news perspectives. In: Proceedings of the ACM/IEEE-CS Joint Conference on Digital Libraries (JCDL) (2017)
5. Hamborg, F., et al.: Identification and analysis of media bias in news articles. In: Gaede, M., Trkulja, V., Petra, V. (eds.) Proceedings of the 15th International Symposium of Information Science, Berlin, pp. 224–236, March 2017
6. Henriksson, A., et al.: Synonym extraction and abbreviation expansion with ensembles of semantic spaces. J. Biomed. Semant. 5(1), 6 (2014)
7. Kristianto, G.Y., Topic, G., Aizawa, A.: Extracting textual descriptions of mathematical expressions in scientific papers. In: D-Lib Magazine (D-Lib 2014), vol. 20, no. 11, p. 9 (2014)

8. Manning, C.D., et al.: The stanford CoreNLP natural language processing toolkit. In: Association for Computational Linguistics (ACL) System Demonstrations (ACL 2014), pp. 55–60 (2014)
9. Pagel, R., Schubotz, M.: Mathematical language processing project. In: England, M., et al. (eds.) Joint Proceedings of the MathUI, OpenMath and ThEdu Workshops and Work in Progress Track at CICM Co-located with Conferences on Intelligent Computer Mathematics (CICM 2014), Coimbra, Portugal, July 7–11, 2014, vol. 1186. CEUR Workshop Proceedings. CEUR-WS.org (2014)
10. Schubotz, M.: Augmenting Mathematical Formulae for More Effective Querying & Efficient Presentation. Epubli Verlag, Berlin (2017). ISBN: 9783745062083
11. Schubotz, M., Veenhuis, D., Cohl, H.S.: Getting the units right. In: Kohlhase, A., et al. (ed.) Joint Proceedings of the FM4M, MathUI, and ThEdu Workshops, Doctoral Program, and Work in Progress at the Conference on Intelligent Computer Mathematics 2016 Co-located with the 9th Conference on Intelligent Computer Mathematics (CICM 2016), Bialystok, Poland, July 25–29, 2016, Vol. 1785. CEUR Workshop Proceedings. CEUR-WS.org (2016)
12. Schubotz, M., et al.: Challenges of mathematical information retrieval in the NTCIR-11 math Wikipedia task. In: Baeza-Yates, R.A., et al. (eds.) Proceedings of the 38th International ACM SIGIR Conference on Research and Development in Information Retrieval (SIGIR 2015), pp. 951–954. ACM, Santiago (2015). ISBN: 978-1-4503-3621-5
13. Schubotz, M., et al.: Semantification of identifiers in mathematics for better math information retrieval. In: Proceedings of the 39th International ACM SIGIR Conference on Research and Development in Information Retrieval (SIGIR 2016), pp. 135–144. ACM, Pisa (2016). ISBN: 978-1-4503-4069-4

Short Papers

Query Expansion for Sentence Retrieval Using Pseudo Relevance Feedback and Word Embedding

Piyush Arora[✉], Jennifer Foster, and Gareth J.F. Jones

School of Computing, ADAPT Centre, Dublin City University, Dublin, Ireland
{parora,jfoster,gjones}@computing.dcu.ie

Abstract. This study investigates the use of query expansion (QE) methods in sentence retrieval for non-factoid queries to address the query-document term mismatch problem. Two alternative QE approaches: i) pseudo relevance feedback (PRF), using Robertson term selection, and ii) word embeddings (WE) of query words, are explored. Experiments are carried out on the WebAP data set developed using the TREC GOV2 collection. Experimental results using P@10, NDCG@10 and MRR show that QE using PRF achieves a statistically significant improvement over baseline retrieval models, but that while WE also improves over the baseline, this is not statistically significant. A method combining PRF and WE expansion performs consistently better than using only the PRF method.

Keywords: Query expansion · Pseudo relevance feedback · Word embeddings · Sentence retrieval

1 Introduction

Sentence retrieval is a challenging information retrieval (IR) task which is useful in question answering and summarization systems. Retrieval of sentences relevant to an answer is a difficult task due to short length of the target items which are more likely to suffer from vocabulary mismatch with respect to the query. We focus on the new task of answer passage retrieval for non-factoid queries [3,8]. In this paper we investigate two unsupervised query expansion (QE) methods which seek to address the query-document term mismatch issue. First, we use Robertson's standard Okapi QE relevance feedback method [7]. Second, we propose a word embedding (WE) method [6] to expand the query using words similar to the query based on vector similarity. Finally, we explore the combination of these alternative sources of evidence for QE. We investigate the following research questions:

1. How do unsupervised approaches using traditional retrieval techniques with QE perform for the task of sentence retrieval from a target corpus?
2. Can we leverage word embeddings to improve retrieval effectiveness in a sentence retrieval task?

© Springer International Publishing AG 2017
G.J.F. Jones et al. (Eds.): CLEF 2017, LNCS 10456, pp. 97–103, 2017.
DOI: 10.1007/978-3-319-65813-1_8

2   Related Work

Finding relevant sentences from a given document for a given query is a common task in applications which involve generating summaries and abstracts from individual documents [5]. Our work differs from this since we are focused on retrieval of relevant sentences from a *complete target document collection* for a given query to find answer sentences for non-factoid queries, similar to the work reported in [8].

A related study [1] explored different retrieval models and QE techniques to find novel information from a document collection. Our work is focused on relevancy rather than novelty but we also explore QE models. Supervised approaches such as Learning-to-Rank (L2R) have been used for answer sentence retrieval [3,5,8]. Our work investigates whether we can achieve similar performance using QE techniques as compared to L2R techniques.

Word embedding is a method for forming low-dimensional vector representations of words based on co-occurrence in a reference corpus, thereby providing a means of semantic comparison between words. WE is of increasing interest among the IR community, and there are a number of examples of recent work exploring the use of WE for document-based QE [2,4,9]. [4] investigated different methods for document retrieval, [2] compared different types of embeddings, and [9] explored effective ways of combining embeddings and co-occurrence based statistics for document retrieval. Our work is similar but we focus on the task of sentence retrieval.

3   Experimental Methodology

As our baseline models, we perform sentence retrieval using: (i) a language modeling (LM) method that uses Jelinek-Mercer smoothing [10], and (ii) the BM25 model [11]. We then perform QE using the following methods:

1. **Pseudo Relevance Feedback (PRF):** We use a standard PRF method in which an initial retrieval run is carried out using the selected baseline retrieval system. A number of the top ranked documents are assumed to be relevant, and potential QE terms from within these documents are ranked using the Robertson Offer Weight [7]. A fixed number of these terms are then added to the original query.
2. **Semantic expansion, using word embeddings (WE):** We explore two novel approaches to using WE for QE. The embedding of each word is computed using the *Word2Vec* [6] method, which learns vector representation using a feed-forward neural network by predicting a word given its context (the cbow model).
   For a given query Q consisting of n terms $q_1, \ldots, q_n$, for each term q_i we generate a pool of potential candidate expansion terms c_i, such that $c_i =$ the top z most similar terms $t^1, \ldots, t^z$ for each term q_i. Thus the complete

set of potential expansion candidates for query Q is $C_i^j = c_1, \ldots, c_n$, where i varies from 1 to n and j varies from 1 to z. The most similar terms for each term q_i are obtained by calculating the cosine similarity between its vector representation and the terms in the document corpus. We then select the top k expansion terms for Q from the potential candidates, using one of two methods:

(a) **QueryWord approach**: We sort all the terms in C_i^j, based on the cosine similarity score between each term t_i^j and the corresponding query term q_i. We select the top k terms from this sorted list as the expansion terms. The use of the cosine similarity in this way biases this list towards expansion terms which are closely related to terms in Q, these terms may be synonyms or words with a close semantic relationship to terms appearing in Q.

(b) **Centroid approach**: We form a centroid vector (CV) of the query, by summing the vectors of all the query terms q_i in Q. We then sort all the terms in C_i^j, based on the cosine similarity score between each term t_i^j and CV. We select the top k ranked terms as the expansion terms which are most similar to CV. The main goal is to retrieve terms which are related to all the terms in Q as a unit.

3. **Combining WE and PRF**: PRF seeks expansion terms in the top ranked documents following an actual search which is hoped will have retrieved relevant items, while WE seeks to identify words that are similar in meaning to the query words independent of their use in the query. PRF and WE thus provide different sources of information which may be exploited in combination in the QE process. To explore the potential for this combination, we examine an approach which linearly combines terms expanded using PRF and WE calculated as follows,

$$COW = (w_t) * WE_{EQT} + (1 - w_t) * PRF_{EQT} \tag{1}$$

where EQT stands for expanded query terms.

4 Experimental Setup

We use the WebAP dataset [8], which was developed using the TREC Gov2 collection, and has 82 queries with a total of 6,399 documents consisting of 991,233 sentences which are marked at the sentence level on a 5 level scale of topical relevance: *4*: perfect, *3*: excellent, *2*: good, *1*: fair, *0*: none. Following [8], we used precision (P@10), normalized discounted cumulative gain (NDCG@10) and mean reciprocal rank (MRR) to compare the performance of our methods.

The Lucene toolkit[1] was used to perform sentence retrieval. We performed stemming and stopword removal using the Lucene EnglishAnalyzer. Sentence retrieval used lucene's implementation of LM or BM25. We implemented our own

---

[1] https://lucene.apache.org/core/4_4_0/core/overview-summary.html.

version of Robertson QE. The Gensim tookit was used to learn and incorporate *word embeddings* for use in QE as described above. We used two different types of embeddings:

1. *Global embeddings*: Embeddings trained on Google news, consisting of about 3 million 300 dimension English word vectors which are released for research. [2]
2. *Local embeddings*: Embeddings learnt using subcollection of the WebAP dataset consisting of 6,399 documents, with parameter settings as follows– method: continuous bag-of-words (cbow), embedding size=200, window for training = 10, iterations set for learning = 20, the rest of the parameters were kept as default. [3]

5   Results and Analysis

Baselines: We explored optimization of parameters by varying λ in the LM retrieval model in the range of $[0.05, 1.0]$, and performed grid search in the range of $[0.1, 1.5]$ with an increment of 0.1 for the b and k parameters for the BM25 model. Our best results were obtained for $\lambda = 0.45$ for LM and $b = 0.4$ and $k = 0.4$ for BM25, as shown in Table 2, these values were then fixed for subsequent experiments. Table 1 also presents the best result reported in [8] for their L2R model. Our baseline results are comparable to their results using LM. However, note that they used Dirchlet smoothing techniques whereas our model uses Jelinek-Mercer smoothing, which might be the reason for small variation in the results.

Table 1. Our baseline results and previous results as reported in [8].

	P@10	Ndcg@10	MRR	P@10	Ndcg@10	MRR
	LM retrieval model					
Results from [8]	0.145	0.134	0.339			
Best model from [8]	LearningToRank (only sentence level features)			LearningToRank (features from neighboring sentences)		
Best results from [8]	**0.174**	**0.159**	**0.344**	**0.194**	**0.180**	**0.403**
Our implementation	LM retrieval model			BM25 model		
Baseline	0.149	0.127	0.293	0.158	0.142	0.330

---

[2] https://github.com/mmihaltz/word2vec-GoogleNews-vectors.
[3] We learnt different embeddings by varying the training method, dimension size, window size, no. of iterations in internal development experiments, but the results obtained showed little variation in performance.

Table 2. PRF based expansion results, the best scores are in boldface. + and * indicate statistically significant improvement over the baseline with p < 0.05 and p < 0.1 respectively, using student's t-test

	P@10	Ndcg@10	MRR	P@10	Ndcg@10	MRR
	LM retrieval model			BM25 model		
Baseline	0.149	0.127	0.293	0.158	0.142	0.330
Weight, w_t, 0.6	**0.153**	**0.140**	0.319	0.170	0.160	**0.352**
Weight, w_t, 0.7	**0.153**	**0.140**	**0.328**	0.173	**0.162***	0.352
Weight, w_t, 0.8	0.150	0.137	0.322	**0.174***	**0.162$^+$**	0.347
Weight, w_t, 0.9	0.149	0.133	0.317	0.164	**0.153***	0.346

Pseudo Relevance Feedback (PRF): We varied the number of assumed relevant documents R, and expansion query terms (EQT) in the range of $\{5, 10, 15, 20\}$. We linearly varied the weight of initial query terms Q and expansion terms EQT.

$$Expanded\ Query = w_t * Q + (1 - w_t) * EQT \tag{2}$$

where w_t varies in range $[0.5, 1]$, with an increment of 0.1.

For both LM and BM25, the best PRF results were obtained using $R = 10$ and $EQT = 10$, as shown in Table 2, with varying weights between initial and expanded query terms. Based on these results, we fixed values $R = 10$ and $EQT = 10$ for further experiments using PRF. PRF shows significant improvement over both baselines, similar effects were reported in earlier work on sentence retrieval by Allan et al. [1]. The BM25 model performed consistently better than LM, thus we report further results using only BM25.

Word Embeddings (WE): We used the value $z = 10$ for all the experiments, where z determines the number of terms entered into the pool of potential candidate expansion terms for each individual query term as described earlier in Sect. 3. The overall number of expansion terms k selected from the pool was varied as $\{5, 10, 15\}$. We linearly varied the weight of query terms and expansion terms as shown in Eq. 2. For each combination of embedding (Global and Local) and the expansion technique (QueryWord and Centroid) the best results (based on P@10) are shown for BM25 in Table 3.

Combining WE and PRF: Table 3 shows results using the combined expansion approach. The number of expanded terms k for expansion using WE was varied as $\{5, 10, 15\}$ while $R=10$ and $EQT=5$ was used for performing PRF. As described in Eq. 1, w_t varies in range $[0.1, 1]$ with an increment of 0.1. For each combination of embedding (Global and Local) and the expansion technique (QueryWord and Centroid) the best results (based on P@10) are presented for BM25. The combined expansion approach WEPRF, performs consistently better than using either the WE and PRF approaches.

Table 3. Embedding based WE and WEPRF approach for sentence retrieval, best scores are in boldface. * indicates that the difference in the results compared to the baseline is statistically significant with p < 0.1, using student's t-test. NDCG scores are calculated at rank 10.

	Word embedding (WE)			Combined approach (WEPRF)		
	BM25 model			BM25 model		
	P@10	NDCG	MRR	P@10	NDCG	MRR
Baseline	0.158	0.142	0.330	0.158	0.142	0.330
Best PRF result	**0.174***	**0.162$^+$**	0.347	**0.174***	**0.162$^+$**	0.347
QueryWord approach for semantic expansion: using global embeddings						
Best result	0.158	0.144	0.340	**0.179***	**0.166***	0.357
QueryWord approach for semantic expansion: using local embeddings						
Best result	**0.168**	**0.151**	**0.350**	0.165	0.157	0.352
Centroid approach for semantic expansion: using global embeddings						
Best result	0.165	0.144	0.320	0.173	0.162	0.354
Centroid approach for semantic expansion: using local embeddings						
Best result	0.159	0.142	0.316	0.168	0.160	**0.361**

Analysis: We explored sentence retrieval with QE, best results obtained using WEPRF approach are better than Learning-to-Rank results reported in [8] using only sentence level features, but are slightly lower than the best results using information from neighbourhood sentences. However, the simpler unsupervised technique WEPRF is still quite good and can be used for retrieval problems where data is not sufficient to train effective Learning-to-Rank models.

Using only WE techniques shows that the QueryWord approach trained on *Local embeddings* performs relatively better for QE than using *Global embeddings* or the Centroid approach. *Local embeddings* tend to generate better expanded terms which are more related to the query terms and the corpus. In the combined approach (WEPRF), where the PRF method provides potentially in-context expanded terms using top potential relevant sentences, the best results are obtained using QueryWord approach with *Global embeddings*, which generates more diverse expanded terms (number of expanded terms = 5) to improve the retrieval effectiveness.

We performed manual analysis to analyze QE using WE. For TopicId: 704 *goals green party political views*, both Global and Local embeddings were able to expand and identify most of the common terms such as ("democratic, republic, caucus, candidacy, viewpoints, opinions, ideology, politicians"), and are thus able to capture context which improve the task of sentence retrieval. Further, for TopicId: 741 ("artificial intelligence"), *Global embeddings* learnt using Google ngrams drifted towards "intelligence in security, intelligence agencies and counter-terrorism" aspects, whereas *Local embeddings* were able to capture aspects related to the main query more effectively ("neural, bayesian, combinatorial, acm, aaai, icml etc"). Further analysis indicates that WE techniques are complementary to PRF techniques, and the combination approach performs better as shown in Table 3.

6   Conclusions and Further Investigation

We explored query expansion techniques for sentence retrieval from a corpus of documents, achieving our best performance for an approach that combines the use of word embeddings and pseudo-relevance feedback. We plan to perform further analysis to learn more about how to exploit semantic representations for this task.

Acknowledgments. We thank the reviewers for their feedback and comments. This research is supported by Science Foundation Ireland (SFI) as a part of the ADAPT Centre at Dublin City University (Grant No: 12/CE/I2267).

References

1. Allan, J., Wade, C., Bolivar, A.: Retrieval and novelty detection at the sentence level. In: Proceedings of SIGIR 2003, pp. 314–321 (2003)
2. Diaz, F., Mitra, B., Craswell, N.: Query expansion with locally-trained word embeddings (2016). arXiv preprint arXiv:1605.07891
3. Keikha, M., Park, J.H., Croft, W.B., Sanderson, M.: Retrieving passages and finding answers. In: Proceedings of the 2014 Australasian Document Computing Symposium, p. 81 (2014)
4. Kuzi, S., Shtok, A., Kurland, O.: Query expansion using word embeddings. In: Proceedings of CIKM 2016, pp. 1929–1932 (2016)
5. Metzler, D., Kanungo, T.: Machine learned sentence selection strategies for query-biased summarization. In: SIGIR Learning to Rank Workshop, pp. 40–47 (2008)
6. Mikolov, T., Chen, K., Corrado, G., Dean, J.: Efficient estimation of word representations in vector space. CoRR, abs/1301.3781 (2013)
7. Robertson, S.E.: On term selection for query expansion. J. Documentation **46**(4), 359–364 (1990)
8. Yang, L., et al.: Beyond factoid QA: effective methods for non-factoid answer sentence retrieval. In: Ferro, N., et al. (eds.) ECIR 2016. LNCS, vol. 9626, pp. 115–128. Springer, Cham (2016). doi:10.1007/978-3-319-30671-1_9
9. Roy, D., Ganguly, D., Mitra, M., Jones, G.J.F.: Word vector compositionality based relevance feedback using kernel density estimation. In: Proceedings of CIKM 2016, pp. 1281–1290 (2016)
10. Ponte, J.M., Croft, W.B.: A language modeling approach to information retrieval. In: Proceedings of SIGIR 1998, pp. 275–281 (1998)
11. Robertson, S., Zaragoza, H., et al.: The probabilistic relevance framework: Bm25 and beyond. Found. Trends® Inf. Retrieval **3**(4), 333–389 (2009)

WebShodh: A Code Mixed Factoid Question Answering System for Web

Khyathi Raghavi Chandu[1(✉)], Manoj Chinnakotla[2], Alan W. Black[1],
and Manish Shrivastava[3]

[1] Carnegie Mellon University, Pittsburgh, USA
{kchandu,awb}@cs.cmu.edu
[2] Microsoft India, Hyderabad, India
manojc@microsoft.com
[3] IIIT Hyderabad, Hyderabad, India
m.shrivastava@iiit.ac.in

Abstract. Code-Mixing (CM) is a natural phenomenon observed in many multilingual societies and is becoming the preferred medium of expression and communication in online and social media fora. In spite of this, current Question Answering (QA) systems do not support CM and are only designed to work with a single interaction language. This assumption makes it inconvenient for multi-lingual users to interact naturally with the QA system especially in scenarios where they do not know the right word in the target language. In this paper, we present *Web-Shodh* - an end-end web-based Factoid QA system for CM languages. We demonstrate our system with two CM language pairs: *Hinglish* (Matrix language: Hindi, Embedded language: English) and *Tenglish* (Matrix language: Telugu, Embedded language: English). Lack of language resources such as annotated corpora, POS taggers or parsers for CM languages poses a huge challenge for automated processing and analysis. In view of this resource scarcity, we only assume the existence of bi-lingual dictionaries from the matrix languages to English and use it for lexically translating the question into English. Later, we use this loosely translated question for our downstream analysis such as Answer Type(AType) prediction, answer retrieval and ranking. Evaluation of our system reveals that we achieve an MRR of 0.37 and 0.32 for Hinglish and Tenglish respectively. We hosted this system online and plan to leverage it for collecting more CM questions and answers data for further improvement.

1 Introduction

CM is the phenomenon of "embedding of linguistic units such as phrases, words and morphemes of one language into an utterance of another language" [1]. The lexicon and syntactic formulations from both the languages are mixed to form a single coherent sentence. Some of such mixtures are known as Spanglish, Hinglish, Tenglish, Portunol and Franponaisor[1]. CM usually prevails in a multilingual

---

[1] Mixing of Spanish-English, Hindi-English, Telugu-English, Portugese-Spanish and French-Japanese language pairs respectively.

© Springer International Publishing AG 2017
G.J.F. Jones et al. (Eds.): CLEF 2017, LNCS 10456, pp. 104–111, 2017.
DOI: 10.1007/978-3-319-65813-1_9

configuration with speakers having more than one common language. Moreover, anglicization of languages is also a very common phenomenon these days, which leads to the representation of native words in English letters phonetically. The study on cross script code mixing is essential mainly because of the prominent usage of English keyboards in countries like India. Studies on statistical usage of code-switching among facebookers found that there is about 33% of intra-sentential switching [2]. This work also showed that 45% of switching is due to real lexical need, which is a considerably high percentage. The increasing use of CM is also driven by the ease and speed of communication mainly facilitated by the easier choice of words and a richer set of expressions to choose from. In spite of this, current QA systems [3,4] only support interaction in a single language. This severely hampers the ability of a multi-lingual user to interact naturally with the QA system. This is especially true in scenarios involving technical and scientific terminology. For example, when a native Telugu speaker wants to know the director of the movie *Heart Attack*, he is more likely to express it as *"heart attack cinema ni direct chesindi evaru?" (Translation: who directed the movie heart attack)* where the words *heart attack, direct, cinema* are all English words. Hence, to increase the reach, impact and effectiveness of QA in multi-lingual societies [5], it is imperative to support QA in CM languages [6]. However, any automated analysis and processing of CM text poses serious challenges due to lack of normalized representations adhering to standard syntactic and phonetic rules. The problem is further compounded by the unavailability of language resources such as annotated corpora, language analysis tools such as POS taggers, parsers *etc.*

In this paper, we present *WebShodh* - an end-to-end open domain factoid Question Answering (QA) system for Web which provides a *ranked list of potential answers* to a CM question. We demonstrate our system using CM in two dominantly spoken languages in India; Hindi and Telugu[2]. In view of resource-scarcity, we only assume the existence of bi-lingual dictionaries from the source language to English. Our system performs a lexical level language identification and translation into English. We use this high-level loosely translated question to classify and infer the expected answer type. We also fire the entire loosely translated English query to Google using their Search API and retrieve the top 10 search results from the web along with their titles and snippets, which are then processed to identify potential candidates for answers based on the hints offered by AType. Finally, we rank these candidate answers based on various features to finally output a ranked list of answers. We evaluated our system on both these CM languages and share the quantitative and qualitative analysis of our results. Overall, our system achieves an MRR of 0.37 and 0.32 for Hinglish and Tenglish respectively. We hosted our system *WebShodh* online (http://128.2.208.89/webshodh/cmqa.php) and intend to use it for collecting more

---

[2] Hindi is one of the most spoken languages in India, with 370 million native speakers and is an official language along with English. Telugu is the most spoken Dravidian language in South India with about 70 million native speakers.

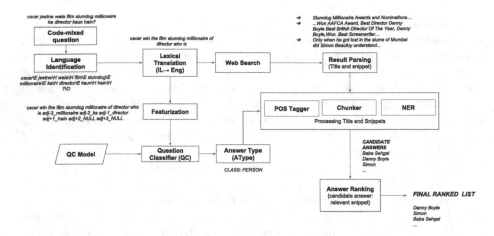

Fig. 1. Architecture of *WebShodh*: a web based factoid QA system for code-mixed languages

QA data for CM languages - an important step forward if we want to try out more data-intensive techniques such as deep learning.

The paper is organized as follows: in Sect. 2, we discuss the related previous work in this area. Section 3 describes the overall system architecture and delves into each of the steps in the pipeline. In Sect. 4, we present the experimental setup including data creation, experimental results and qualitative error analysis. Section 5 discusses the conclusions and future scope of the work.

2 Related Work

Linguistic and conversational motives for CM have been studied in [7–9]. [10] describes the grammatical contexts in which CM has taken place in student interactions. The recent years have shown rapid upsurge in understanding and analyzing these languages as they are among the most prominently used languages on social media. The intuitive first step towards tackling this domain is lexical language identification, which has been addressed in EMNLP[3] and FIRE[4] in 2014. The challenges of this non-trivial task have been presented by [11]. [12] have studied POS tagging in code-mixed social content and have concluded that the tasks of language identification and transliteration still stand as major challenges. Question Classification (QC) and Question Answering (QA) systems have been well studied for monolingual settings previously by [13–16]. [17] have introduced the space of QC in code-mixed languages. This work used an SVM based QC technique and presented results for coarse and fine grained categorizations, based on ontology of question hierarchy described by [16]. While this work was mainly done for Hindi-English pair, it was later studied for Bengali-English by

---

[3] http://emnlp2014.org/workshops/CodeSwitch/call.html.

[4] http://fire.irsi.res.in/fire/home.

[18]. [19] have presented an approach to mine the ever growing content on social media for generating a CM QA corpus in Bengali-English which contains both CM questions and answers and also proposed an evaluation strategy using the corpus.

To the best of our knowledge, our system is the first end-end factoid QA system designed specifically for CM questions.

3 Web Based Code-Mixed QA System

In this section, we describe the details of our system - *WebShodh*. This system is hosted at http://128.2.208.89/webshodh/cmqa.php and is currently supporting Hinglish and Tenglish. A video demonstration of the working of *WebShodh* is available at https://www.youtube.com/watch?v=aVsZVfere5w[5]. Figure 1 presents the architecture of *WebShodh* along with an illustrated example Hinglish CM question *"Oscar jeetne wala Slumdog Millionaire film ka director kaun hain?"* *(Translation: who is the director of the oscar award winning film "Slumdog Millionaire"?)*. Given a natural language question expressed in CM, it was passed through a language identification module from [20]. The principal idea is to lexically translate this question into English so that - (a) we can leverage monolingual resources in English, which is a resource rich language for subsequent processing (b) quality of web search in English is better compared to that in the matrix languages.

Question Classification: The complexities of identifying the Answer Type (AType) for CM questions are discussed in [17] and they also propose a technique for SVM based AType classification. Given the translated CM questions, they use a featurizer to create a bag of features consisting of lexical level features along with the adjacent words of 'Wh-' word, for representing the query. This is passed through an SVM based Question Classifier (QC) which classifies the CM question into one of the given types such as - HUMAN, LOCATION, ENTITY, ABBREVIATION, DESCRIPTION and NUMERIC, the type hierarchy defined by [16]. In this work, we just consider the coarse-grained categories for AType classification since the training data is too sparse in fine-grained category.

In this work, we extended the work done by [17] by including additional class of features to the SVM model. To improve the generalization capability, POS tags features of the words from the respective languages that are identified lexically are used. In addition, pre-trained embeddings from Google news vectors for each of the lexically translated words are used. We considered 10 representative samples from each AType. Later, we compute the centroids for each AType in this 300-dimensional space. For each of the adjacent words on both sides of 'Wh-' word, we get their word2vec embedding, calculate distance with the AType centroids and find out the closest AType to include that as a feature. Besides this, we performed a five-fold cross validation with a grid search for tuning the kernel

---

[5] This video is recorded in real time frame to demonstrate the speed of the system for practical purposes.

and C parameters in SVM. We used an RBF kernel with *gamma* value set to the inverse of feature vector size for better performance. Due to the above changes, we improved the overall accuracy of the QC system across the 6 categories from 63% to 71.96%.

Retrieval of Web Results: We submit the loosely translated English question as a search query to Google using Search API and retrieve the top 10 relevant documents along with their titles, URLs and snippets. "Snippet-tolerant property" [21] is leveraged to arrive at the answer by exploiting the information present in the relevant snippets, as processing the entire document is computationally expensive and time consuming.

Candidate Answer Generation: We run POS tagging, chunking and Named Entity Recognizer tools on the retrieved snippets and titles. The categories of NER are mapped to QC categories, based on which the relevant candidate answers are retrieved. We filter only the words and phrases whose NER tags map to the given AType and pass them to the next phase as candidate answers.

Answer Ranking: For each candidate answer, its relevance score is computed by adding the cosine similarity between the translated CM question and all congregated titles and snippets where the candidate answer occurs. The final list of answers is displayed in a ranked order according to the above relevance score. Redundancy of the correct answer, which occurs in multiple relevant documents, potentially improves its ranking score. But NER on huge text introduces latency in the pipeline. Hence we need to decide on an appropriate trade-off between them.

4   Evaluation Dataset and Results

We used *WebShodh* - our end-end open domain CM QA system for also collecting the evaluation data. We took the help of 10 native speaker volunteers each for Hindi and Telugu languages. All of them were bi-lingual speakers who were also fluent in English. We gave them access to the web interface of our system and requested them to try out at least 10 factoid questions of their choice.

A maximum of 10 ranked answers for each of the CM factoid question are displayed. As a part of the feedback process, the user was asked to select the correct answer and submit it to the system. Through this, we are collecting the data corresponding to a question, its answer along with the answer rank. In this way we have collected 100 questions for each language pair. The details of this evaluation dataset is given in Table 1. The user feedback on question category was purposefully omitted from the interface as there is certain domain knowledge involved in annotating question types, which the users may not be aware of. The data obtained through this platform offers a huge potential to improve CM QA further and hence the system is hosted online. Language Mix Ratio (LMR) is the ratio of the number of words from Embedded language to the total number of words in the sentence. From Table 1, we can observe that on an average, LMR is 0.3937 and 0.3973 respectively for Hinglish and Tenglish CM questions.

QUESTION (HINDI-ENGLISH CM)	QC LABEL	ANSWER	ANALYSIS
Question: oscar jeetne wala film slumdog millionaire ka director kaun hain? **Gloss Translation:** oscar won of film slumdog millionaire of director who is ? **Meaning:** Who is the director of the oscar winning film Slumdog Millionaire?	**Predicted:** Human **Correct:** Human	**Given answer:** Danny Boyle **Correct answer:** Danny Boyle	'Danny Boyle' was provided redundantly in the candidate sentences and was tagged appropriately as PERSON in NER and hence is the highest scored answer. Entity Normalization could further increase this score.
Question: world war 1 kis saal mein shuru hua hain? **Gloss Translation:** world war 1 which year in begin happen is ? **Meaning:** In which year did World War 1 begin?	**Predicted:** Entity **Correct:** Numeric	**Given answer:** 1914 **Correct answer:** 1914	Misclassification as entity leads to candidate answers as all noun phrases and the year '1914' had a POS tag of NP and hence retrieved as highest ranked answer..
Question: acetyl salicylic acid ka doosra naam kya hain? **Gloss Translation:** acetyl salicylic acid of second name what is ? **Meaning:** What is another name of acetyl salicylic acid ?	**Predicted:** Human **Correct:** Entity	**Given answer:** Not found **Correct answer:** aspirin	Most of the examples in training data annotated with 'HUMAN' has adjacent word as 'name'. The misclassification lead to not identifying an answer.
Question: cheap thrills gana kis album se hain? **Gloss Translation:** cheap thrills song which album from is? **Meaning:** Which album does cheap thrills song belong to?	**Predicted:** Entity **Correct:** Entity	**Given answer:** Sia **Correct answer:** This is Acting	Correctly classified as entity. Goes through all noun phrases and explicit mention of the word 'album' is not present in the candidate sentences that increases the score of 'This is Acting'.

QUESTION (TELUGU-ENGLISH CM)	QC LABEL	ANSWER	ANALYSIS
Question: Dan Brown rasina modati pustakam lo protagonist evaru? **Gloss Translation:** Dan Brown written first book in protagonist who? **Meaning:** Who is the protagonist in Dan Brown's first book?	**Predicted:** Human **Correct:** Human	**Given answer:** Robert Langdon **Correct answer:** Robert Langdon	The question is correctly classified and the redundancy of the exact answer 'Robert Langdon' increased its score.
Question: Amnesty International yokka headquarters ekkada undi ? **Gloss Translation:** Amnesty International of headquarters where is? **Meaning:** Where is the headquarters of Amnesty International?	**Predicted:** Location **Correct:** Location	**Given answer:** Uganda **Correct answer:** London	'yokka' has been incorrectly classified as an English word. As a result of this it was not lexically translated to English. Hence accurate set of documents were not retrieved by the query.
Question: ee 19th century painter Marquesas Islands lo chanipoyaru? **Gloss Translation:** What 19th century painter Marquesas Islands in died? **Meaning:** Which 19th century painter died in Marquesas Islands?	**Predicted:** Entity **Correct:** Human	**Given answer:** Paul Gauguin **Correct answer:** Paul Gauguin	The question is misclassified as entity. Candidate answers include all Noun Phrases in this scenario and Paul Gauguin is an NP which is ranked highest.
Question: Japan lo highest point ekkada undi? **Gloss Translation:** Japan in highest point where is? **Meaning:** Where is the highest point in Japan?	**Predicted:** Location **Correct:** Location	**Given answer:** Shizuoka **Correct answer:** Mount Fuji	The correct answer is in the third position in the ranked list. Mount Fuji is in the border of Shizuoka. The extent of granularity of answer varies according to questions.

Fig. 2. Qualitative analysis of results with representative positive and negative examples

Table 1. CM QA Evaluation Dataset Details

Distribution parameters	Hinglish	Tenglish
Number of questions	100	100
Total number of words	833	667
Percentage of English words	39.37%	39.73%
Percentage of native words	60.62%	60.26%
Avg. CM words per question	5	4
Avg. length of questions	8	6

Table 2. Results of end to end WebShodh QA system

Metric	Hinglish	Tenglish
Precision at 1	0.37	0.32
Precision at 3	0.58	0.55
Precision at 5	0.67	0.65
Precision at 10	0.73	0.71
MRR	0.37	0.32

This section presents the quantitative and qualitative analysis of end-to end CM QA system. We use standard evaluation metrics such as precision at various ranks and Mean Reciprocal Rank (MRR) for measuring the effectiveness of our QA system. Table 2 shows the precision at 1, 3, 5 and 10 for both the language pairs along with their corresponding MRR. Figure 2 provides a qualitative analysis of our results for both the language pairs. This analysis is based on the following categories: (a) Both QC label and answer are correct and correct answer is present at rank 1 (b) QC label is incorrect but the predicted answer is correct (c) QC label and answer are incorrect (d) QC label is correct but the answer predicted is incorrect.

5 Discussion and Conclusion

An accurate one to one mapping of alphabet does not exist across most languages that belong to different language families. This raises the issues of spelling variations while romanizing. This problem is commonly observed in the case of 'th' and 't'. Romanized Hindi and Telugu do not have specific environmental conditions or rules to use these letters and are often used interchangeably for wx notations of 't', 'T', 'w' and 'W'. The same problem is observed in the case of other hard and soft consonants. Similarly inconsistencies in representing long and short vowels usually cause errors in transliteration and thus the error is sent downstream to the task of translation. Consider the code-mixed question *'phata poster nikla hero movie lo protogonist evaru?'* (meaning: who is the protagonist in phata poster nikla hero movie?). Though the question itself is in Tenglish, *'phata poster nikla hero'* itself is a Hinglish code-mixed entity, which is the name of a movie. So the non-English words within the entity should be not be lexically translated to get the correct answer. Such entities need to be identified to avoid lexical translation.

Telugu is an agglutinative language which combines multiple morphemes to form a single word. *Sandhi* is the phenomenon of interplay of sounds at the boundaries of adjacent words leading to fusion and alteration of sounds, commonly observed in this language. For example, consider the word *'perenti'* in Telugu (meaning: what is the name) which is a frequently occurring word in Tenglish question dataset. It is a combination of two words *'peru'* (meaning: name) and *'enti'* (meaning: what) based on certain phonetic *sandhi* rules. It depends on the idiolect of the person on choosing to type *'peru enti'* or *'perenti'*. Hence the problem of dealing with code-mixing is compounded with noisy text. We plan to work on these issues further so that we can maximize the benefit of reaping bilingual dictionaries. We also plan to extend the system to Spanglish (code-mixing of Spanish and English) by building a cross script bilingual dictionary and language identification system. Unlike a pidgin, Spanglish could be the primary language of some people, mostly in the areas of Puerto Rico.

In conclusion, today's linguistically pluralistic societies need tools which support interaction in CM languages. In this paper, we presented *WebShodh* - an end-end web-based Factoid QA system for CM languages. We demonstrated our system with two pairs of CM languages - Hinglish and Tenglish. In view of resource scarcity, our system used very few resources such as bi-lingual dictionaries for these languages. We use Google Search API for retrieving the web results along with their snippets and titles. Evaluation of our system reveals that we achieve an MRR of 0.37 and 0.32 for Hinglish and Tenglish respectively. We hosted the system *WebShodh* online to collect more questions in order to understand the intricate variations of these newly formed languages in real world and leverage it for collecting more CM question/answer data which is critical for future research and further improvement of the system.

References

1. Myers-Scotton, C., Linguistics, C.: Bilingual Encounters and Grammatical Outcomes. Oxford University Press, Oxford (2002)
2. Hidayat, T.: An Analysis of Code Switching used by Facebookers (2008)
3. Brill, E., Dumais, S., Banko, M.: An analysis of the AskMSR question-answering system. In: EMNLP-Volume 10 (2002)
4. Zhang, D., Lee, W.S.: A web-based question answering system (2003)
5. Magnini, B., et al.: Overview of the CLEF 2004 multilingual question answering track. In: Peters, C., Clough, P., Gonzalo, J., Jones, G.J.F., Kluck, M., Magnini, B. (eds.) CLEF 2004. LNCS, vol. 3491, pp. 371–391. Springer, Heidelberg (2005). doi:10.1007/11519645_38
6. Tay, M.W.J.: Code switching and code mixing as a communicative strategy in multilingual discourse. World Englishes **8**(3), 407–417 (1989)
7. Lesley, M., Pieter, M.: One Speaker, Two Languages: Cross-Disciplinary Perspectives on Code-Switching. Cambridge University Press, Cambridge (1995)
8. Beatrice, A.: Automatic Detection of English Inclusions in Mixed-lingual Data with an Application to Parsing. Dissertation, University of Edinburgh (2007)
9. Auer, P.: Code-Switching in Conversation: Language, Interaction and Identity (2013)
10. Dey, A., Fung, P.: A hindi-english code-switching corpus. In: LREC, pp. 2410–2413 (2014)
11. Barman, U., Das, A., Wagner, J., Foster, J.: Code mixing: a challenge for language identification in the language of social media. In: EMNLP (2014)
12. Vyas, Y., et al.: POS tagging of english-hindi code-mixed social media content. In: EMNLP, vol. 14, pp. 974–979 (2014)
13. Ferrucci, D., et al.: Building watson: an overview of the DeepQA project. AI Mag. **31**(3), 59–79 (2010)
14. Moschitti, A., et al.: Using syntactic and semantic structural kernels for classifying definition questions in Jeopardy! In: EMNLP, pp. 712–724 (2011)
15. Xu, J., Zhou, Y., Wang, Y.: A classification of questions using SVM and semantic similarity analysis. In: ICICSE, pp. 31–34 (2012)
16. Li, X., Roth, D.: Learning question classifiers. In: International Conference on Computational Linguistics-Volume 1, pp. 1–7 (2002)
17. Chandu, K.R., Chinnakotla, M., Shrivastava, M.: Answer ka type kya he? Learning to classify questions in code-mixed language. In: International Conference on World Wide Web, pp. 853–858. ACM (2015)
18. Majumder, G., Pakray, P.: NLP-NITMZ@ MSIR 2016 system for CodeMixed crossScript question classification. In: ECIR, pp. 7–10 (2016)
19. Banerjee, S., et al.: The first cross-script code-mixed question answering corpus. In: ECIR (2016)
20. Bhat, I.A., et al.: IIIT-H system submission for FIRE 2014 shared task on transliterated search. In: FIRE, pp. 48–53 (2014)
21. Zhang, D., Lee, W.S.: Question classification using support vector machines. In: International ACM SIGIR Conference on Research and Development in Informaion Retrieval, pp. 26–32 (2003)

Extracting Graph Topological Information and Users' Opinion

Mirko Lai[1,2]($\boxtimes$), Marcella Tambuscio[1], Viviana Patti[1], Giancarlo Ruffo[1], and Paolo Rosso[2]

[1] Dipartimento di Informatica, Università degli Studi di Torino,
C.so Svizzera 185, 10149 Torino, Italy
{lai,tambusci,patti,ruffo}@di.unito.it
[2] PRHLT Research Center, Universitat Politècnica de València,
Camino de Vera, s/n, 46022 Valencia, Spain
prosso@dsic.upv.es

Abstract. This paper focuses on the role of social relations within social media in the formation of public opinion. We propose to combine the detection of the users' stance towards BREXIT, carried out by content analysis of Twitter messages, and the exploration of their social relations, by relying on social network analysis. The analysis of a novel Twitter corpus on the BREXIT debate, developed for our purposes, shows that like-minded individuals (sharing the same opinion towards the specific issue) are likely belonging to the same social network community. Moreover, opinion driven homophily is exhibited among neighbours. Interestingly, users' stance shows diachronic evolution.

Keywords: Stance detection · BREXIT · Community detection

1 Introduction

The political public debate is radically changed after the increasing usage of social media in last years. Politicians use them in order to conduct their political campaigns, and to engage users. On the other hand, users interact each other sharing their opinions and beliefs about political agenda or public administration. In this domain, techniques to study and analyse social media users' activity have been gaining importance in recent years, and (now more than ever) automatic approaches are needed in order to deal with this enormous amount of users' generated content. For instance, interest is growing in opinion mining, considered an important task to classify and monitor users' sentiment polarity [8], and in Stance Detection (SD), a finer grained task where the focus is on detecting the orientation *pro* or *con* that users assume within debates towards specific target entity, e.g., a controversial issue [7]. SD could be very useful to probe the citizens' perspective towards particular national and international political issues. Many recent works also suggest the exploitation of users' social community to develop features helping to detect their opinions [2,9]. To learn more

© Springer International Publishing AG 2017
G.J.F. Jones et al. (Eds.): CLEF 2017, LNCS 10456, pp. 112–118, 2017.
DOI: 10.1007/978-3-319-65813-1_10

about the role of social relations in the formation of public opinion we address two research questions: first, if individuals that share the same opinion towards a specific issue are likely to belong to the same community [6]; second, if link formation can be better understood in term of homophily (i.e., users with the same opinion are more likely to be connected to each other). We also explore the possibility to have a diachronic evolution in stance, e.g., people changing their stance after some particular events, happening when the debate is still active [3]. Here, we analysed the political discussion in United Kingdom (UK) about the European Union membership referendum, held on June 23rd 2016, commonly known as BREXIT, on Twitter. We showed that our hypotheses are supported by the analysis of real data proposing a new SD annotation scheme that takes into account temporal evolution, and a method for SD based on SVM in order to label the stance of users involved in the discussion.

2   Dataset

Data collection. In order to explore social relations and temporal evolution of users' stance, we collected about 5M of English tweets containing the hashtag #brexit using the Twitter Stream API, during the time span between June 22nd and 30th. First, we grouped tweets according to three time intervals, corresponding to relevant clear-cut events related to the referendum, in a short and highly focused time window:

- *"Referendum Day"* - the 24 hours preceding the polling stations closing (between June 22nd at 10:00 p.m. and June 23nd at 10:00 p.m.);
- *"Outcome Day"* - the 24 hours following the formalisation of referendum outcome (between June 24nd at 8:00 a.m. and June 25nd at 8:00 a.m.);
- *"After Pound Falls"* - the 24 hours after the financial markets' turbulence that followed the referendum (between June 28nd at 12:00 p.m. and June 29nd at 12:00 p.m.).

Then, we selected a random sample of 600 users from 5,148 that wrote at least 3 tweets in each time interval. We defined a *triplet* as a collection of three random tweets written by the same user in a given time interval. Finally, we created the TW-BREXIT corpus that consists of 1,800 triplets.

Manual annotation. We employed CrowdFlower[1] to annotate the so-obtained corpus. We asked the human contributors to annotate the user's stance on the target *BREXIT* (i.e. UK exit from EU). In particular, given a triplet posted by an user, they had to infer the user's stance, by choosing between three options:

- *Leave*: if they think that the user is *in favour* of the UK exit from EU;
- *Remain*: if they think that the user supports staying within the EU (i.e. the user is *against* BREXIT);

---

[1] http://www.crowdflower.com.

– *None*: if they could not infer user's stance on BREXIT (e.g., all the messages are unintelligible, or the user do not express any opinion about the target, or the user expresses opinion about the target, but the stance is unclear).

The final TW-BREXIT corpus contains 1,760 labelled triplets in agreement (majority voting)[2].

Social Network. By the *friends/list* Twitter API, we collected the follower list for the 4,548 available[3] users over 5,148 that wrote at least 3 tweets in each interval in order to explore users' social network. We obtained a graph where a node represents a user and an edge between two users will exist if one follows the other. The graph consists in 4,114,523 nodes connected by 13,189,524 edges. We then extracted a sub-graph consisting in 198,419 nodes connected by 6,604,298 edges after removing friends having less than 10 relations in order to reduce computational issues.

3   Content and Network Analysis

Diachronic evolution of stance. In order to provide insights on temporal evolution, we analysed the label distribution in TW-BREXIT over the three temporal intervals. Not surprisingly, we observe an unbalanced distribution for stance as shown in Table 1. We used the hashtag #brexit for collecting data: despite it is apparently a neutral hashtag, a recent study [4] shows that most of tweets containing #brexit were posted by people that expressed stance in favour of Brexit, but since we are not interested in predicting the referendum outcome this bias is not crucial for the next analysis. It is more important to notice that label distribution changes over the time, in particular between "Outcome Day" and "After Pound Falls" phases. Then, we considered the point of view of a single user exploring if her/his own stance changes over time. We found that 57,66% of the users was labelled with the same stance in all the three temporal intervals (37,16% Leave, 15,5% None, 5% Remain). Very interestingly, 42,33% of users' labelled stance changes across different temporal intervals. In particular, 9,5% of users' stance varies from *Leave* (L) to *None* (N) (7% L → L → N; 2,5% L → N → N). From these results we cannot infer that users effectively changed opinion, but for sure they express their stance in their tweets in a different way depending on the phase of the political discussion. This is an argument in favour of the hypothesis that stance should be analysed not in isolation but also in a diachronic perspective, which will be matter of future deeper investigations.

Automatic content analysis: stance detection. We aim to automatically estimate the stance of all users of our dataset in order to explore how the stance is distributed in the social network. Then, we propose a machine learning supervised approach using SVM to annotate the stance s of the remaining 3,948

---

[2] Inter-Annotator Agreement: 65.48. The corpus is available for research purposes.

[3] Some users set privacy in order to hide profile information, while others shut down their profile after the referendum.

Table 1. Label distribution over the time

Time span	Leave	Remain	None
Average	961 (51%)	236 (14%)	563 (35%)
Referendum day	55.67%	13.67%	30.67%
Outcome day	55.67%	14%	30.33%
After Pound falls	**50%**	13.67%	**36.33%**

users, using the following five features computed over a triplet: bag of words (BoW), structural-based (structural), sentiment-based (sentiment) (described in [5]), community-based (community), and temporal-based (temporal). The community feature returns the community of the user who wrote the triple, while the temporal one, the given time interval of the triplet. The F-Measure $\frac{F_{leave} + F_{remain}}{2}$ obtained by SVM using all the mentioned features is 67% and it overcomes the performance of SVM trained with unigrams (58.25%) and unigrams plus n-grams (60.14%) (baselines proposed by [7]).

Community Detection. Subsequently, we analysed the network topology. Figure 1(a) shows the graph plotted by the software Gephi[4] coloured by user's community. The users community's membership was assigned by the Louvain Modularity method [1]. Figure 1(b) shows the graph where users have been coloured according the annotated stance computed with SVM. Table 2 highlights that the percentage of users' stance in community D is evidently biased towards the stance "Remain"; in communities B, E, and F towards the stance "Leave"; in communities A and C towards the stance "None". The existences of communities so defined in term of stance could allow filter bubble phenomena to occur.

Neighbourhood Overlap. Lastly, we evaluated the stance similarity among couples of connected nodes. Then, we defined users *agreement* as a measure of the likelihood that two users i and j have the same stance (i.e., $s(i) = s(j)$) in the same time interval, and then we explored how the agreement between two users changes depending on the rate of the common neighbours. *Neighbourhood Overlap* (NO) is defined as the number of neighbours that nodes i and j have in common divided by the sum of neighbours of both i and j (not counting i and j themselves):

$$NO(i,j) = \left(\frac{|\{N_i \cap N_j\}|}{|\{N_i \cup N_j\} \setminus \{i,j\}|} \right) \tag{1}$$

where N_i and N_j are the sets of neighbours of nodes i and j respectively. Table 3 shows how to compute the agreement score $A_{i,j}$ between i and j. Considering $E_l = \{(i,j) \in E | \wedge NO(i,j) = l\}$ as the subset of edges that are incidents to neighbours of both i and j, when $NO(i,j)$ value is exactly equal to l, we computed the *agreement* A_l related to the NO level l as it follows:

---

[4] http://gephi.org.

$$A_l(i,j) = \sum_{i,j|(i,j)\in E_l} \frac{A_{i,j}}{|E_l|}. \tag{2}$$

Roughly speaking, through this measure we want to explore if the opinion agreement among users changes accordingly the rate of common neighbourhood. We computed users' agreement in our dataset for each time interval and then we took the averaged value for different values of neighbourhood overlap. Figure 2 shows that the agreement between two users increases depending on the percentage of friends. Results showed the tendency of users to associate with similar others according to opinion driven homophily.

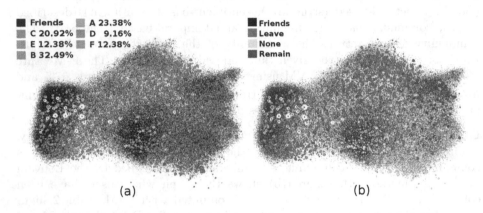

(a)   (b)

Fig. 1. In (a), each node is coloured depending on assigned community. Otherwise, in (b), they are coloured according to the annotated stance of the user's triplet by SVM (red for *Leave*, yellow for *None*, blue for *Remain*, mixed colours when stance changes over time). Followers and the remaining users are black-coloured (Color figure online).

Table 2. Users' stance distribution over communities. The percentage shows the average users' distribution in communities over the three temporal phases.

Community	A	B	C	D	E	F
Leave	29.63%	**84.61%**	26.31%	18.96%	**85.6%**	**75%**
Remain	11.11%	0.37%	17.02%	**57.47%**	2.37%	0%
None	**56.79%**	14.28%	**54.38%**	18.39%	10.06%	22.92%

4   Discussion

In this paper we have shown that users having the same stance towards a particular issue tend to belong to the same social network community. Moreover, we found evidences that the neighbours are more likely to have similar opinions. The obtained results show that stance verified by human annotators over

Table 3. The table shows the agreement score for couple of users (i, j) over the temporal phases. The maximum value is 1 in the case i and j agree $(s(i) = s(j))$ in all the three temporal phases, 0 if one or both users have label "None" and -1 otherwise.

	Agreement	One or both none	Disagreement
Referendum day	0.33	0	-0.33
Outcome day	0.33	0	-0.33
After Pound falls	0.33	0	-0.33
	1	0	-1

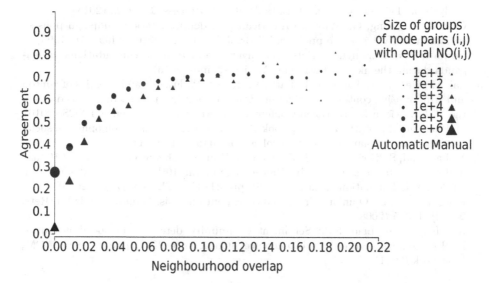

Fig. 2. A shape (circle or triangle) represents a group of node pairs (i, j) with equal $NO(i, j)$ (rounded to two decimal points). Shape size is proportional to the size of such groups. The agreement score $A_{i,j}$ was computed with manual annotation stance (triangle) and with user's stance computed by SVM (circle). We noted that the *affinity* among two users increases depending on the rate of NO.

the same user varies over time, even though we exclusively focused on three 24-hours time slots in a time span of only 8 days. This suggests that stance should be studied considering the diachronic evolution of the debate. We are planning to combine the diachronic evolution of users' stance with the dynamic social network perspective and to explore this methodology on other political corpora. In our future research we would also like to understand the role that influencers could have on the stance change. Moreover, we would like to investigate the use of irony within polarised communities in order to figure out if social network relations influence the use of this figurative language.

Acknowledgments. The work of the last author has been partially funded by the Spanish Ministry of Economy, Industry and Competitiveness (MINECO) under the research project SomEMBED TIN2015-71147-C2-1-P and by the Generalitat Valenciana under the grant ALMAMATER (PrometeoII/2014/030).

References

1. Blondel, V.D., Guillaume, J.L., Lambiotte, R., Lefebvre, E.: Fast unfolding of communities in large networks. J. Stat. Mech. Theory Exp. (2008). 10
2. Deitrick, W., Hu, W.: Mutually enhancing community detection and sentiment analysis on Twitter networks. J. Data Anal. Inf. Process. **1**, 19–29 (2013)
3. Gelman, A., King, G.: Why are American presidential election campaign polls so variable when votes are so predictable? Br. J. Polit. Sci. **23**(04), 409–451 (1993)
4. Howard, P.N., Kollanyi, B.: Bots, #strongerin, and #brexit: computational propaganda during the uk-eu referendum. ArXiv e-prints (2016)
5. Lai, M., Hernández Farías, D.I., Patti, V., Rosso, P.: Friends and enemies of clinton and trump: using context for detecting stance in political tweets. In: Proceedings of the 15th Mexican International Conference on Artificial Intelligence. LNCS (2016)
6. McPherson, M., Smith-Lovin, L., Cook, J.M.: Birds of a feather: homophily in social networks. In: Annual Review of Sociology, pp. 415–444 (2001)
7. Mohammad, S., Kiritchenko, S., Sobhani, P., Zhu, X., Cherry, C.: Semeval-2016 task 6: detecting stance in tweets. In: Proceedings of the 10th International Workshop on Semantic Evaluation (SemEval-2016), pp. 31–41. ACL, San Diego (2016)
8. Pang, B., Lee, L.: Opinion mining and sentiment analysis. Found. Trends Inf. Retr. **2**(1–2), 1–135 (2008)
9. Xu, K., Li, J., Liao, S.S.: Sentiment community detection in social networks. In: Proceedings of the 2011 iConference, iConference 2011, pp. 804–805. ACM, New York (2011)

Estimating the Similarities Between Texts of Right-Handed and Left-Handed Males and Females

Tatiana Litvinova[1,3(✉)] ⓘD, Pavel Seredin[1,2,3] ⓘD, Olga Litvinova[1] ⓘD,
and Ekaterina Ryzhkova[1] ⓘD

[1] RusProfiling Lab, Voronezh 394043, Russia
centr_rus_yaz@mail.ru
[2] Voronezh State University, Voronezh 394018, Russia
[3] The Kurchatov Institute, Moscow 123098, Russia

Abstract. Identifying the characteristics of text authors is of critical importance for marketing, security, etc., and there has been a growing interest in this issue recently. A major feature to be researched using text analysis has been gender. Despite a lot of studies that have obviously contributed to the progress in the field, identification of gender with text authors so far remains challenging and daunting. One of the reasons is that current research shows no consideration of the mutual influences of various individual characteristics including gender and laterality. In this paper, using the material of a specially designed corpus of Russian texts named *RusNeuroPsych*, including the neuropsychological data of the authors, we calculated the distance between texts written by right-handed and left-handed males and females (4 classes). For this study we have chosen handedness as one of the most important laterality measures. In order to calculate the distance between the classes, a formula measuring the Wave-Hedges distance was employed. The text parameters were topic-independent and frequent (the indices of lexical diversity, a variety of parts of speech ratios, etc.). It was shown that texts by authors of different genders but with an identical type of handedness are more similar linguistically than those by individuals of the same gender but with a different type of manual preference. We suppose that it could be useful to build a classifier for classes "gender + handedness" instead of predicting gender itself.

Keywords: Gender attribution · Authorship profiling · Handedness · Laterality · Distance measures · Russian language

1 Problem Statement

Author gender identification is one of the most challenging issues that has been facing authorship profiling over the last 15 years [2, 5, 15, 16]. PAN's participants are frequently confronted with it as well [13, 14]. Despite considerable progress, text-based gender identification is still a complex issue particularly in cross-genre settings (in cases of gender identification for English at PAN 2016, most systems obtained accuracies below the baseline, see [14] for details). There are several factors contributing to the complexity of the issue, and it is primarily the fact that apart from gender, each individual is inherently characterised by different personality traits, emotional states, etc. which in

© Springer International Publishing AG 2017
G.J.F. Jones et al. (Eds.): CLEF 2017, LNCS 10456, pp. 119–124, 2017.
DOI: 10.1007/978-3-319-65813-1_11

combination affect one's writing style and are thus to be investigated as one. For example, Schler et al. [17] showed mutual influences of gender and age: writing style grows increasingly "male" with age, pronouns and assent/negation become scarcer, while prepositions and determiners become more frequent. As of now, however, little is known on how gender interacts with other individual characteristics with respect to authorship profiling task.

In the current research we are addressing mutual interactions of author gender with one of the major neuropsychological features of humans that reflects individual differences during the collaboration of two cerebral hemispheres, i.e. laterality [3]. It is the primary integral characteristics of the work of the brain that are central to general regulatory and cognitive processes. We made use of one of the most important and frequently used behavioral indicator for cerebral laterality – degree of handedness [11]. As a result of a plethora of works, the correlations between the degree of handedness and cognitive, as well as psychological, characteristics of individuals have been established [1, 3, 11] considering gender as well [10]. Morton [8] showed that individuals of the same hemisity but opposite sex had more personality traits in common than those of the same sex but different hemisity (hemisity is a term that refers to inherent opposite right or left brain-oriented differences in thinking and behavioral styles. Right and left brain-oriented groups were shown to reveal neuroanatomical differences [8]).

A lot of research has been conducted on the impact of handedness on language processing [9]. Although the degree of lateralisation of language function between the hemispheres is known to vary in left-handed and right-handed populations, linguistic characteristics of texts by authors with different types of manual asymmetry have not been looked into in depth. In one of the few papers on the topic [6] texts by individuals with different types of motor and sensory asymmetry were shown to have statistically significant differences.

The objective of the paper is to identify the similarity measurements of written texts by males and females with different degrees of handedness. To the best of our knowledge, it is the first time this issue has been addressed.

2   Materials and Methods

The study was performed using a specially designed corpus of Russian written texts named RusNeuroPsych that, apart from texts, contains rich metadata [6] [1]. Each text was written in the presence of the experimenter. Topics of the texts are letter to a friend and description of a picture. Each respondent wrote one or two texts. The collection is divided into two parts: "Children" (texts written by school children aged from 12 to 17) and "Adult" (texts written by peoples from 18 to 35, mostly students). For this study we use subcorpus "Adults" which contains 392 texts by 209 peoples as well as rich metadata (gender, year of birth, education, Big 5 scores, neuropsychological testing scores, scores on HADS test). An average text contains 153 words. The index of handedness of the

---

[1] The corpus is freely available at http://en.rusprofilinglab.ru/korpus-tekstov/rusneuropsych-corpus/.

respondents was calculated as the difference between the number of "right", "left" and "symmetrical" answers divided into the number of tests (7). As a result, all the respondents were divided into three subgroups according to their handedness: left $(-1, -0.33)$, medium $(-0.33, 0.33)$, right $(0.33, 1)$ levels. Respondents with the medium level of handedness were then removed. As a result, we had 128 texts by right-handed and 30 texts by left-handed females and 79 texts by right-handed and 21 by left-handed males. The comparisons were performed for the following classes of texts: (1) written by right-handed females and left-handed females; (2) written by left-handed males and right-handed females; (3) written by right-handed males and left-handed females; (4) written by right-handed males and left-handed males; (5) written by right-handed males and right-handed females; (6) written by left-handed males and left-handed females.

In order to tackle these issues, all the texts were linguistically labelled using the morphological analyser Pymorphy2 and the online service istio.com, as well as a the LIWC software [12], including dictionaries we designed. Therefore indices of lexical diversity of texts, frequencies of different parts of speech (POS) and their ratios, frequency of punctuation marks, frequencies of some vocabulary groups (emotion and perception), individual parts of speech, etc. (235 in total) were obtained (cf. [7]). At the next stage the parameters with the values equaling zero in more than 50% of the texts were excluded from the total list of the parameters (156 parameters). Then we tested the hypothesis assuming that there are statistically significant differences between the same text parameters from the selections we had designed. In order to establish whether there was a statistically significant difference in two groups of the text parameters with normal data distribution, the independent-samples t-test was used. When the data was not normally distributed, the non-parametric Mann-Whitney criterion was employed. In both cases, in order to test the hypothesis, assuming that there are statistically significant differences between the groups of parameters, we calculated the achieved level of the significance of a statistical test (t-test and Mann-Whitney test), i.e. the acceptable one for the first-order error, and then compared it with the specified level for this study $p < 0.05$.

3   Results and Discussion

Statistically significant differences for groups of texts were identified in the following linguistic parameters:

- *Right-handed females and left-handed females*: The proportion of function words (FW) in the text; TTR (100) (type/token ratio in the first 100 words of the text); proportion of words from the list of 100 most frequent words in Russian; proportion of the particle "NOT" («HE»); proportion of deictic words; number of FW/number of punctuation marks;
- *Left-handed males and right-handed females:* Proportion of FW in the text; Proportion of quantitative words (numerals + pronominal adverbs); proportion of words describing perception;
- *Right-handed males and left-handed females:* proportion of words from the list of 100 most frequent words in Russian; proportion of the preposition ON («HA»);

proportion of the preposition "BY" («У»); proportion of words describing emotions; number of FW/number of commas; number of FW/number of punctuation marks; proportion of the total number of punctuation;
- *Right-handed males and left-handed males:* Proportion of function words + pronouns in the text; Proportion of function words in the text; percentage of 5 most frequent words excluding function words; Proportion of function words in 5 most frequent words; Proportion of quantitative words (numerals + pronominal adverbs)/total number of words; Proportion of perception words; Number of FW/number of commas;
- *Right-handed males and right-handed females:* TTR(100); proportion of 5 most frequent words including FW; proportion of all punctuation marks;
- *Left-handed males and left-handed females:* proportion of words describing perception.

At the final stage of the analysis, we calculated the distance measure for six classes of individuals depending on their gender and degree of handedness. In order to do that, using the average values of the text parameters where there are statistically significant differences of the corresponding selections, weighted centroids were designed, $[A_1 \ldots A_n]$ and $[B_1 \ldots B_n]$, that were based on n elements, which are the average values of the selected text parameters of the compared selections; n is the number of the elements of the centroid that equals the number of text parameters for the two compared classes (selections) and existing statistically significant differences.

The distance measure of the classes A and B was calculated based on the formula for the Wave-Hedges distance [4]

$$S(A, B) = \sum_i^n \frac{|A_i - B_i|}{\max(A_i B_i)} \tag{1}$$

Table 1 lists the values of the function of the distance measure S(A, B) calculated according to (1) for the corresponding classes, as well as the ratio of the distance measure to the number of elements in the centroid S(A, B)/n in order to account for the effect of the number of elements (text parameters with some differences) in the class. It should be noted that even though the arrangement of the compared classes using only one of the calculated distance measures S(A, B) or S(A, B)/n is in good agreement, in order to get a good result we had to arrange the class considering both distance S(A, B) and S(A, B)/n. As the rank goes up (from 1 to 6), there is less similarity between the compared classes.

The analysis of the material revealed that texts by right-handed males and females respectively were found to be the most similar quantitatively, while texts by males and particularly females with different manual preferences vary the most, which enables us to argue that handedness has a considerable effect on the quantitative text parameters. This is to be considered while developing the methods of identifying text author gender. The obtained results prove that it is necessary that psychophysiological parameters of individuals and their effect on writing are investigated not in isolation but in combination.

Table 1. Distance measures for texts by males and females with different types of handedness

Gender_Handedness	Distance measure S(A, B)	Ratio of distance measures and number of parameters in centroid S(A, B)/n	Rank of the class
F_R/F_L	1.52	0.25	6
M_L/F_R	0.96	0.24	3
M_R/F_L	1.28	0.21	3
M_R/M_L	0.72	0.12	2
M_R/F_R	0.17	0.08	1
M_L/F_L	0.26	0.26	3

4   Conclusion and Future Work

The study showed that neuropsychological characteristics of text authors are to be employed in identification of their gender, even though it has to be said that the results are preliminary. There are currently plans to increase the number of participants in order to test the results, as well as to carry out a similar study using available texts of "children's" subcorpus that is part of RusNeuroPsych and designed by students from grades 6 to 10. Besides, in this light we are going to examine not only the degree of handedness but also the other indices of functional asymmetry (motor, sensory, cognitive) as well as personality traits, trying to link such tasks in order to build a common framework to allow us to have a better understanding of how people use language. There are plans to design a classifier for "gender + handedness". In future we are planning to build classifiers for classes "gender + handedness" instead of predicting gender itself. We argue that studies of how individual characteristics manifest themselves in writing and interact allow the accuracy of prognostic models to be increased and make an overall contribution to authorship profiling.

Acknowledgments. This research is financially supported by the Russian Science Foundation, project No. 16-18-10050, "Identifying the Gender and Age of Online Chatters Using Formal Parameters of their Texts".

References

1. Beratis, I., Rabavilas, A., Kyprianou, M., Papadimitriou, G., Papageorgiou, C.: Investigation of the link between higher order cognitive functions and handedness. J. Clin. Exp. Neuropsychol. **35**(4), 393–403 (2013)
2. Company, J., Wanner, L.: How to use less features and reach better performance in author gender identification. In: Proceedings of the 9th International Conference on Language Resources and Evaluation (LREC), pp. 1315–1319. Reykjavik, Iceland (2014)
3. Khomskaia, E.D., Efimova, I.V., Budyka, Ye.V., Yenikolopova, Ye.V.: Neyropsikhologiya individual'nykh razlichiy [Neuropsychology of Individual Differences]. Russian Pedagogical Agency, Moscow (1997) (in Russian)

4. Kocher, M., Savoy, J.: Distance measures in author profiling. Inf. Process. Manag. **53**(5), 1103–1119 (2017)
5. Koppel, M., Argamon, S., Shimoni, A.R.: Automatically categorizing written texts by author gender. Literary Linguist. Comput. **17**(4), 401–412 (2002)
6. Litvinova, T., Ryzhkova, E., Litvinova, O.: Features a written speech production of people with different profiles of the lateral brain organization (on the Basis of a New Type RusNeuroPsych Corpus). In: Proceedings of the 7th Tutorial and Research Workshop on Experimental Linguistics (ExLing 2016), pp. 103–107. International Speech Communication Association, Saint Petersburg, Russia (2016)
7. Mikros, G.: Systematic stylometric differences in men and women authors: a corpus-based study. In: Köhler, R. Altmann, G. (eds.) Issues in Quantitative Linguistics 3. Dedicated to Karl-Heinz Best on the Occasion of His 70th Birthday, pp. 206–223. RAM – Verlag (2013)
8. Morton, B.E.: Behavioral laterality of the brain: support for the binary construct of hemisity. Front. Psychol. **4**, 683 (2013)
9. Newman, S., Malaia, E., Seo, R.: Does degree of handedness in a group of right-handed individuals affect language comprehension? Brain Cogn. **86**, 98–103 (2014)
10. Nicholls, M.E.R., Forbes, S.: Handedness and its association with gender-related psychological and physiological characteristics. J. Clin. Exp. Neuropsychol. **18**, 905–910 (1996)
11. Papadatou-Pastou, M., Tomprou, D.: Intelligence and handedness: meta-analyses of studies on intellectually disabled, typically developing, and gifted individuals. Neurosci. Biobehav. Rev. **56**, 151–165 (2015)
12. Pennebaker, J., Booth, R., Boyd, R., Francis, M.: Linguistic Inquiry and Word Count: LIWC2015. Pennebaker Conglomerates, Austin (2015)
13. Rangel, F., Celli, F., Rosso, P., Potthast, M., Stein, B., Daelemans, W.: Overview of the 3rd author profiling task at PAN 2015. CLEF 2015 Labs and Workshops, Notebook Papers, CEUR-WS.org. Toulouse, France (2015)
14. Rangel, F., Rosso, P., Verhoeven, B., Daelemans, W., Potthast, M., Stein, B.: Overview of the 4th author profiling task at PAN 2016: Cross-Genre Evaluations. Working Notes Papers of the CLEF 2016 Evaluation Labs. CEUR-WS.org. Évora, Portugal (2016)
15. Sarawgi, R., Gajulapalli, K., Choi, Y.: Gender attribution: tracing stylometric evidence beyond topic and genre. In: Proceedings of the Fifteenth Conference on Computational Natural Language Learning, pp. 78–86. Association for Computational Linguistics, Portland (2011)
16. Sboev, A., Litvinova, T., Gudovskikh, D., Rybka, R., Moloshnikov, I.: Machine learning models of text categorization by author gender using topic-independent features. Procedia Comput. Sci. **101**, 135–142 (2016)
17. Schler, J., Koppel, M., Argamon, S., Pennebaker, J.W.: Effects of age and gender on blogging. In: AAAI Spring Symposium: Computational Approaches to Analyzing Weblogs, vol. 6, pp. 199–205 (2006)

Evaluation of Hierarchical Clustering via Markov Decision Processes for Efficient Navigation and Search

Raul Moreno[✉], Wĕipéng Huáng, Arjumand Younus, Michael O'Mahony, and Neil J. Hurley

Insight Centre for Data Analytics, University College Dublin, Dublin, Ireland
{raul.moreno,weipeng.huang,arjumand.younus,
michael.omahony,neil.hurley}@insight-centre.org

Abstract. In this paper, we propose a new evaluation measure to assess the quality of a hierarchy in supporting search queries to content collections. The evaluation measure models the scenario of a searcher seeking a particular target item in the hierarchy. It takes into account the structure of the hierarchy by measuring the cognitive challenge of determining the correct path in the hierarchy as well as the reduction in search time afforded by hierarchy. The goal is to propose a general-purpose measure that can be applied in different application contexts, allowing different hierarchical arrangements of content to be quantitatively assessed.

1 Introduction

Content collections are commonly arranged in hierarchical taxonomies. For example, a commercial retailing site in the clothing sector may arrange its catalogue in categories such as "Leisurewear", which may contain a sub-category "Sportswear". A customer with a particular product in mind can then search through the category hierarchy in order to zone in on the sub-set of the collection that contains the required target product. Generally, there is more than one way to organise a hierarchy. For example, a book such as "Harry Potter" might be stored under a high-level category "Children's Books" and sub-category "Fantasy", or indeed "Fantasy" might be the higher-level category with "Children's Books" beneath it. Or a different taxonomy might instead use a category of "Books about Wizards". Given a particular hierarchical organisation of content, we address the question of how to evaluate its usefulness for supporting content search queries. Such a measure would facilitate the quantitative comparison of one hierarchy with another, without the need for a ground-truth hierarchy.

A range of methods for measuring the coherence of sub-categories within a hierarchy have been proposed [1]. However, we seek a measure that evaluates the *structure* of the hierarchy as well as the coherence of the hierarchical clusters. In particular, we seek a measure that addresses the question of how easy it is to navigate through a hierarchy from its root to some target content, accounting

G.J.F. Jones et al. (Eds.): CLEF 2017, LNCS 10456, pp. 125–131, 2017.
DOI: 10.1007/978-3-319-65813-1_12

for the cognitive cost of choosing a correct path at each branch of the hierarchy. A few hierarchical clustering measures address this issue [2,3]. Johnson et al. [3] evaluate hierarchy structure but their approach relies on the availability of ground-truth clusters. Cigarran et al. approach [2] proposes a goal-oriented evaluation measure of the hierarchical clustering quality, which considers the content of the cluster, the hierarchical arrangement and the navigation cost. Notably, the approach utilizes the idea of a Minimal Browsing Area (MBA)[1] to measure navigation cost. We differ from these approaches in that our focus is on evaluation without ground-truth and, based on Markov decision processes [4,5] our novel evaluation measure is designed to model the behaviour of a searcher navigating a hierarchy.

2 Markov Decision Processes

Markov decision processes (MDPs) [4,5] provide an appropriate mathematical framework for modelling decision making where outcomes are partly random and partly under the control of a decision maker. An MDP is a discrete time process, in which at each time step, the process is in some state s and the decision-maker must choose an available action a to move to a new state s'. Each move has an associated reward $R_a(s, s')$, which provides the motivation for choosing particular decisions. The goal of an MDP may be stated in terms of seeking a *policy* for choosing the action at each step that earns the maximum expected cumulative reward over time.

2.1 Navigating a Hierarchy as an MDP

In our scenario, the searcher must make a decision at each visited node in the hierarchy. There are two possible actions at each state (or node in the hierarchy):

1. Stay at the current node and examine the documents stored under it
2. Navigate to a child node

Assuming that the target document is stored under the current node i, the reward of choosing to remain at this node may be written as the overall reduction in the number of documents (n_i) that need to be examined, compared with the full document set (n). Navigating to the child containing the target accumulates reward 0 at the current step, however opens the possibility of an enhanced reward at a subsequent step, when the user finally decides to explore the documents at a lower-level node. Once the target document is examined, the search process ends. If an incorrect child is chosen, then a fixed reward of -1 is obtained regardless of subsequent steps, which can never reach the target. Hence, it is unnecessary to examine the search any further after a bad choice is made. In summary, the

---

[1] An MBA for the target is the minimal part of the hierarchy that, starting from the top node, a user must explore in order to reach all relevant items.

search process consists of visits to a set of states $\{s_0, s_1, \dots\}$ containing the target document with the following rewards assigned to each choice:

$$R_{\text{remain}}(s_i) = 1 - \frac{n_i}{n} \quad R_{\text{good}}(s_i, s_{i+1}) = 0 \quad R_{\text{bad}}(s_i, s_{i+1}) = -1.$$

Effectively, the search ends once a non-zero reward is obtained.

The expected reward depends on the navigation policy π that the searcher follows in order to decide between the alternatives at each step in the process. Given a particular policy, π, we can evaluate the probabilities associated with the different possible state transitions. In particular, given a target document t, we write the probabilities at node i of the hierarchy as:

- $p_i(t)$ = probability of choosing the correct path to the target
- $q_i(t)$ = probability of choosing an incorrect path
- $r_i(t)$ = probability of choosing to examine all documents under the node i

We have that $p_i(t) + q_i(t) + r_i(t) = 1$. Also, $q_i(t)$ is the sum of the probabilities to navigate to all incorrect nodes at the level, which indicates the tree structure is not necessarily binary.

Let $\{s_0, s_1, \dots, s_{d(t)}\}$ be the set of nodes on the path from the root ($= s_0$) to the leaf node, $s_{d(t)}$, that contains the target t. Finally, we propose to define the *hierarchy score* given target document t as the expected reward, i.e.,

$$\text{HQ}(L, \{t\}, \pi) = \mathrm{E}[R(\pi)] = \sum_{i=1}^{d(t)} \left(\prod_{j=0}^{i-1} p_j(t) r_i(t) (1 - \frac{n_i}{n}) - \prod_{j=0}^{i-1} p_j(t) q_i(t) \right),$$

where $p_0 = 1$. The value of the hierarchy for searching a document set D using policy π is

$$\text{HQ}(L, \pi) = \sum_{t \in D} w_t \, \text{HQ}(L, \{t\}, \pi),$$

where w_t is the importance weight of document t such that $\sum_t w_t = 1$ (e.g. w_t may be taken as the relative frequency of a search for t).

As an illustration of the formula, consider a hierarchy in which the number of documents is reduced by a factor of ϵ at each level of the hierarchy. Thus, the reward if the target is found at level i, is $1 - \epsilon^i$. We examine the formula in the limit as $i \to \infty$, for different choices of the three decision probabilities.

In particular, considering that $p_i = \gamma$, $r_i = \alpha(1 - \gamma)$ and $q_i = (1 - \alpha)(1 - \gamma)$ for some $\alpha \in [0, 1], \gamma \in [0, 1]$, we have

$$E[R] = \sum_{i=0}^{\infty} \gamma^i \alpha (1-\gamma)(1-\epsilon^i) - \sum_{i=0}^{\infty} \gamma^i (1-\alpha)(1-\gamma)$$

$$= (1-\gamma)\left(\sum_{i=0}^{\infty} (\alpha - (1-\alpha))\gamma^i - \alpha \sum_{i=0}^{\infty} (\gamma \epsilon)^i \right)$$

$$= (1-\gamma)\left((2\alpha - 1) \sum_{i=0}^{\infty} \gamma^i - \alpha \sum_{i=0}^{\infty} (\gamma \epsilon)^i \right)$$

$$= (1-\gamma)\left(\frac{2\alpha - 1}{1-\gamma} - \frac{\alpha}{1-\gamma \epsilon} \right)$$

$$= (2\alpha - 1) - \alpha \frac{1-\gamma}{1-\gamma \epsilon}.$$

With $\gamma > 0$ and $\alpha \to 0$, the searcher always descends through the hierarchy, with a non-zero probability of taking the wrong branch at each descent, so that $R \to -1$. As $\alpha \to 1$, $E[R] \to 1 - (1-\gamma)/(1-\gamma \epsilon)$. The searcher is always inclined to take the correct branch and the overall benefit of the hierarchy depends on the rate ϵ at which number of documents reduces as the searcher descends the hierarchy, with a maximum reward of 1 in the limit as $\epsilon \to 1$.

Consider the special case where $\gamma = 1/3$ and $\alpha = 1/2$, then the searcher has no guidance on which path to choose at each level in the tree and $p_i = q_i = s_i = 1/3$. Now $E[R] = -1/(3 - \epsilon)$. A negative reward in $-[\frac{1}{2}, \frac{1}{3}]$ illustrates that unless the hierarchy provides some useful guidance to the user, it is better to exhaustively search all documents.

3   Choosing a Navigation Policy

The navigation policy depends on the information that is available to determine which path to take. Depending on the application, different types of node summaries might be available. If the documents are organised using topic modelling, for example, there may be a set of words that describe each node. Alternatively a set of prototype examples of the contents of each child might be presented.

For the evaluation measure, we assume that, given a target t, it is possible to compute the distance between t and a cluster of documents, C, and write this distance as $d(t, C)$. In many applications, a pairwise distance between documents is available: $d(t_1, t_2)$, $\forall t_1 t_2 \in D$, and the cluster distance can be based on these distances e.g.

$$d(t, C) = \frac{1}{|C|} \sum_{t' \in C} d(t, t').$$

Other possibilities include defining a distance based on the word distribution in each cluster (e.g., see [6,7]), when topic modelling has been used to construct the hierarchy. In any case, we develop the navigation policy under the assumption the information available to the searcher is the set the values $d(t, C)$ available for each child node, C that can be reached from a parent node.

3.1 Policy: Greedy Descent to Threshold Depth

To develop a navigation policy, a criterion is required to decide whether to end the descent into the hierarchy at the current node. Furthermore, if a decision to explore deeper is made, a criterion to determine which child to explore is required. A practical real scenario is that the searcher has some search budget in mind, and is willing to examine a cluster of documents when the number of documents in the cluster is below a certain threshold. This results in the following probability:

$$r_i(\pi_1) = \begin{cases} 1, & \text{if } n_i \leq thr \\ 0, & \text{otherwise} \end{cases}$$

Also, the choice of child depends deterministically on $d(t, C)$, with the searcher choosing the child that is closest to the target. Let $\{C_1, \ldots, C_k\}$ be the available child nodes at a particular decision point, let $\mathbb{1}()$ correspond to the indicator function which is 1 when its Boolean argument is true and zero otherwise and let $\ell = \arg\min_j d(t, C_j)$. Then

$$p_i(\pi_1) = (1 - r_i(\pi))\mathbb{1}(t \in C_\ell)$$
$$q_i(\pi_1) = (1 - r_i(\pi))\mathbb{1}(t \notin C_\ell).$$

This policy represents a reasonable approximation to how searchers explore real-world hierarchies and therefore it is the primary policy that we propose to use in our measure. A crisp hierarchy in which the $d(t, C_j)$ are distinct and items tend to be organised in coherent clusters will evaluate highly in our measure with this policy. Outlying items that are equally distant from several clusters are likely to yield negative reward, as the searcher is likely to eventually choose an incorrect path, the deeper the descent into the hierarchy. The value of the threshold depth should be chosen based on the application context.

4   Toy Example and Conclusion

As an example, we generate $M = 10,000$ random points arranged in 8 clusters with samples spread around each cluster center as shown in Fig. 1a. Figure 1b illustrates the binary tree that is obtained on application of agglomerative clustering using the pairwise distance between samples. Every node contains all items which appear in the lower levels, therefore we can decide in each step to: examine all the documents in the current node to retrieve our target, navigate to the child where the target is located, or choose the wrong path to others.

We apply our hierarchical evaluation following the policy described in 3.1 with a threshold of $n/8$. A score per item is obtained, measuring the quality of the hierarchy for this specific retrieved target.

As an overall evaluation we compute the mean value of these scores, obtaining a hierarchy quality $HQ = 0.52$. Examining the distribution of the quality reveals

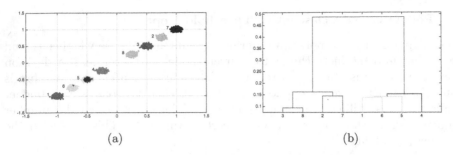

Fig. 1. Figure (a) shows the 10000 random samples generated in 8 clusters, and targets with negative evaluation in the black triangles. Figure (b) represents the hierarchy obtained with agglomerative clustering and the flaws in the dendrogram.

that 79% of the values are above 0.8 quality, according to the expected 1/8 reduction in search complexity when the searcher succeeds in finding the target. A minority of samples get a negative reward, as the hierarchy leads them to the incorrect node/child, reducing the overall quality to 0.52. Let's examine the targets with negative evaluation (see Fig. 1a), these points belong to clusters 5 and 6. As Fig. 1b shows, all points in cluster 5 and some in cluster 6 are closer on average to cluster 4 than they are to the full cluster formed by 1, 5, 6. Therefore, when the navigation reaches that decision point in the hierarchy, it chooses the wrong path to cluster 4 and hence misses the target in cluster 5. So, although the hierarchy has chosen the right cluster in its leaf nodes, it contains a flaw at that internal node. Our proposed measure captures that flaw successfully.

Finally, we can conclude the proposed measure provides a reliable measure of the hierarchical quality and successfully detects flaws in the dendrogram. Further analysis should be done in a wide range of situations and different policies should be applied.

Acknowledgments. This project has been funded by Science Foundation Ireland under Grant No. SFI/12/RC/2289.

References

1. Liu, Y.C., Li, Z.M., Xiong, H., Gao, X.D., Wu, J.J., Wu, S.: Understanding and enhancement of internal clustering validation measures. IEEE Trans. Cybern. **43**(3), 982–994 (2013)
2. Cigarran, J.M., Peñas, A., Gonzalo, J., Verdejo, F.: Evaluating hierarchical clustering of search results. In: Consens, M., Navarro, G. (eds.) SPIRE 2005. LNCS, vol. 3772, pp. 49–54. Springer, Heidelberg (2005). doi:10.1007/11575832_7
3. Johnson, D.M., Xiong, C., Gao, J., Corso, J.J.: Comprehensive cross-hierarchy cluster agreement evaluation. In: AAAI (Late-Breaking Developments) (2013)
4. Gosavi, A.: Reinforcement learning: a tutorial survey and recent advances. INFORMS J. Comput. **21**(2), 178–192 (2009)
5. Sutton, R.S., Barto, A.G.: Reinforcement Learning: An Introduction, vol. 1. MIT press Cambridge, Cambridge (1998)

6. Kuang, D., Park, H.: Fast rank-2 nonnegative matrix factorization for hierarchical document clustering. In: Proceedings of the 19th ACM SIGKDD International Conference on Knowledge Discovery and Data Mining, pp. 739–747. ACM (2013)
7. Blei, D.M., Jordan, M.I., Griffiths, T.L., Tenenbaum, J.B.: Hierarchical topic models and the nested chinese restaurant process. In: NIPS (2003)

Plausibility Testing for Lexical Resources

Magdalena Parks[1(⊠)], Jussi Karlgren[2], and Sara Stymne[1]

[1] Uppsala University, Uppsala, Sweden
magdalena.f.parks@gmail.com
[2] Gavagai, Stockholm, Sweden

Abstract. This paper describes principles for evaluation metrics for lexical components and an implementation of them based on requirements from practical information systems.

1 Evaluating Information System Components

The performance of a component in a complex processing pipeline can influence the function of downstream components, meaning that end-to-end testing also must be performed on entire systems, using approaches based on use cases with target notions that validate the function of the system for the purpose it is built, such as many of the evaluation measures formulated in workshops at CLEF. But a task-based evaluation does not reveal the performance of individual components. Evaluation of knowledge-based components in an information system should be done systematically, ideally in ways which are similar to unit tests done for other technical components, motivated by the need for a development and maintenance team to:

1. know that the component does what is expected of it;
2. support modular design with clean interfaces between modules and simplify refactoring and other development tasks;
3. support cross-project portability of modules;
4. enable bug tracking, optimization, and error analyses in case of system failure or performance dips.

Evaluating learning lexical resources is challenging for several reasons [1].

Firstly, operationalizable intrinsic measures for knowledge based models risk being irrelevant for system performance or measure outcome, rather than learning process. Secondly, hands-off evaluation which takes a general view of component performance and disregards the body of knowledge which it has been designed to address risks missing important aspects of what a component is intended to do. Thirdly, a purely declarative test may be applied in several different ways, some of which may not provide guidance for system improvement.

Typical evaluation resources are based on a list of selected probe terms and target terms which are in some identified relation to those target terms. Examples of synonym tests are the TOEFL test [2], WordSimilarity-353 [3], SIMLEX 999 [4], ConceptSim [5] and so forth, with a more fine-grained approach for probing several semantic relations in BLESS [6]. These resources are reasonable for

© Springer International Publishing AG 2017
G.J.F. Jones et al. (Eds.): CLEF 2017, LNCS 10456, pp. 132–137, 2017.
DOI: 10.1007/978-3-319-65813-1_13

testing a static resource and are designed to be a sample of general language usage.

2   Requirements for a Habitable Evaluation Scheme

The reasons an information system needs a learning lexical resource (rather than a stable dictionary) can be varied. One central motivation is to help an information system select the most reasonable term from a set of candidates. This is useful in many tasks: in generation, speech recognition, and various variant suggestion situations, both to rank alternatives in the face of choice and to provide defaults in the absence of information. To do this well a lexical resource must have some sense of context. A tentatively useful approach is to use the Cloze procedure, and our suggested test shares many of the starting points with it [7]. In this paper we suggest an approach for testing a lexical resource which is based on *semantic coherence* and allows for topic or domain tailoring.

An evaluation method must not be biased towards any particular kinds of representation since the ultimate goal of an evaluation would be to fairly compare the different knowledge representations. However, the form of the semantic representation can vary greatly from system to system, making a more direct evaluation very difficult. A more universal method of evaluating semantic representations is to design a task which can only be accurately performed given an accurate knowledge of word semantics. Such a task must meet certain criteria.

First, it is essential that the task should be able to be reliably performed using semantic knowledge. Next, it is important that the task should not be able to be reliably performed without that semantic knowledge. More specifically, syntax should not play a key role in a system's ability to perform the task. While it may be impossible to fully eliminate the role of all non-semantic information on the solvability of the task, it is necessary to minimize the impact as much as possible.

Last, it is necessary for the task to have a solution that can be confirmed without the use of semantic knowledge. Using existing semantic knowledge to confirm the solution would bias the test towards systems similar to the semantic knowledge system used. While using a human-generated answer key as a gold standard would be an acceptable solution in terms of correctness, it is highly inefficient if we ever hope to expand the test to various domains or to include different vocabulary, as we would need to have humans carefully go through and create a new gold standard for any new test cases. Ideally, we would like to use a task which can be programmatically generated, including an answer key, while still requiring semantic knowledge to solve.

3   Plausible Utterances

The proposed task is a *plausible utterances* task. To generate this task, we create a set of sentences containing a number of sentences which occurred naturally and have remained unaltered, as well as a number of sentences which occurred naturally but have had one or more words within them swapped for some other word. We call this set a *coconut*, as per examples given by Karlgren et al. [1]. The task is then to sort these sentences in order of likelihood of having occurred naturally. We can then evaluate the ordering of these results to see whether the system reliably ranks the real sentences higher than the fake sentences.

The plausible utterances test can meet the criteria stated above. By replacing words in the sentence, we are creating a sentence which contains words that are, in all likelihood, in discord with the rest of the sentence. As long as semantics plays some role in this, semantic knowledge can be used to perform the task, satisfying our first criteria. Additionally, we can satisfy the second criteria as long as we carefully select word replacements in such a way as to minimize syntactic incongruity. This test approaches language in a more natural setting than e.g. the *word intrusion* task suggested by Chang et al. [8], where human assessors were asked to pick out a randomly introduced topical term in a set of topically consistent terms, and is related to the notion of *perplexity*, used to evaluate consistency and coverage language models for e.g. speech recognition [9]; in contrast with perplexity measures which are intrinsic to a data set, our suggested coconut items are transparent and comprehensible to any human language user for e.g. error analysis purposes.

Finally, since we have programmatically generated the coconut, we know which sentences have been manipulated, and which ones have not, so we naturally have an answer key. While it is possible that some manipulated sentences may, by chance, be reasonably plausible, generating and testing with multiple different coconuts should reduce the impact of these false positives.

4   Method of Fake Sentence Generation

The key method involved in generating this plausible utterances task is the generation of the fake sentences. We do this by selecting words from naturally occurring sentences and replacing them with other, syntactically similar words.

It is necessary that our swapping methods minimize the syntactic incongruity of the sentence. To this end, we make use of POS tags as part of our swapping methodology. Specifically, we elect to only replace words which have the same POS. This easily results in sentences which have minimal syntactic incongruity, while adding only the additional requirement of using tagged data.

We use the Brown corpus [10] as our source of naturally occurring sentences, and limit our experiments to swapping only singular nouns, with the tag NN.

While the vast majority of words are simply swapped with any word of the same POS, we find that some specific instances are not covered by the POS tagset. For instance, if we take the natural noun phrase a dog, we might end up with a apple, which is syntactically incorrect. We correct for this by altering the determiner based on the first letter of the replacement word, a simple correction that works in most cases.

5   Swapping Strategies

Choices concerning the selection of words to be replaced, as well as their replacements, can have a profound impact on the difficulty level of the coconut and can present different challenges to the semantic knowledge systems being tested. Two variants of the coconut are proposed here.

5.1   Word Coconut

The first variant involves the selection of a probe word. A number of sentences are selected which naturally contain the probe word, and a number of sentences are altered to contain the word. To generate these fake sentences, we randomly select real sentences from the Brown corpus [10] which contain some word that has the same part of speech as our probe word. We then swap that word with our probe word. Probe words are randomly selected out of all words that we see occur in more than one sentence, to avoid unique words from showing up in our tests. In this variation, the sentences each contain the same word in different contexts, and the task is to determine which contexts are the most likely to have occurred naturally.

Examples (1) show an example of a word coconut that could be generated using the probe word dog using the POS tag NN, such that one real sentence is selected and one fake sentence is generated. Note that while both sentences syntactically work, the first sentence would make sense semantically while the second sentence is nonsensical.

(1) a. The woman walked the dog to the park.
    b. He sipped his dog as he read the newspaper.

Examples (2), on the other hand, show a coconut which may be generated using the same parameters, but is semantically ambiguous. While our methodology may generate such coconuts, we can expect that in most cases there will be relatively few ambiguous sentence pairs within a coconut. For this reason, we should not expect even the best system to be able to solve this task perfectly in all cases, but we can expect it to perform comparatively better given the same coconuts.

(2) a. The dog sat on the bed.
    b. The old woman saw a dog.

5.2   Sentence Coconut

The second variant involves the selection of a sentence and a target word in that sentence. That target word is then replaced with various words, and each variant is used as part of the test set. We randomly select a sentence, similar to our process for the word coconut. In this variation, the sentences are all identical except for the target word. The task is to determine which target word is most likely to have occurred naturally. Examples (3) show a sentence coconut.

(3) a. The woman walked the dog to the park.
     b. The woman walked the idea to the park.

Ideally, we wish to be able to compare a system's performance on the different coconut types relative to each other. For this reason, we choose to generate coconuts of both types containing an equal number of sentences. In our experiments, we use eight sentences in each coconut.

Next, we observe that the sentence coconut generation method results in only a single real sentence, while word coconuts may have more. Again, in the interest of comparability, we choose to make word coconuts conform to the sentence coconut, only retrieving a single real usage of the given word for use in the coconut. In the end, we have both word coconuts and sentence coconuts each with a total of eight sentences, one of which is a natural utterance.

6   Evaluating the Evaluation

To test the evaluation we created two naïve n-gram language models from the Reuters corpus [11] using the SRILM toolkit [12], using an additive smoothing of 0.1. First, we use the n-gram data to measure the likelihood of an entire probe sentence. Secondly, more specifically, we take a window around the swap word of the probe sentence, based on the range on the n-gram model in question. To test a 3-gram model, we use a 5 word window, centered on the swapped word. The idea here is not to test the sequences in the probe sentence we know are parts of real utterances: by using this context window, we are isolating the n-grams involved in the part of the sentence that we are unsure is a real utterance. We do not expect either of these implementations to perform our task particularly well, but intend to use them to demonstrate certain characteristics of our coconut test.

We run our experiment using, as described above, coconuts of size 8, each with only a single unmodified sentence. We use 2-gram, 3-gram, and 4-gram models for each test. We generate 46 coconuts of each type, and for each type run our experiment using the full context and the context window. For each coconut, we find the ranking of the only unmodified sentence, with the best being 1 and the worst being 8. We calculate the average rank over all coconuts for our final results, shown in Table 1.

Table 1. Results of Coconut size 8, using ngram models, with different implementations

	Context window word coconut	Full context word coconut	Context window sentence coconut	Full context sentence coconut
2-gram	3.93	3.43	4.85	5.15
3-gram	3.85	3.37	5.24	4.83
4-gram	3.83	3.37	4.91	4.72

7   Conclusion

We see that sentence coconuts are more resilient to changes in the methodology used to perform the plausible utterances task when compared to word coconuts. While the results for the word coconuts quickly settle on results which favor the full context implementation, we see that the sentence coconuts do not reliably demonstrate such a pattern. This is an indication that the result is less dependent on the task implementation than the language model itself. Contrary to expectation, we see that the implementation trained on a narrow and thus more specific context window is outperformed on the word coconuts by the implementation trained on entire sentences as context. This is most likely due to data scarcity, and illustrates well that word coconuts, being a less exact model of performance, are more sensitive to implementation details than sentence coconuts. We find that sentence coconuts are recommended as an evaluation metric, in that they provide a better resolution between test models. The coconut tests are currently being introduced in an industrial setting as part of a test suite for testing a learning lexical resource deployed in numerous languages.

References

1. Karlgren, J., et al.: Evaluating learning language representations. In: Mothe, J., Savoy, J., Kamps, J., Pinel-Sauvagnat, K., Jones, G.J.F., SanJuan, E., Cappellato, L., Ferro, N. (eds.) CLEF 2015. LNCS, vol. 9283, pp. 254–260. Springer, Cham (2015). doi:10.1007/978-3-319-24027-5_25
2. Landauer, T., Dumais, S.: A solution to Plato's problem: the latent semantic analysis theory of acquisition, induction and representation of knowledge. Psychol. Rev. **104**(2), 211–240 (1997)
3. Finkelstein, L., Gabrilovich, E., Matias, Y., Rivlin, E., Solan, Z., Wolfman, G., Ruppin, E.: Placing search in context: The concept revisited. In: Proceedings of the 10th international conference on World Wide Web. ACM (2001)
4. Hill, F., Reichart, R., Korhonen, A.: Simlex-999: evaluating semantic models with (genuine) similarity estimation. arXiv preprint arXiv:1408.3456 (2014)
5. Schwartz, H.A., Gomez, F.: Evaluating semantic metrics on tasks of concept similarity. In: Proceedings of FLAIRS (2011)
6. Baroni, M., Lenci, A.: How we BLESSed distributional semantic evaluation. In: Proceedings of the 2011 Workshop on GEometrical Models of Natural Language Semantics, pp. 1–10. ACL (2011)
7. Taylor, W.L.: Cloze procedure: a new tool for measuring readability. Journalism Q. **30**(4), 415–433 (1953)
8. Chang, J., Boyd-Graber, J.L., Gerrish, S., Wang, C., Blei, D.M.: Reading tea leaves: how humans interpret topic models. In: Proceedings of NIPS (2009)
9. Katz, S.: Estimation of probabilities from sparse data for the language model component of a speech recognizer. IEEE Trans. Acoust. Speech Signal Process. **35**(3), 400–401 (1987)
10. Francis, W.N., Kucera, H.: Brown Corpus Manual, p. 15. Brown University (1979)
11. Lewis, D.D.: Reuters-21578, distribution 1.0 (1997)
12. Stolcke, A., et al.: SRILM - an extensible language modeling toolkit. In: Interspeech, vol. 2002, p. 2002 (2002)

An Improved *Impostors* Method
for Authorship Verification

Nektaria Potha$^{(\boxtimes)}$ and Efstathios Stamatatos

University of the Aegean, 83200 Karlovassi, Greece
{nekpotha,stamatatos}@aegean.gr

Abstract. Authorship verification has gained a lot of attention during
the last years mainly due to the focus of PAN@CLEF shared tasks. A verification method called *Impostors*, based on a set of external (impostor)
documents and a random subspace ensemble, is one of the most successful
approaches. Variations of this method gained top-performing positions in
recent PAN evaluation campaigns. In this paper, we propose a modification of the *Impostors* method that focuses on both appropriate selection
of impostor documents and enhanced comparison of impostor documents
with the documents under investigation. Our approach achieves competitive performance on PAN corpora, outperforming previous versions of
the *Impostors* method.

Keywords: Authorship analysis · Authorship verification · Text categorization

1 Introduction

Authorship verification is the task of examining whether two (or more) documents are written by the same author [10,11,13]. It is a fundamental task in
authorship analysis since any authorship attribution problem can be decomposed
into a series of verification problems [9]. In comparison to closed-set attribution,
the verification task is more challenging since it focuses on whether the candidate
author and the text under investigation have a *similar enough* style rather than
what candidate author is *the most similar*. On the other hand, an advantage
of verification over closed-set attribution is that the performance of a verification method is affected by less factors since the candidate set size is always
singleton and the distribution of training texts over the authors is not so important. Authorship verification methods have been applied in several applications
in humanities [7,18] and forensics [5]. Recently, a series of related PAN@CLEF
shared tasks were organized attracting multiple submissions [16,17].

The *Impostors* method was introduced by Koppel and Winter [11] and so far,
it is one of the most successful approaches. Variations of this method won first
places in PAN-2013 and PAN-2014 shared tasks in authorship verification [8,14].
This method uses a set of external documents by other authors (with respect to
the ones under investigation) and builds a simple random subspace ensemble.

© Springer International Publishing AG 2017
G.J.F. Jones et al. (Eds.): CLEF 2017, LNCS 10456, pp. 138–144, 2017.
DOI: 10.1007/978-3-319-65813-1_14

Essentially, it attempts to transform the verification problem from a one-class classification task to a binary classification task since it calculates whether the texts by the candidate author or the impostors are closer to the disputed texts.

In this paper, we propose a modified version of the *Impostors* method that enhances its performance. Rather than selecting the impostor texts randomly, we propose to use the texts with the highest min-max similarity to the texts under investigation. To use a metaphor, in a police lineup it doesn't make sense to draw the suspects from the general population. Rather, all the suspects should have similar characteristics. Another weakness of the original method is that it disregards cases where at least one impostor text is found more similar to the disputed text in comparison to the texts of the candidate author. To compensate, we propose to rank similarities in decreasing order and take into account the position of the candidate author's text. The proposed approach is evaluated on several PAN corpora and achieves very competitive results.

2   Previous Work

There are two main paradigms in authorship verification. *Intrinsic methods* perform analysis only on the documents under investigation and handle the verification problem as a one-class classification task. They are robust since they do not require external resources and fast since they analyse a few documents [4,6,13]. On the other hand, *extrinsic methods* analyse an additional set of external documents and transform the verification problem to a binary classification task [1,11,19]. They are usually more effective [16,17]. From another perspective, a set of verification approaches consider verification problems as instances of a binary classification task and attempt to train a classifier that can distinguish between positive (same-author) and negative (different-author) problems [2,12]. Such methods heavily depend on the properties of the training corpus.

A modified version of the *Impostors* method, called *General Impostors* (GI) was introduced by Seidman [14]. Since the original method only handles pairs of documents, GI considers the case where multiple documents by the candidate author are available (following the guidelines of PAN shared tasks). Another modification is proposed by Khonji and Iraqi [8]. They focus on the GI weakness of disregarding cases where at least one impostor document is found more similar to the disputed text than the candidate author's text and they utilise the similarity information from those cases. However, the absolute similarity score may significantly differ when different sets of documents are used. Based on that, in this paper we introduce the use of the ranking information. Yet another modification of the *Impostors* method is described by Gutierrez et al. [3]. They propose an aggregate function that iterates over document pairs and applies homotopy-based classification.

3   The Proposed Method

The GI method accepts as input data a set of documents by the same author (known documents) and exactly one document of disputed authorship (unknown document) and provides a score in [0, 1] indicating whether the unknown and known documents are by the same author [14]. This score can be viewed as the probability of a positive answer and can be transformed to a binary answer given an appropriate threshold. GI requires a set of external documents by other authors (with respect to the ones included in the documents under investigation). It randomly selects a subset of these external documents to serve as impostors. Then, it builds a random subspace ensemble by selecting randomly in each repetition a subset of features and a subset of impostors and calculates the similarity of (both known and unknown) documents with impostors [14]. The main idea is that if the known and unknown documents are by the same author, then the known documents will outperform impostors in terms of similarity with the unknown document.

In this paper, we propose a modification of GI (see Algorithm 1) that attempts to improve the following points:

1. *Impostor selection*: Instead of selecting the impostor documents for each verification problem randomly, we propose to select the external documents with the highest min-max similarity [11] score with respect to the known documents. That way, we increase the probability to consider challenging impostor documents that have at thematic or stylistic similarity with the known documents.
2. *Ranking information*: We only compare the impostor document with the unknown document (rather than with both the known and unknown documents). We consider the impostor document as a direct competitor of the known document and therefore we want to know what of these two is more similar to the unknown document. Moreover, instead of taking into account only the cases where the known document is found more similar to the unknown document than all the impostors, we rank (decreasingly) the similarities of both known document and the impostors and consider the ranking position of the known document. For example, if a known document is found to be more similar to the unknown document than all but one impostor across all repetitions (not necessarily the same impostor each time), the original GI method will return a score of 0 while our method will provide a score of 0.5.

4   Experiments

4.1   Setup

To evaluate the proposed approach and compare it with the original GI method and its most important variations, we use the corpora developed at PAN evaluation campaign on authorship verification in 2014 and 2015. These corpora include multiple verification problems in four languages (Dutch, English, Greek, and Spanish) and cover several genres (newspaper articles, essays, reviews, literary texts, etc.) Separate training and evaluation parts are provided for each corpus. In PAN-2014, known and unknown documents within a verification problem have thematic similarities and belong to the same genre. On the other hand,

Data: D_{known}, $d_{unknown}$, $D_{external}$
Parameters: $repetitions$,$|Impostors_{problem}|$,$|Impostors_{repetition}|$,$rate$
Result: $FinalScore$
for *each* $d_{known} \in D_{known}$ **do**
    for *each* $impostor \in D_{external}$ **do**
      | $MinMax(impostor) = minmaxSimilarity(impostor, d_{known})$;
    end
    Select $Impostors_{problem} \subset D_{external}$ with highest $MinMax(:)$;
    /* Select $Impostors_{problem} \subset D_{external}$ randomly   */
    Set $Score(d_{known}) = 0$;
    repeat *repetitions* times
        Select $Impostors_{repetition} \subset Impostors_{problem}$ randomly;
        Select $rate\%$ of features randomly;
        for *each impostor* $\in Impostors_{repetition}$ **do**
            $Sim(impostor) = similarity(impostor, d_{unknown})$;
            /* $Sim(impostor) =$
                     $similarity(impostor, d_{known}) * similarity(impostor, d_{unknown})$ */
    end
    $Sim_{known} = similarity(d_{known}, d_{unknown})$;
    /* $Sim_{known} = similarity(d_{known}, d_{unknown})^2$ */
    Rank $S = Sim(:) \cup Sim_{known}$ in decreasing order;
    pos = position of Sim_{known} in S;
    $Score(d_{known}) = Score(d_{known}) + 1/(repetitions * pos)$;
    /* if $Sim_{known} > max(Sim(:))$ then
        | $Score(d_{known}) = Score(d_{known}) + 1/repetitions$;
    end
                                                 */
    end;
    $FinalScore = aggregate(Score(:))$;
end

Algorithm 1. The proposed method. Changes with respect to the original GI are shown in blue. Original GI is shown in comments.

in PAN-2015, known and unknown documents within a problem may belong to different genres and their thematic areas may be distinct which make the task even harder. More details about these corpora as well as evaluation results of PAN participants are provided in [16,17].

The GI method and the proposed variation have several parameters that need to be set. Previous studies attempted to fine-tune these parameters separately [8,14]. To simplify this process, we focus on fine-tuning parameter $a = |Impostors_{problem}|$ and then use $repetitions = a/5$ and $|Impostors_{repetition}| = a/10$. Moreover, we use character 5-grams as features, a fix $rate = 0.5$ and the *min-max similarity* function. The *aggregate* function is selected among min, max, and average for each training corpus separately. Most of the times, average is selected [14]. Since GI is a stochastic algorithm, each experiment is repeated five times and we report average Area Under the ROC curve (AUROC) measures as used at PAN-2014 and PAN-2015 evaluation campaigns. The set of external documents ($D_{external}$) is constructed for each corpus separately. We submit queries in Bing search engine using significant (with highest *tf-idf*) words from the set of known documents of the training corpus and download the first results. More than 1,000 documents per corpus were downloaded and html tags were stripped off. No further pre-processing is performed.

4.2   Results

For each one of the PAN-2014 and PAN-2015 authorship verification corpora, we report the performance of our implementation of the original GI method and the proposed variation. To study the contribution of each proposed change described in Sect. 3 separately, we also report performances of taking into account only the impostor selection change (Proposed-1) and only the ranking information change (Proposed-2). Additionally, we include the performance of other variations of the *Impostors* method as described by Khonji and Iraqi [8] and Gutierrez et al. [3]. The AUROC evaluation results are presented in Table 1. As can be seen, the proposed approach outperforms in all but one case (PAN15-EN) the original GI method, in most of the cases by a large margin. The proposed method is also very competitive with respect to Khonji and Iraqi [8], the overall winner of PAN-2014. There is a mixed picture as concerns the contribution of the impostor selection change and the ranking information change and it is not clear which one of them is most important. However, their combination (proposed-full) is better than each one of them in all but one case (PAN14-DR).

Table 1. AUROC results of the proposed approach and other variations of the *Impostors* method.

	PAN14-DE	PAN14-DR	PAN14-EE	PAN14-EN	PAN14-GR	PAN14-SP	PAN15-DU	PAN15-EN	PAN15-GR	PAN15-SP
Khonji and Iraqi (2014)	0.913	**0.736**	0.590	0.750	0.889	**0.898**				
Gutierrez et al. (2015)							0.592	0.739	0.802	0.755
Original GI	0.947	0.660	0.618	0.649	0.772	0.604	0.667	**0.803**	0.656	0.785
Proposed-1	0.970	0.704	0.565	0.738	0.520	0.540	0.662	0.765	0.811	0.825
Proposed-2	0.901	0.698	0.655	0.634	0.860	0.772	0.595	0.786	0.742	0.802
Proposed-full	**0.976**	0.685	**0.762**	**0.767**	**0.929**	0.878	**0.709**	0.798	**0.844**	**0.851**

5   Conclusion

Two main changes of the *Impostors* method are proposed in this paper. The first change makes the selection of impostor documents per verification problem a deterministic procedure ensuring that impostors will have similar characteristics with the candidate author's texts. The second change attempts to enrich the information that is kept in each repetition of the random subspace ensemble. Experiments in several authorship verification corpora demonstrate that the combination of these changes significantly enhance the performance and the proposed approach is competitive, if not better, than another variation of GI that won the first place in PAN-2014 evaluation campaign [8]. The presented results further attest the effectiveness of the *Impostors* method and future work can more thoroughly examine the use of alternative text representation schemes and the profile-based paradigm [15].

References

1. Bagnall, D.: Author identification using multi-headed recurrent neural networks. In: Cappellato, L., Ferro, N., Gareth, J., San Juan, E. (eds.) Working Notes Papers of the CLEF 2015 Evaluation Labs (2015)
2. Fréry, J., Largeron, C., Juganaru-Mathieu, M.: UJM at CLEF in author identification. In: CLEF 2014 Labs and Workshops, Notebook Papers. CLEF and CEUR-WS.org (2014)
3. Gutierrez, J., Casillas, J., Ledesma, P., Fuentes, G., Meza, I.: Homotopy based classification for author verification task-notebook for PAN at CLEF 2015. In: Cappellato, L., Ferro, N., Gareth, J., San Juan, E. (eds.) Working Notes Papers of the CLEF 2015 Evaluation Labs (2015)
4. Halvani, O., Winter, C., Pflug, A.: Authorship verification for different languages, genres and topics. Digit. Investig. **16**, S33–S43 (2016)
5. Iqbal, F., Khan, L.A., Fung, B.C.M., Debbabi, M.: E-mail authorship verification for forensic investigation. In: Proceedings of the 2010 ACM Symposium on Applied Computing, pp. 1591–1598. ACM (2010)
6. Jankowska, M., Milios, E.E., Keselj, V.: Author verification using common n-gram profiles of text documents. In: Proceedings of COLING, 25th International Conference on Computational Linguistics, pp. 387–397 (2014)
7. Kestemont, M., Stover, J.A., Koppel, M., Karsdorp, F., Daelemans, W.: Authenticating the writings of Julius Caesar. Expert Syst. Appl. **63**, 86–96 (2016)
8. Khonji, M., Iraqi, Y.: A slightly-modified GI-based author-verifier with lots of features (ASGALF). In: CLEF 2014 Labs and Workshops, Notebook Papers. CLEF and CEUR-WS.org (2014)
9. Koppel, M., Schler, J., Argamon, S., Winter, Y.: The fundamental problem of authorship attribution. Engl. Stud. **93**(3), 284–291 (2012)
10. Koppel, M., Schler, J., Bonchek-Dokow, E.: Measuring differentiability: unmasking pseudonymous authors. J. Mach. Learn. Res. **8**, 1261–1276 (2007)
11. Koppel, M., Winter, Y.: Determining if two documents are written by the same author. J. Am. Soc. Inf. Sci. Technol. **65**(1), 178–187 (2014)
12. Pacheco, M., Fernandes, K., Porco, A.: Random forest with increased generalization: a universal background approach for authorship verification. In: Cappellato, L., Ferro, N., Jones, G., San Juan, E. (eds.) CLEF 2015 Evaluation Labs and Workshop - Working Notes Papers. CEUR-WS.org (2015)
13. Potha, N., Stamatatos, E.: A profile-based method for authorship verification. In: Likas, A., Blekas, K., Kalles, D. (eds.) SETN 2014. LNCS (LNAI), vol. 8445, pp. 313–326. Springer, Cham (2014). doi:10.1007/978-3-319-07064-3_25
14. Seidman, S.: Authorship verification using the impostors method. In: Forner, P., Navigli, R., Tufis, D. (eds.) CLEF 2013 Evaluation Labs and Workshop - Working Notes Papers (2013)
15. Stamatatos, E.: A survey of modern authorship attribution methods. J. Am. Soc. Inf. Sci. Technol. **60**, 538–556 (2009)
16. Stamatatos, E., Daelemans, W., Verhoeven, B., Juola, P., López-López, A., Potthast, M., Stein, B.: Overview of the author identification task at PAN 2015. In: Working Notes of CLEF 2015 - Conference and Labs of the Evaluation Forum (2015)

17. Stamatatos, E., Daelemans, W., Verhoeven, B., Stein, B., Potthast, M., Juola, P., Sánchez-Pérez, M.A., Barrón-Cedeño, A.: Overview of the author identification task at PAN 2014. In: Working Notes for CLEF 2014 Conference, pp. 877–897 (2014)
18. Stover, J.A., Winter, Y., Koppel, M., Kestemont, M.: Computational authorship verification method attributes a new work to a major 2nd century African author. J. Am. Soc. Inf. Sci. Technol. 67(1), 239–242 (2016)
19. Veenman, C., Li, Z.: Authorship verification with compression features. In: Forner, P., Navigli, R., Tufis, D. (eds.) CLEF 2013 Evaluation Labs and Workshop - Working Notes Papers (2013)

Comparison of Character n-grams and Lexical Features on Author, Gender, and Language Variety Identification on the Same Spanish News Corpus

Miguel A. Sanchez-Perez[✉], Ilia Markov,
Helena Gómez-Adorno, and Grigori Sidorov

Center for Computing Research, Instituto Politécnico Nacional, Mexico City, Mexico
miguel.sanchez.nan@gmail.com, imarkov@nlp.cic.ipn.mx,
helena.adorno@gmail.com, sidorov@cic.ipn.mx

Abstract. We compare the performance of character n-gram features ($n = 3-8$) and lexical features (unigrams and bigrams of words), as well as their combinations, on the tasks of authorship attribution, author profiling, and discriminating between similar languages. We developed a single multi-labeled corpus for the three aforementioned tasks, composed of news articles in different varieties of Spanish. We used the same machine-learning algorithm, Liblinear SVM, in order to find out which features are more predictive and for which task. Our experiments show that higher-order character n-grams ($n = 5-8$) outperform lower-order character n-grams, and the combination of all word and character n-grams of different orders ($n = 1-2$ for words and $n = 3-8$ for characters) usually outperforms smaller subsets of such features. We also evaluate the performance of character n-grams, lexical features, and their combinations when reducing all named entities to a single symbol "NE" to avoid topic-dependent features.

Keywords: Feature selection · Authorship attribution · Author profiling · Discriminating between similar languages · Lexical features · Character n-grams

1 Introduction

This paper focuses on three natural language processing (NLP) tasks that have experienced an increase in interest in recent years: authorship attribution (AA), author profiling (AP), and discriminating between similar languages (DSL). Authorship attribution (AA) is the task that aims at automatically identifying the author of a text [1], when author profiling (AP) aims at identifying profiling aspects of an author, such as age, gender, or native language based solely on a sample of his or her writing.[1] Discriminating between similar languages (DSL) is the task of predicting the language variety in which a given text was written.

---

[1] In this paper, we only address gender identification.

© Springer International Publishing AG 2017
G.J.F. Jones et al. (Eds.): CLEF 2017, LNCS 10456, pp. 145–151, 2017.
DOI: 10.1007/978-3-319-65813-1_15

From the machine-learning perspective, all the three tasks can be viewed as a multi-class, single-label classification problem, where automatic methods have to assign class labels (e.g., author's name (AA); author's gender (AP); language variety (DSL)) to objects (text samples). Practical applications of these tasks vary from electronic commerce and forensics to machine translation and information retrieval systems.

Character n-grams and lexical features (unigrams and bigrams of words), as well as their combinations, have proved to be predictive for these tasks, including when the Spanish language or its varieties are concerned [2,3]. Thus the research question addressed in this work is to examine which features and feature combinations are the best predictive for author, gender, and language variety identification when evaluated on the same corpus in Spanish. Moreover, we evaluate the impact of NEs on these tasks.

2 Related Work

In this section, we will focus on the best approaches for the Spanish language published in the most recent editions of two widely known workshops: PAN[2] and VarDial[3]. These workshops provide a common platform for researchers interested in evaluating and comparing their systems' performance on the authorship identification-related and discriminating between similar languages tasks, respectively.

In the 2014 edition of the PAN Authorship Attribution (AA) competition [4], the task consisted in identifying the author of a text on a corpus composed of newspaper opinion articles. The winner approach for Spanish [5] used a modification of the *Impostors* method [6]. The author identification (author verification) task in PAN 2015 [7] focused on a cross-genre scenario, that is, when training and test sets are on different genre (e.g., tweets vs. news articles). The best approach for Spanish [8] relied on a variety of features, including character n-grams, words, POS tags, and sentence length.

In the 2015 edition of the PAN Author Profiling (AP) task [9], the winning approach [10] for gender identification on the Spanish tweets corpus was based on second order attributes technique. In 2016 [2], the shared task focused on cross-gender AP conditions. The best approach [11] in identifying the gender on the Spanish dataset relied on words, sentiment and topic derivation, and stylistic features.

The 2016 edition of the VarDial workshop for discriminating between similar languages (DSL) [12] used a corpus of short excerpts of news texts, covering Argentine, Castilian, and Mexican Spanish. The overall winner [13] employed character n-gram features ($n = 1-7$). This year edition [3] included Argentinian, Peruvian, and Peninsular Spanish. The winner [14] used character n-grams ($n = 1-4$) for predicting the language group and character n-grams of different order,

---

[2] http://pan.webis.de.
[3] http://ttg.uni-saarland.de/vardial2017/sharedtask2017.html.

POS n-grams, and proportions of capitalized letters, punctuation marks, and spaces for identifying the language varieties within the group.

The results for the DSL task are usually higher than those for AA or AP. For instance, the best performing system [14] in the VarDial 2017 workshop [3] achieved 92.74% of accuracy, while the results for AA and AP under single-genre conditions are usually around 80% [2,7]. As can be seen, the state-of-the-art approaches in both shared tasks employed character n-gram and lexical features. Therefore, it makes it important to evaluate the performance of these features on the same corpus for the three tasks.

3   Corpus

There are numerous works that tackle the evaluation of character n-gram and lexical features' performance for the English language, since for English there is a large number of corpora and lexical resources. However, for Spanish, the availability of corpora is scarce, which limits the amount of research done for this language. For the evaluation of character n-gram and lexical features, we built a corpus composed of news articles in eight varieties of the Spanish language: Argentinian, Mexican, Colombian, Chilean, Venezuelan, Panamanian, Guatemalan, and Peninsular Spanish.

The corpus includes only the news with a minimum size of 750 characters. We removed all the news with distributed authorship, e.g., *AP, La prensa, Editorial*, etc. Overall, between 10 and 40 texts (news articles) were selected for each author in the corpus; these ranges were set so that the corpus is not highly unbalanced with respect to the number of documents per author. Additionally, we manually checked each news content and deleted names of authors, places, emails, and any other information that may help to reveal the authorship of a text. Finally, during the manual inspection of the corpus, we labeled each text with author's gender (male or female).

The corpus is composed of 5,187 news articles written by 232 different authors (2,968 articles written by male authors and 2,219 by female authors) and includes eight varieties of Spanish distributed as follows: Argentina: 449, Venezuela: 828, Colombia: 929, Guatemala: 598, Spain: 908, Mexico: 682, Panama: 418 and Chile: 375. The Spanish News Corpus is freely available on our website[4], where you will find more information about the corpus statistics.

4   Experimental Settings and Results

We evaluated the performance of character n-grams and lexical features, as well as some of their combinations. Character n-grams vary in order from 3 to 8, while lexical features include unigrams and bigrams of words. Each model was evaluated by measuring classification accuracy on the entire corpus under stratified 10-fold cross-validation. Following previous research [15], we removed features

---

[4] http://www.cic.ipn.mx/~sidorov/SpanishNewsCorpus.zip.

Table 1. Accuracy results (%) for lexical features, character (char.) n-grams, and their combinations in the AA, AP, and DSL tasks, before and after reducing all NEs to a single symbol.

Word unigrams	Word bigrams	Char. 3-grams	Char. 4-grams	Char. 5-grams	Char. 6-grams	Char. 7-grams	Char. 8-grams	Accuracy with NEs (%)				Accuracy without NEs (%)			
Features								**AA**	**AP**	**DSL**	**N**	**AA**	**AP**	**DSL**	**N**
✓								73.74	73.99	92.92	38,360	68.77	70.79	86.70	31,884
	✓							70.04	73.13	91.05	94,501	63.97	70.43	84.36	89,448
		✓						74.01	69.87	91.50	25,631	70.43	66.96	86.97	19,209
			✓					76.60	72.80	93.75	83,917	74.15	70.10	90.57	62,514
				✓				76.71	73.92	93.75	189,240	74.55	71.43	91.34	148,026
					✓			76.13	74.94	94.04	336,422	73.85	72.45	91.65	285,365
						✓		75.30	75.11	94.04	498,014	73.16	73.55	91.77	445,546
							✓	74.17	**75.61**	93.64	628,180	72.12	**73.90**	91.61	579,349
Combinations								**AA**	**AP**	**DSL**	**N**	**AA**	**AP**	**DSL**	**N**
✓	✓							74.82	74.78	92.94	132,861	70.08	72.43	87.97	121,332
✓	✓	✓						75.32	71.68	92.60	158,492	72.53	70.50	89.47	140,541
✓	✓	✓	✓					76.46	72.97	93.14	242,409	73.97	71.16	90.17	203,055
✓	✓	✓	✓	✓				77.31	73.51	93.45	431,649	74.63	71.74	90.73	351,081
✓	✓	✓	✓	✓	✓			77.91	74.15	93.70	768,071	**75.13**	72.14	91.19	636,446
✓	✓	✓	✓	✓	✓	✓		**77.96**	74.57	93.99	1,266,085	*	72.86	92.30	1,081,992
✓	✓	✓	✓	✓	✓	✓	✓	77.93	75.28	**94.16**	1,894,265	*	73.05	**92.52**	1,661,341

with a frequency less than 5 in the entire corpus, which significantly reduces the size of the feature set (on average by approximately 80%). As machine-learning algorithm, we selected Support Vector Machines (SVM); it was the classifier of choice of the majority of the teams in the previous editions of the PAN and VarDial competitions [2,12]. Given that the number of features is much larger than the number of instances, we used Crammer and Singer's linear kernel algorithm with default parameters implemented in the WEKA's [16] Liblinear [17] package. Following the practice of the VarDial workshop [18], we conducted additional experiments reducing all named entities (NEs) to a single symbol ($\#NE\#$) in order to evaluate their impact on these tasks.

Table 1 shows the obtained results for the AA, AP (gender identification), and DSL tasks in terms of accuracy (%) before and after replacing the NEs with a symbol. For each experiment, the number of features (N) is provided. The top accuracy values for each task are shown in bold typeface. The asterisks correspond to experiments that did not finish on time. We believe that the number of classes (238) for AA leads to a high computational cost for this SVM kernel. It is worth mentioning that when reducing the NEs this algorithm takes much more time to converge (about 6 times more).

As one can see from Table 1, higher-order character n-grams ($n = 5-8$) outperform both lower-order character n-grams and n-grams of words for all the three tasks when evaluated in isolation. The combination of all word and character n-grams provides the best results for two out of three considered tasks, AA and DSL, which is in line with the previous research [19]. These results are consistent with and without replacing NEs.

Moreover, it can be seen that the results continue to improve when adding higher-order character n-grams to the combination of features. However, higher-order character n-gram features significantly increase the size of the feature set, especially when used in combinations with each other, and consequently, the computational cost of the training process, while the accuracy improvement is only marginal. Therefore, we limited our experiments with the maximum order of 8 for character n-grams.

The best model for the AP and DSL tasks slightly outperforms the BOW approach when NE's are present (1.62% and 1.24%, respectively). However, when NEs are reduced the difference becomes higher (3.11% and 5.82%, respectively). The average drop in accuracy after reducing NEs is 2.52% for character n-grams and approximately 5% for lexical features. This confirms that lexical features are more topic-specific, which sometimes leads to unintended extraction of topic or domain information [20].

The average accuracy drop after reducing NEs is 3.52% for AA, 2.29% for AP, and 3.47% for DSL. For the AA task, the accuracy drop of 3.52% on our corpus is lower than the one of 5%–20% reported in [21], when for DSL the drop of 3.47% is higher than the one of around 2% reported in the VarDial workshop proceedings [18]. One of the possible explanations is the nature of our corpus, which contains shorter texts than the fiction novels corpus used for AA in [21], but much longer texts than the VarDail corpus of excerpts of journalistic texts.

5 Conclusions

In this paper, we examined the performance of character n-grams ($n = 3-8$), lexical features (unigrams and bigrams of words), and their combinations on the tasks of authorship attribution (AA), author profiling (AP) (only gender identification), and discriminating between similar languages (DSL) on a developed multi-labeled corpus of news articles in different varieties of Spanish.

The obtained results indicate that higher-order character n-grams outperform lower-order character n-grams for all the three tasks and provide the best results for gender identification when used in isolation (75.61% of accuracy). The combination of all word and character n-grams of different orders ($n = 1-2$ for words and $n = 3-8$ for characters) outperforms other combinations of such features and provides the best results for author and language variety identification (77.96% and 94.16%, respectively). We also evaluated the impact of named entities on these tasks. Our experiments showed that reducing them all to a single symbol "NE" to avoid topic-dependent features decreases accuracy by around 2.5%–3.4%, depending on the task. This work serves as a baseline for more complex methods based on dimensionality reduction or deep learning.

Acknowledgments. This work was partially supported by the Mexican Government (CONACYT projects 240844, SNI, COFAA-IPN, SIP-IPN 20171813, 20171344, 20172008) and CONACYT under the Thematic Networks program (Language Technologies Thematic Network Projects 260178 and 271622).

References

1. Stamatatos, E.: A survey of modern authorship attribution methods. J. Am. Soc. Inf. Sci. Technol. **60**, 538–556 (2009)
2. Rangel, F., Rosso, P., Verhoeven, B., Daelemans, W., Potthast, M., Stein, B.: Overview of the 4th author profiling task at PAN 2016: cross-genre evaluations. In: Working Notes Papers of the CLEF 2016 Evaluation Labs. CLEF and CEUR-WS.org (2016)
3. Zampieri, M., Malmasi, S., Ljubešić, N., Nakov, P., Ali, A., Tiedemann, J., Scherrer, Y., Aepli, N.: Findings of the vardial evaluation campaign 2017. In: Proceedings of the 4th Workshop on NLP for Similar Languages, Varieties and Dialects (VarDial 2017) (2017)
4. Stamatatos, E., Daelemans, W., Verhoeven, B., Stein, B., Potthast, M., Juola, P., Sánchez-Pérez, M.A., Barrón-Cedeño, A.: Overview of the author identification task at PAN 2014. In: Working Notes of CLEF 2014, pp. 877–897 (2014)
5. Khonji, M., Iraqi, Y.: A slightly-modified GI-based author-verifier with lots of features. In: Working Notes of CLEF 2014 (2014)
6. Koppel, M., Winter, Y.: Determining if two documents are by the same author. J. Am. Soc. Inf. Sci. Technol. **65**, 178–187 (2014)
7. Stamatatos, E., Daelemans, W., Verhoeven, B., Juola, P., López-López, A., Potthast, M., Stein, B.: Overview of the author identification task at PAN 2015. In: Working Notes of CLEF 2015 (2015)
8. Bartoli, A., Dagri, A., Lorenzo, A.D., Medvet, E., Tarlao, F.: An author verification approach based on differential features. In: Working Notes of CLEF 2015 (2015)
9. Rangel, F., Celli, F., Rosso, P., Pottast, M., Stein, B., Daelemans, W.: Overview of the 3rd author profiling task at PAN 2015. In: CLEF 2015 Labs and Workshops, Notebook Papers, vol. 1391. CEUR (2015)
10. Álvarez-Carmona, M.A., López-Monroy, A.P., Montes-y-Gómez, M., Villaseor-Pineda, L., Jair-Escalante, H.: INAOE's participation at PAN'15: Author profiling task. In: Working Notes Papers of the CLEF 2015 Evaluation Labs, vol. 1391. CEUR (2015)
11. Gencheva, P., Boyanov, M., Deneva, E., Nakov, P., Georgiev, G., Kiprov, Y., Koychev, I.: PANcakes team: a composite system of genre-agnostic features for author profiling. In: Working Notes Papers of the CLEF 2016 Evaluation Labs. CLEF and CEUR-WS.org (2016)
12. Malmasi, S., Zampieri, M., Ljubešić, N., Nakov, P., Ali, A., Tiedemann, J.: Discriminating between similar languages and arabic dialect identification: a report on the third DSL shared task. In: Proceedings of the 3rd Workshop on Language Technology for Closely Related Languages, Varieties and Dialects (VarDial 2016), pp. 1–14 (2016)
13. Çöltekin, C., Rama, T.: Discriminating similar languages: experiments with linear SVMs and neural networks. In: Proceedings of the 3rd Workshop on NLP for Similar Languages, Varieties and Dialects (VarDial 2016), pp. 15–24 (2016)

14. Bestgen, Y.: Improving the character n-gram model for the DSL task with BM25 weighting and less frequently used feature sets. In: Proceedings of the 4th Workshop on NLP for Similar Languages, Varieties and Dialects (VarDial 2017) (2017)
15. Markov, I., Stamatatos, E., Sidorov, G.: Improving cross-topic authorship attribution: the role of pre-processing. In: Proceedings of the 18th International Conference on Computational Linguistics and Intelligent Text Processing (CICLing 2017) (2017)
16. Witten, I., Frank, E., Hall, M., Pal, C.: Data mining: practical machine learning tools and techniques, 4th edn. Morgan Kaufmann (2016)
17. Fan, R.E., Chang, K.W., Hsieh, C.J., Wang, X.R., Lin, C.J.: LIBLINEAR: a library for large linear classification. J. Mach. Learn. Res. **9**, 1871–1874 (2008)
18. Zampieri, M., Tan, L., Ljubešić, N., Tiedemann, J., Nakov, P.: Overview of the DSL shared task 2015. In: Proceedings of the Joint Workshop on Language Technology for Closely Related Languages, Varieties and Dialects (LT4VarDial 2015), pp. 1–9 (2015)
19. Gómez-Adorno, H., Markov, I., Baptista, J., Sidorov, G., Pinto, D.: Discriminating between similar languages using a combination of typed and untyped character n-grams and words. In: Proceedings of the 4th Workshop on NLP for Similar Languages, Varieties and Dialects (VarDial 2017) (2017)
20. Tsur, O., Rappoport, A.: Using classifier features for studying the effect of native language on the choice of written second language words. In: Proceedings of the Workshop on Cognitive Aspects of Computational Language Acquisition (CACLA 2007), pp. 9–16. ACL (2007)
21. Ríos-Toledo, G., Sidorov, G., Castro-Sánchez, N.A., Nava-Zea, A., Chanona-Hernández, L.: Relevance of named entities in authorship attribution. In: Proceedings of the 15th Mexican International Conference on Artificial Intelligence (MICAI 2016). LNAI. Springer (2017). http://www.micai.org/2016/pre-print/LNAI/paper_148.pdf

Enriching Existing Test Collections with OXPath

Philipp Schaer$^{(\boxtimes)}$ and Mandy Neumann

TH Köln (University of Applied Sciences), Cologne, Germany
{philipp.schaer,mandy.neumann}@th-koeln.de

Abstract. Extending TREC-style test collections by incorporating external resources is a time consuming and challenging task. Making use of freely available web data requires technical skills to work with APIs or to create a web scraping program specifically tailored to the task at hand. We present a light-weight alternative that employs the web data extraction language OXPath to harvest data to be added to an existing test collection from web resources. We demonstrate this by creating an extended version of GIRT4 called GIRT4-XT with additional metadata fields harvested via OXPath from the social sciences portal Sowiport. This allows the re-use of this collection for other evaluation purposes like bibliometrics-enhanced retrieval. The demonstrated method can be applied to a variety of similar scenarios and is not limited to extending existing collections but can also be used to create completely new ones with little effort.

Keywords: Test collections · Metadata enrichment · GIRT · OXPath · Harvesting of metadata · Scholarly retrieval

1 Introduction

Building TREC-style test collections for information retrieval evaluation is a costly activity. It involves at least three main tasks: (1) setting up an appropriate set of documents, (2) generating a list of topics (50 or even more, as suggested by Voorhees [9]), and (3) obtaining relevance assessments (most of the time by employing domain experts or search specialists as assessors). All three tasks combined sum up and make the generation of new test collection or the redesign and extension of existing test collections a time consuming and challenging task. Generating a completely new test collections is the most complex scenario. Therefore we would like to focus on the enrichment of existing test collections, especially the set of documents. This would allow the reuse of documents, topics and relevance assessments while enabling the old test collection to be reused in other evaluation contexts like scholarly search or bibliometrics-enhanced retrieval [6].

Previous projects like EFIREval[1] already focused on adapting test collections to new environments or incorporated richer information about the different

---

[1] https://sites.google.com/site/ekanoulas/grants/EFIREval.

© Springer International Publishing AG 2017
G.J.F. Jones et al. (Eds.): CLEF 2017, LNCS 10456, pp. 152–158, 2017.
DOI: 10.1007/978-3-319-65813-1_16

retrieval scenarios and searchers' activities but only few tried to augment the document collection, which is why we would like to focus on this desideratum.

Research question. How can we enrich parts of existing test collections, like the document collection, by incorporating external resources like digital libraries or other freely available web data sets with as little effort as possible?

Approach. We propose a light-weight method for extending and augmenting the documents sets in test collections by incorporating the web extraction language OXPath. This language that derived from XPath is capable of extracting huge sets of information from large web corpora. It is used by scholarly literature portals like dblp to build up their data sets.

Contributions. We show the feasibility of our approach by extending the GIRT4 collection that was used in the Domain-Specific Track of CLEF with freely available data from the social sciences portal Sowiport[2]. After harvesting the additional data we created an extended collection called GIRT4-XT that augments the original GIRT4 documents with additional attributes like ISSN codes. This way the rather old test collection that initially was used to do cross-lingual and domain-specific retrieval evaluations can be used for other evaluation purposes like bibliometrics-enhanced retrieval.

2   Related Work

Re-using existing test collections for other purposes in general is not a new idea. Berendsen et al. [2] were using the previously mentioned GIRT collection to generate a so-called pseudo test collection that is automatically generated. The relatively spare data of the GIRT collection (content bearing metadata only being the title and a rather short abstract) comes with a rich set of annotations (see Table 1). These annotations were used to generate pseudo topics and relevance assessments. This pseudo test collection provided training material for learning to rank methods.

A similar approach was used by Roy, Ray, and Mitra [8] who used the CiteSeerX collection to generate a test collection for citation recommendation services. They extracted the textual part of a citation context to form a query. The cited references were taken to be the relevant documents for that query. This way 2,826 queries were obtained but most queries (contexts) have only one relevant citation, making this test collection rather sparse.

Larsen and Lioma [5] described different strategies to generate a scholarly IDEAL test collection. While they came up with some new ideas and strategies of gathering and curating a document collection they rely on manually crafted topics and relevance assessments to complete the test collections. As they outline, the scholars that are the sources of topics and relevance assessments are notoriously busy, hard to engage and unlikely to be crowdsourced. They named

---

[2] http://sowiport.gesis.org.

INEX as a role model of community effort in collecting relevance judgments from its participants and encouraged to follow that road.

The reuse of document and test collections is common practice by adding new topics and relevance assessments or by transferring them to new application domains (e.g. from IR evaluation to recommender systems). Both approaches most often rely on manual work and judgments. Another approach for building up test collections was presented in the Social Book Search [4] track of CLEF. They built their task on top of the INEX Amazon/LibraryThing collection [1] and enriched it with content from forum discussions on the LibraryThing website to extract topics and relevance assessments. This is a rather technical methodology to obtain this crucial part of a test collection which involved the generation of custom web crawlers for this single purpose.

3   Materials and Methods

As suggested by some of the related work (e.g. Social Book Search), test collections can be created or enhanced with freely available web data. But web pages are meant to be displayed to a human user, as opposed to APIs that provide a means for software applications to gather the structured data that makes up the content of those web pages. Thus for compiling a corpus from web data, one would have to either have access to such an API, or work directly with the human-oriented HTML interface. The former would definitely require some programming/scripting skills, while the latter would either require extensive programming skills for scraping the web page content, or a lot of human effort to collect the desired information manually. In the past, several attempts have been made to ease the process of acquiring web data for non-technical users, by providing web data extraction tools.

OXPath is an open-source language focusing on deep web crawling that takes a declarative approach to the problem [3]. Based on the XML query language XPath, it enables the simulation of user interaction with a web page and the extraction of information in the course of these interactions. To achieve this, OXPath extends the capabilities of XPath with five new elements: (1) actions like clicking and form filling, (2) interactions with the visual appearance of a page, (3) means of identifying nodes by multiple relations, (4) extraction markers to yield hierarchical records of sought-after information, and (5) the Kleene star to enable navigation of paginated content. With these means, it is possible to craft an expression to harvest a lot of data with just a few lines of code.

OXPath can be used e.g. for harvesting bibliographic metadata for digital libraries like dblp, as presented by Michels et al. [7]. In contrast to other tools made for extracting bulk data from web pages, OXPath proves to be particularly memory-efficient as shown by Furche et al. [3].

Regarding document sets in test collections, OXPath can also be used to extract additional information from such digital libraries to extend the test collection with new attributes. Taking the social sciences portal Sowiport as an example, we created a light-weight OXPath wrapper that is able to harvest targeted information from a specific set of records and save the extracted data in a

Listing 1. Sample OXPath wrapper for harvesting the SOLIS database of Sowiport and extracting title, language and published attributes for each result.

```
 1  doc('http://sowiport.gesis.org/')
 2    /descendant::field()[2]/{click /}
 3      //div[@id='facets']//a[contains(.,'SOLIS')]/{click /}
 4      //*[@id='limit']/option[contains(., '100')]/{click /}
 5      /(//*[contains(@title, 'next')]/{click /})*
 6        //*[contains(@class,'record')]:<record>
 7        [.//a[@class~='title'][./b:<title=normalize-space(.)>]/{click /}
 8          /. [? .//*[@id='detailed_view_metadata']//table//td
 9           [ ./preceding-sibling::td[contains(.,'Editor:')]]]//a
10             :<editor=normalize-space(.)>]
11          [? .//*[@class='recordsubcontent']//table//td
12           [ ./preceding-sibling::th[contains(.,'Database:')]]]
13             :<id=substring-after(normalize-space(.), 'Acquis. id: ')>]
14        ]
```

hierarchically structured form. Listing 1 demonstrates a sample OXPath wrapper that is able to interact with the web page of Sowiport[3]. It narrows down the list of presented items to those from a specific database (in this case the social science literature database SOLIS, that GIRT4 is based on) and navigates through the result list in a loop (lines 3–5). By clicking the title of each record element (line 7), the element's detail view is opened where additional data can be found. For example, in lines 8–10 the editor field is located in the page and each listed editor extracted separately. In a similar vein, the acquisition id ("Acquis. id") is extracted from a different location on the same page (lines 11–13). The extracted data is hierarchical in nature and can be serialized e.g. in XML or CSV format for further processing.

4   Results

By harvesting additional data from the SOLIS database in Sowiport using a relatively simple declarative expression, we were able to extend the original GIRT4 data with additional information, such as ISSN/ISBN codes or editor, publisher and location information (see Table 1 for an overview). The items from the GIRT collection were matched with the harvested data via their id which was both present in the harvested SOLIS data (acquisition id) and the GIRT4 data set (DOCID without the GIRT prefix).

Of a total of 151.319 documents in GIRT4 we extended 135.214 documents with data from SOLIS/Sowiport. Note that only the documents on social science literature were extended while the social science project descriptions also included in GIRT were ignored. The new test collection is called GIRT4-XT and includes a total of six new metadata fields that were not included in the original

---

[3] Note that we replaced all German terms from the Sowiport portal with English equivalencies in this listing.

data set (editor, ISSN, ISBN, location, publisher, and page numbers). Some of the SOLIS records include links to full texts but as most of them are behind publisher pay walls we were not able to extract them.

Table 1. Overview on the included fields of the original GIRT4 corpus, the available SOLIS data from the Sowiport portal and the combined GIRT4-XT corpus. Three different states are marked in the table: – = field data not available; ○ = available in unstructured form; • = available in structured form.

Corpus	id	author	editor	title	source	issn	isbn	pubyear	keywords	class.	abstract	full text	method	location	publisher	pages	language	country
GIRT4	•	•	–	○	○	–	–	•	•	•	•	–	•	–	–	–	•	•
SOLIS	•	•	•	○	•	•	•	•	•	•	•	○	•	○	○	•	•	•
GIRT4-XT	•	•	•	○	○	•	•	•	•	•	•	–	•	○	○	•	•	•

5 Discussion and Conclusion

We showed how to extend and enrich existing information retrieval test collections by harvesting freely available metadata from digital library systems by employing the web extraction language OXPath. This method allows us to reuse existing test collections (especially their topics and relevance assessments) in different domains by adding new metadata to the existing documents in the collection.

We demonstrated the feasibility of the process by extending GIRT4 with additional document annotations like editor names, ISSN codes of the related journal or page numbers. This way new kinds of experiments are possible like those discussed in the bibliometrics-enhanced IR community, but the proposed methods and techniques are not limited to this domain. Another use case for our test collection enrichment strategy might be the TREC Genomics Track test collections[4]. As suggested by Larsen and Lioma [5] these collections can be augmented by references extracted from PubMed, a scenario more than suitable for OXPath.

The proposed approach heavily relies on the usage of OXPath as it is an easy-to-learn, light-weight, and all-in-one rapid development technology to gather the additional (meta-)data from web resources like digital libraries. Although the advantages outweigh the disadvantages we would like to point out some shortcomings of OXPath that have to be considered. First of all OXPath is not tuned for speed which results in rather moderate processing times. Internally the whole web page has to be rendered and processed to allow a human-comparable

---

[4] http://skynet.ohsu.edu/trec-gen/.

extraction mechanism. When processing many hundred thousand web pages the harvesting process can take many days. There are ways to distribute the whole process on parallel threads but this is not a built-in feature. Another point is that there are relatively few tools to support the development process[5]. In spite of these limitations, OXPath is still a powerful and useful tool for harvesting semi-structured data from web resources.

In the future, we want to employ OXPath not only for the enhancement of existing test collections, but also for the creation of completely new ones, were all the data necessary should be extracted from web resources. One of our role models for this is the Social Book Search collection.

Acknowledgements. This work was supported by Deutsche Forschungsgemeinschaft (DFG), grant no. SCHA 1961/1-2.

References

1. Beckers, T., Fuhr, N., Pharo, N., Nordlie, R., Fachry, K.N.: Overview and results of the INEX 2009 interactive track. In: Lalmas, M., Jose, J., Rauber, A., Sebastiani, F., Frommholz, I. (eds.) ECDL 2010. LNCS, vol. 6273, pp. 409–412. Springer, Heidelberg (2010). doi:10.1007/978-3-642-15464-5_44
2. Berendsen, R., Tsagkias, M., Rijke, M., Meij, E.: Generating pseudo test collections for learning to rank scientific articles. In: Catarci, T., Forner, P., Hiemstra, D., Peñas, A., Santucci, G. (eds.) CLEF 2012. LNCS, vol. 7488, pp. 42–53. Springer, Heidelberg (2012). doi:10.1007/978-3-642-33247-0_6
3. Furche, T., Gottlob, G., Grasso, G., Schallhart, C., Sellers, A.: Oxpath: a language for scalable data extraction, automation, and crawling on the deep web. VLDB J. **22**(1), 47–72 (2013). doi:10.1007/s00778-012-0286-6
4. Koolen, M., Kazai, G., Preminger, M., Doucet, A.: Overview of the INEX 2013 social book search track. In: Information Access Evaluation meets Multilinguality, Multimodality, and Visualization - Fourth International Conference of the Cross-Language Evaluation Forum (CLEF 2013), p. 26, Valencia, Spain (2013). https://hal.archives-ouvertes.fr/hal-01073644
5. Larsen, B., Lioma, C.: On the need for and provision for an "ideal" scholarly information retrieval test collection. In: Proceedings of the 3rd Workshop on Bibliometric enhanced Information Retrieval (BIR2016), pp. 73–81 (2016). http://ceur-ws.org/Vol-1567/paper8.pdf
6. Mayr, P., Scharnhorst, A., Larsen, B., Schaer, P., Mutschke, P.: Bibliometric-enhanced information retrieval. In: Rijke, M., Kenter, T., Vries, A.P., Zhai, C.X., Jong, F., Radinsky, K., Hofmann, K. (eds.) ECIR 2014. LNCS, vol. 8416, pp. 798–801. Springer, Cham (2014). doi:10.1007/978-3-319-06028-6_99
7. Michels, C., Fayzrakhmanov, R.R., Ley, M., Sallinger, E., Schenkel, R.: Oxpath-based data acquisition for DBLP. In: Proceedings of the 17th ACM/IEEE-CS on Joint Conference on Digital Libraries (JCDL 2017), pp. 319–320. NY, USA. ACM, New York (2017)

---

[5] We developed an extension for the text editor Atom ourselves, see https://atom.io/packages/language-oxpath.

8. Roy, D., Ray, K., Mitra, M.: From a scholarly big dataset to a test collection for bibliographic citation recommendation. In: Workshops at the Thirtieth AAAI Conference on Artificial Intelligence (2016). http://www.aaai.org/ocs/index.php/WS/AAAIW16/paper/view/12635

9. Voorhees, E.M.: Topic set size redux. In: SIGIR 2009, pp. 806–807. ACM (2009). doi:10.1145/1571941.1572138

Best of the Labs

A Highly Available Real-Time News Recommender Based on Apache Spark

Jaschar Domann$^{(\boxtimes)}$ and Andreas Lommatzsch

Agent Technologies in Business Applications and Telecommunication Group (AOT),
Institute Technische Universität Berlin, Ernst-Reuter-Platz 7, 10587 Berlin, Germany
{jaschar.domann,andreas.lommatzsch}@campus.tu-berlin.de
http://www.aot.tu-berlin.de

Abstract. Recommending news articles is a challenging task due to the continuous changes in the set of available news articles and the context-dependent preferences of users. In addition, news recommenders must fulfill high requirements with respect to response time and scalability. Traditional recommender approaches are optimized for the analysis of static data sets. In news recommendation scenarios, characterized by continuous changes, high volume of messages, and tight time constraints, alternative approaches are needed. In this work we present a highly scalable recommender system optimized for the processing of streams. We evaluate the system in the CLEF NewsREEL challenge. Our system is built on APACHE SPARK enabling the distributed processing of recommendation requests ensuring the scalability of our approach. The evaluation of the implemented system shows that our approach is suitable for the news recommendation scenario and provides high-quality results while satisfying the tight time constraints.

Keywords: Apache Spark · Stream recommender · Distributed algorithms · Real-time recommendation · Scalable machine learning

1 Motivation

In the recent years, there has been an immense growth in the popularity of online news. One downside of this huge amount and oversupply of published news is that it becomes very difficult to find relevant articles. Therefore, services are needed supporting users in browsing news and guiding them to the most interesting news item taking into account the user preferences. In addition, recommender systems should consider the specific context (e.g. day of week or news categories) and relevant events. In order to support users in finding relevant items in large data collections, recommendation algorithms have been developed. Traditionally, these approaches are based on complex models computed on static data sets. But in the field of news recommendation, the sets of items and the user preferences change continuously: On the one hand the popularity of items follows a Zipfian distribution (a very low fraction of items receives a very large number of impressions) and on the other hand the number of article impressions

© Springer International Publishing AG 2017
G.J.F. Jones et al. (Eds.): CLEF 2017, LNCS 10456, pp. 161–172, 2017.
DOI: 10.1007/978-3-319-65813-1_17

highly depends on the hour of the day, the day of week, and the number of pub-lished news messages in specific domains [6]. So it becomes essential to gather changes in real-time and to adapt the recommender models continuously.

The main objective of this work is to implement a highly scalable and stream-based system which provides online news recommendations. This system must be capable of handling large data streams and answering recommendation requests within the customer-defined time constraints.

In this work we present our recommender system. This system makes use of a dynamically adapting most-popular algorithm, which keeps track of the currently most frequently read news articles. For ensuring scalability the system is implemented based on the APACHE SPARK framework[1]. The APACHE SPARK framework enables the distributed processing of recommendations and assures high scalability.

We evaluate this approach in the CLEF NEWSREEL[2] challenge. The chal-lenge offers two tasks: First, the Living Lab evaluation (NEWSREEL Task 1) uses the Open Recommendation Platform[3] (hosted by PLISTA). The performance of the recommender systems is evaluated based on live user feedback. Second, the Evaluation Lab (NEWSREEL Task 2) focuses on the offline evaluation that is based on re-playing a data stream recorded in the Living Lab evaluation. The re-played dataset consists of the published news items and the user-item inter-actions collected in four weeks. Task 2 enables the reproducible analysis of the recommender performance and the simulation of different load scenarios.

The remaining paper is structured as follows: Sect. 2 describes the scenario in detail and points out the specific challenges. Related work is discussed in Sect. 3. In Sect. 4, we present our approach and explain the applied methods. The evaluation results are discussed in Sect. 5. Finally, a conclusion and an outlook to future work are given in Sect. 6.

2   Problem Description

The CLEF NEWSREEL challenge gives researchers the possibility to evalu-ate news recommender algorithms under realistic conditions [4]. The challenge enables the evaluation both on live user feedback ("online") and based on a recorded stream of data ("offline").

The underlying use case is shown in Fig. 1. On a web-based news portal below each news article up to six slots are reserved for news recommendations. The task in the CLEF NEWSREEL is filling these slots with recommendations referring to items on the same portal.

The challenge distincts between *impressions, requests* and *clicks*. An *impres-sion* describes the event, that a user reads an item. A *request* is a message asking for up to 6 recommendations that will be embedded in a given news page. A *click* denotes that a user clicks on a recommendation. A click results in an impression.

---

[1] http://spark.apache.org.

[2] http://www.clef-newsreel.org.

[3] https://orp.plista.com.

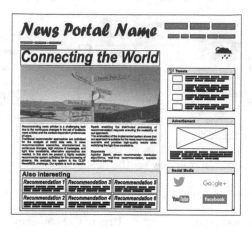

Fig. 1. The figure visualizes the NEWSREEL use case.

Due to the very small click rate, the recommendations have only a low influence on the impressions. The dwell time is not explicitly tracked in the NEWSREEL challenge.

The performance of the recommender algorithm is measured based on the CLICK-THROUGH RATE (CTR) which describes the ratio of clicks on a given recommendation to total number of offered recommendations. This ratio is calculated as follows.

$$CTR_{\text{online}} = \frac{\#\text{clicks}}{\#\text{recommendation requests}}$$

In contrast to the Online Evaluation setting the Offline Evaluation cannot make use of direct user feedback ("clicks"). Thus, an alternative evaluation approach must be applied. The task in the offline challenge is predicting future user impressions. The future user impressions are known to the evaluation component since the offline evaluation is based on a recorded data stream. A recommendation is counted as correct, if the user reads the recommended article ("impression") within a 5 min time window after the request. The impression-based offline CLICK-THROUGH RATE (CTR_{offline}) is calculated as follows.

$$CTR_{\text{offline}} = \frac{\#\text{correct recommendations}}{\#\text{requested recommendations}}$$

The communication between the attendees of the contest is handled by an API provided by the ONLINE RECOMMENDATION PLATFORM (ORP). This API provides all relevant data for recommendations and sends requests to the recommendation systems of the participating teams. In case a user requests an article of an online news portal the ORP sends the impression event to all teams and delegates the recommendation request to a randomly selected team. Figure 2 shows the architecture of the challenge and the interacting components.

The recommendation systems (participating in the NEWSREEL challenge) must be able to handle the load of all messages. Furthermore, they must also

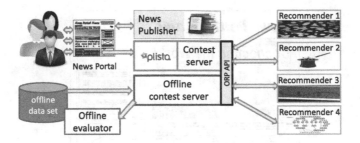

Fig. 2. The graphic gives an overview on the design of the NEWSREEL challenge.

Table 1. Classification of requests with different levels of urgency and frequency.

Request type	Urgency	Frequency
Article update	<1 min	Hourly
User view	<1 s	Every second
Recommendation request	<100 ms	Every minute

meet the requirements for urgency of answering the request classes shown on Table 1. Special attention must be put on efficiency and scalability in order to assure high quality recommendation and to fulfill the technical requirements. Due to the restricted data available in the scenario an exact re-identification of users over different web-sessions is infeasible.

The specific requirements of NEWSREEL are interesting challenges motivating the development of a scenario-optimized solution.

3   Related Work

In this section we briefly review existing recommender algorithms and discuss their specific strengths and weaknesses. We focus on most-popular items methods, user-based collaborative filtering (CF) algorithms as well as item-based CF algorithms.

Most-popular items algorithms suggest the most popular items in the community. The strengths of this approach include that no detailed knowledge about individual user preferences is required and the recommendations can be efficiently pre-computed. The weaknesses of the most-popular items paradigm include that neither context nor individual user preferences are considered, so this may lead to a reduced recommendation precision. In the analyzed scenario, most popular items algorithms seem to be suitable, since all user-item interactions are available enabling the popularity computation in real-time.

User-based CF algorithms recommend items to users by determining users having similar interests to the active user. The active user is here considered as the one who receives the recommendations. If users have overlapping item preferences then they are considered as having similar interests. So a user-based

CF recommender algorithm concludes that an unknown item of similar users is a good suggestion for the active user. Due to the fuzzy user-identification in the NEWSREEL setting, user-based approaches seem to be suboptimal in the analyzed scenario.

Item-based CF algorithms compute the similarity between items based on user feedback. Two items are considered as similar if users tend to assign the same rating for them. So the system suggests an item similar but still unknown to the active user. The strength of this approach is that anonymous user feedback can be used and a valuable explanations can be provided. Due to the continuous changes in the item set in the NEWSREEL setting, the item-based recommender suffers from a cold-start problem that leads to a reduced recommendation performance.

Special challenges in the news recommendation scenario are the continuous changes in the sets of users and items as well as in the user preferences. Use cases characterized by high fluctuating item sets also exist in other domains. Zangerle et al. [9] present an approach to gather music related micro-posts from Twitter as user stream and make use of them for music recommendations. Diaz-Aviles et al. [1] use twitter micro-posts to recommend niche topical news stories. Based on subscribed RSS-feeds and from selected TWITTER messages ("tweets") the recommender engine suggests relevant news. In contrast to these scenarios, the NEWSREEL scenario shows higher requirements with respect to scalability and the number of processed messages. Thus, highly optimized approaches are needed in order to fulfill the tight time constraints.

For the development of recommendation services several frameworks exist, such as MAHOUT [7,8] and LENSKIT [2]. These frameworks also provide solutions to evaluate recommender algorithms. Unfortunately, these frameworks do not completely provide support for stream-based recommendation scenarios, since they are built for static datasets. Since news recommender systems have to be able to handle huge number of requests, the scalability of the systems is an important issue. This includes that in a stream-based scenario the data must be processed continuously in near real-time. Several frameworks have been developed tailored for the efficient processing of big data relying on the distributed processing, such as APACHE FLINK[4], APACHE STORM[5], HERON[6], and APACHE SPARK. Among these frameworks, APACHE SPARK seems to have the largest number of users and the most active community as well as a more comprehensive documentation [5]. Therefore, it fulfills our requirements best in comparison with the other frameworks.

4   Approach

In this section we present the architecture of the developed recommender system and discuss the implemented algorithms in detail.

---

[4] https://flink.apache.org/.
[5] https://storm.apache.org/.
[6] https://twitter.github.io/heron/.

4.1 Recommender Models

The problem analysis has shown that Collaborative Filtering algorithms do not match the scenario's requirements due to the fuzzy identification of users and the sparse information. In order to ensure robust, efficient, and fast adapting algorithms, we focus on most popular algorithms in our recommender system. Since most popular algorithms have the ability of adapting quickly to new items and contexts and do not suffer from fuzzy user identification, these algorithms are well-suited in our scenario. The recommendation strategy is applied separately for each news publisher. Optionally, it can be refined for the news categories ("sub-domains") or for specific contexts or stereotype user groups.

In the NEWSREEL setting, users are not asked to explicitly rate articles. In order to compute the popularity of items, we treat each user-item interaction ("impression", "click") as an indicator that the user is interested in a news item. Since the user proactively clicked on the news article, the interaction can be handled as implicit feedback. In the most-popular approach, the challenge is to determine the most interesting news items at high scale. As the number of articles steadily increases, the model must be updated continuously. In our system, we make use of the Map-Reduce mechanism implemented in SPARK. Map-Reduce is an appropriate method for quickly determining the most popular articles as it allows us to count the number of clicks each article in the stream received in a highly parallelized way.

4.2 Implementation

The NEWSREEL challenge defines a HTTP-based protocol for the interacting between the developed recommender systems and the contest server. Thus, the connector for receiving messages and providing recommendations is a web server. The web server must handle three types of requests having different levels of urgency (cf. Table 1). The architecture of the recommender system must consider the priorities of the messages as well as integrate the APACHE SPARK framework and the web server. The architecture of the developed recommender system is visualized in Fig. 3.

The NEWSREEL contest server sends the messages (impressions, item updates, recommendation requests) to the web server acting as interface of the news recommender system. The web server forwards the messages of a management component that analyzes the received messages. Dependent on the message type (impression, item update, request) and the news portal ("domain") the messages are either used for updating the recommender models or for computing recommendations.

Impression messages and item update messages are delegated to the SPARK component. The SPARK-based component aggregates messages in a buffer and re-calculates the recommender models. The recalculation is done based on the Map-Reduce paradigm enabling the distributed processing of the data. The system runs on a standalone SPARK cluster; this means that the web server is available on the same machine as the cluster master while the Map-Reduce jobs

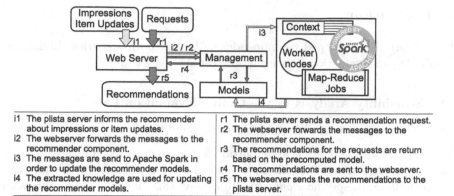

i1 The plista server informs the recommender about impressions or item updates.	r1 The plista server sends a recommendation request.
i2 The webserver forwards the messages to the recommender component.	r2 The webserver forwards the messages to the recommender component.
i3 The messages are send to Apache Spark in order to update the recommender models.	r3 The recommendations for the requests are return based on the precomputed model.
i4 The extracted knowledge are used for updating the recommender models.	r4 The recommendations are sent to the webserver.
	r5 The webserver sends the recommendations to the plista server.

Fig. 3. Integration of the web server and SPARK. The web server delegates recommendation requests to the recommender in order to retrieve results. The web server, the management component, and the SPARK-context are started in the same JVM to ensure fast communication.

are parallelized on the slave nodes. The models build by SPARK are stored in a way that always the most recent recommender models are ready for computing recommendations. The model computation is done continuously and triggered by incoming impression messages.

Recommendation requests must be answered immediately. In order to ensure a minimal processing time, the recommendations are computed based on the models (created by SPARK). Since the models are typically small, requests can be efficiently handled on the master node - no distributed processing is required.

The major part of the implementation is realized in Java. The data processing is implemented using the SPARK JAVA API. The distribution in our scenario causes delays in network that inhibit some benefits of parallel processing. This delays are caused by a setup of multiple *Virtual Machines* connected over the internet per *Virtual Private Network*. According to our own estimation, a setup in a *Local Area Network* would mitigate this negative effect.

4.3   Discussion

The developed system architecture efficiently integrates APACHE SPARK in the web-based recommender system. The system is optimized for the handling of high volumes of messages. The models are continuously updated using the distributed processing capabilities of SPARK. Recommendation requests are handled very fast using the models computed on the SPARK cluster. The combination of distributed model building and direct request handling ensures that at any time recommendations are computed based on fresh models. The distributed model creation builds the basis for applying more complex recommender models, due to the fact that model building and request answering are separated.

5   Evaluation

We evaluate the developed recommender system in both scenarios, Living Lab ("online") and Evaluation Lab scenario ("offline").

5.1   Scalability Analysis in the Offline Evaluation Lab

We evaluate our system with the focus on scalability.

Load Peaks. The scalability is an important aspect in news recommender systems due to the huge number of messages and the high variance in the messages volume induced by specific events. Figure 4 shows that the number of impressions recorded by our system changes significantly over time. The high amplitude in Fig. 4 relates to several big transfers of soccer players. This underlines that a news recommender system must be capable of handling enormous changes in the number of messages and be capable to process load peaks.

Response Time Analysis. To further assess the capabilities of the system, we evaluated the throughput under extreme circumstances. Figure 5 shows the response time histogram on a single machine with 1,000 threads simultaneously requesting recommendations and sending impressions.

The histogram has three peaks. The first peak marks requests that are answered immediately. This case occurs if the incoming JSON-string is corrupted or required information is missing. The second peak represents requests that are cached while the third are normal responses that are handled by the management and recommender components. The mean response time is about 330 ms and considerably larger than the response time constraint of 100 ms. This can be explained by the fact that the environment resided on a single machine with

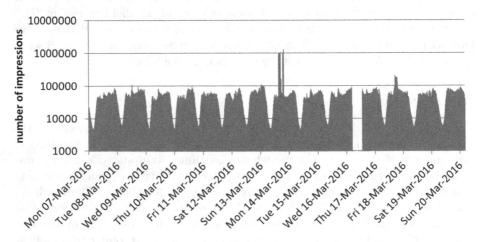

Fig. 4. The graph shows the log-scaled number of impressions that where recorded within two weeks in March. The bin size for this plot is on a 15 min basis.

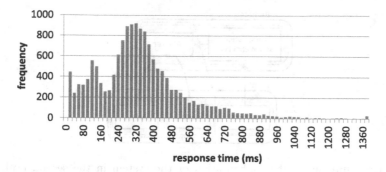

Fig. 5. The graph shows the frequency of response times for 15 k recommendation requests with 1,000 parallel request threads.

maximized throughput. In a production system, the expected mean response time is narrowed accordingly due to an improved parallelism.

Cluster Setup. We analyze the influence of different setups for the SPARK cluster. For our experiments, we built a cluster that consists of a master and two slave nodes (Fig. 6). The slave nodes are deployed on distant virtual machine so that a remarkable network latency within the cluster exists. For the evaluation we replayed the NEWSREEL message stream consisting of the data recorded in May 2016. The collected messages are sent from the local network to our recommender system.

We analyzed the influence of the cluster configuration on the response time. In the first evaluation run one slave node with four workers has been used. In the second evaluation both slave nodes and eight workers have been used. We observed a lower response time in the second evaluation run, due to a lower worker job fail rate. Configurations with more computational power result in more reliable model re-building resulting in a better recommendation performance in high load conditions. Moreover strong differences in CPU and memory resources between the SPARK nodes and long network latencies should be avoided for ensuring a stable working cluster.

5.2 Click-Through Rate Analysis in the Offline Evaluation Lab

For the Evaluation Lab a data record of one week in the period between 7 March and 13 March 2016 was used. This time period has been selected because in this timeframe the Living Lab data stream provided a reliably high volume of data.

In the evaluation process three most popular algorithms, labeled with MOST-POPULAR300, MOSTPOPULAR100 and CATEGORYMP100, have been deployed on a standalone local SPARK cluster running on a single machine. The number in the label describes the size of the operation queue which is used by the algorithm. The replay of the data record is done on the same system. In reference to Fig. 7 a data record analysis shows that the domain with ID 35774 is dominating the number of recommendation requests. Hence, the evaluation results are

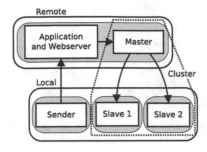

Fig. 6. The figure shows the configuration of the system in the second evaluation. A gray background denotes a distinct virtual machine. The arrows show the path a request takes along the system and the dotted area encloses the cluster components.

Table 2. Comparison of offline CTR from various recommending algorithms.

Algorithm	Description	CTR
MostPopular300	Uses a most-popular algorithm, operates on a queue of items with size 300	2.93 %
CategoryMP100	Uses a most-popular algorithm considering categories (with separate queues for each category), operates on queues of items with size 100	2.84 %
MostPopular100	Uses a most-popular algorithm, operates on a queue of items with size 100	2.83 %
Baseline	Recommends most recent items	0.59 %

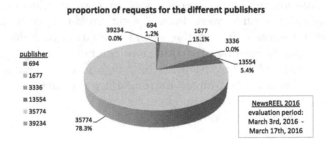

Fig. 7. The figure indicates that requests for the publisher 35774 dominate the News-REEL challenge.

based on an analysis of the impression and requests related to domain 35774. The results are shown in Table 2 and indicate that all three algorithms significantly outperform the CTR_{offline} of the baseline recommender provided by the lab organizers. This verifies that the evaluated algorithms generate good recommendations (measured by a reasonable CTR).

Table 3. Comparison of response rates from the Living Lab results from the 2nd of April to the 1st of June in 2016. The last column shows the number of days the algorithm was active. The table entries are sorted by the minimal response rate column in descending order. The presented system is labeled *Spark* while other participant names are anonymized.

Algorithm label	Mean response rate	Minimal response rate	Days
Spark	99.76%	99.43%	30
6	99.87%	99.43%	28
5	99.87%	99.24%	27
1	98.46%	84.10%	29
7	98.62%	79.82%	30
2	97.40%	75.11%	28
4	95.95%	52.43%	26

5.3 Response Rate Analysis in the Living Lab Evaluation

The implemented system has been evaluated on a single machine in the Living Lab scenario [3]. The performance of the system has been computed based on log files provided by PLISTA. The log files include information about the CTR and the response rate for each algorithm on a daily basis. The latter rate only incorporates responses that did not break any constraints (such as the response time limit of 100 ms). The measured response rates are aggregated in Table 3. The table shows that the system outperforms all other algorithms with respect to the total number of days that the algorithm has been active. Overall, with a response rate of over 99%, we achieved the goal of high availability.

5.4 Discussion

The presented system fulfills the real-time requirements of handling streams with strict time constraints. The Evaluation Lab results confirms the findings. The evaluation shows the high quality of the recommendations and the scalability of the system. Furthermore, the system can be adapted to more complex scenarios by distributing the system among several machines or CPU cores.

6 Conclusion

In this paper we presented a recommender system implementing most popular item strategies. Our system computes the recommendations based on the number of user-item interactions for the most recently read articles. The most popular news articles are suggested to the users. The recommender is based on the idea that articles interesting for many readers are also relevant for new users. The strengths of the approach include the ability to quickly adapt to new trends and events and the good recommendation precision also for new and anonymous users.

The specific requirements with respect to scalability and response time we address by building our recommender system on APACHE SPARK. The use of APACHE SPARK enables us running the recommender as a distributed system in a cluster ensuring the scalability of the approach. We have evaluated the system in a Living Lab ("online") and Evaluation Lab ("offline") scenario. The Living Lab evaluation shows that our recommender system reliably reaches a good CTR while having the highest availability in the NEWSREEL challenge. In the Evaluation Lab we showed that the system can handle huge data streams efficiently.

Future Work. In the future we plan to evaluate additional recommendation models in order to further improve the CTR. Scaling the cluster horizontally makes it possible to use more elaborate algorithms from the SPARK machine learning library while corresponding to the strict time constraints. The usage of APACHE SPARK 2.0 introducing *DataFrames* makes the continuous model generation obsolete and will decrease processing time.

References

1. Diaz-Aviles, E., Drumond, L., Schmidt-Thieme, L., Nejdl, W.: Real-time top-n recommendation in social streams. In: Proceedings of the 6th ACM Conference on Recommender Systems, pp. 59–66. ACM, New York (2012). ISBN: 978-1-4503-1270-7
2. Ekstrand, M.D., Ludwig, M., Konstan, J.A., Riedl, J.T.: Rethinking the recommender research ecosystem: reproducibility, openness, and LensKit. In: Proceedings of the 5th ACM Conference on Recommender Systems, pp. 133–140. ACM (2011)
3. Hopfgartner, F., Kille, B., Lommatzsch, A., Plumbaum, T., Brodt, T., Heintz, T.: Benchmarking news recommendations in a living lab. In: Kanoulas, E., Lupu, M., Clough, P., Sanderson, M., Hall, M., Hanbury, A., Toms, E. (eds.) CLEF 2014. LNCS, vol. 8685, pp. 250–267. Springer, Cham (2014). doi:10.1007/978-3-319-11382-1_21
4. Hopfgartner, F., Brodt, T., Seiler, J., Kille, B., Lommatzsch, A., Larson, M., Turrin, R., Serény, A., Recommendations, B.N.: The CLEF NewsREEL use case. SIGIR Forum **49**(2), 129–136 (2016). doi:10.1145/2888422.2888443. ISSN: 0163–5840
5. Linden, A., Krensky, P., Hare, J., Idoine, C.J., Sicular, S., Vashisth, S.: Magic quadrant for data science platforms. In: Gartner & Forrester & Aragon, collection hiver 2017, pp. 28–29 (2017). https://myleadcorner.files.wordpress.com/2017/04/magic-quadrant-for-data-science-platforms-feb-2017-1.pdf
6. Lommatzsch, A., Albayrak, S.: Real-time recommendations for user-item streams. In: Proceedings of the 30th ACM Symposium on Applied Computing (SAC 2015), pp. 1039–1046. ACM, New York (2015). ISBN: 978-1-4503-3196-8
7. Seminario, C.E., Wilson, D.C.: Case Study evaluation of mahout as a recommender platform. In: RUE @ RecSys, pp. 45–50 (2012)
8. Walunj, S.G., Sadafale,K.: An online recommendation system for e-commerce based on apache mahout framework. In: Proceedings of the 2013 Conference on Computers and People Research, pp. 153–158. ACM (2013)
9. Zangerle, E., Gassler, W., Specht, G.: Exploiting Twitter's collective knowledge for music recommendations. In: Rowe, M., Stankovic, M., Dadzie, A.-S. (eds.) Making Sense of Microposts, pp. 14–17, April 2012. http://ceur-ws.org/Vol-838

The Case for Being Average: A Mediocrity Approach to Style Masking and Author Obfuscation

(Best of the Labs Track at CLEF-2017)

Georgi Karadzhov[1], Tsvetomila Mihaylova[1(✉)], Yasen Kiprov[1],
Georgi Georgiev[1], Ivan Koychev[1], and Preslav Nakov[2]

[1] Faculty of Mathematics and Informatics, Sofia University "St. Kliment Ohridski",
Sofia, Bulgaria
georgi.m.karadjov@gmail.com, tsvetomila.mihaylova@gmail.com,
yasen.kiprov@gmail.com, g.d.georgiev@gmail.com, koychev@fmi.uni-sofia.bg
[2] Qatar Computing Research Institute, HBKU, Doha, Qatar
pnakov@qf.org.qa

Abstract. Users posting online expect to remain anonymous unless they have logged in, which is often needed for them to be able to discuss freely on various topics. Preserving the anonymity of a text's writer can be also important in some other contexts, e.g., in the case of witness protection or anonymity programs. However, each person has his/her own style of writing, which can be analyzed using stylometry, and as a result, the true identity of the author of a piece of text can be revealed even if s/he has tried to hide it. Thus, it could be helpful to design automatic tools that can help a person obfuscate his/her identity when writing text. In particular, here we propose an approach that changes the text, so that it is pushed towards average values for some general stylometric characteristics, thus making the use of these characteristics less discriminative. The approach consists of three main steps: first, we calculate the values for some popular stylometric metrics that can indicate authorship; then we apply various transformations to the text, so that these metrics are adjusted towards the average level, while preserving the semantics and the soundness of the text; and finally, we add random noise. This approach turned out to be very efficient, and yielded the best performance on the Author Obfuscation task at the PAN-2016 competition.

1 Introduction

An important characteristic of the Web nowadays is the perceived anonymity of online activity. For example, a user posting in a forum online would expect to remain anonymous unless the site has asked him/her to share personal information as part of the registration to use their services. This is in theory. The reality is that there is very little anonymity online as Web sites track users in various ways, e.g., by requiring registration, by using cookies, by using third-party services, by linking phone numbers or Google/Facebook accounts to online activity,

G.J.F. Jones et al. (Eds.): CLEF 2017, LNCS 10456, pp. 173–185, 2017.
DOI: 10.1007/978-3-319-65813-1_18

etc. One could try to gain some anonymity by creating a new account or by using a different device or even an anonymous proxy service. Yet, when posting in a forum, the author's anonymity is still potentially at risk, as it is possible to analyze and match his/her posts to those of a known user. This is because each person has his/her own style of writing, which reflects their personality.

Revealing a person's identity requires (i) a hypothesis about who that person might be and (ii) a sufficient sample of text written by that person. Then, stylometric features can be used to predict whether the author of a target piece of text is indeed the one hypothesized. Even without a hypothesis about a target author, stylometry can reveal key demographic characteristics about the author of a piece of text, e.g., his/her gender and age, which is of interest to marketing analysts, political strategists, etc.

Overall, stylometry is a well-established discipline with application to authorship attribution, plagiarism detection, author profiling, etc. However, much less research has been done on the topic of author obfuscation, i.e., helping a person hide his/her own style in order to protect his/her identity and key demographic information. Unlike authorship attribution or author profiling, this is not a simple text classification problem but rather a complex text generation task, where not only the author's style has to be hidden, but the text needs to remain grammatically correct and the original meaning has to be preserved as much as possible.

Below we focus on the task of author obfuscation, i.e., given a document, the goal is to paraphrase it, so that its writing style does not match the style of its original author anymore. We use the task formulation, the data, and the evaluation setup from PAN'2016, the 15th evaluation lab on uncovering plagiarism, authorship, and social software misuse.

A system addressing the task needs to optimize three conflicting objectives simultaneously:

1. **Safety:** forensic analysis should not be able to reveal the original author of the obfuscated text;
2. **Soundness:** the obfuscated text should be semantically equivalent to the original;
3. **Sensibility:** the obfuscated text should be inconspicuous, i.e., it should not raise suspicion that there has been an attempt to obfuscate the author.

The performance of a participating system in the PAN'2016 task is measured as follows:

– *automatically*: using automatic authorship verifiers to measure *safety*; and
– *manually*: using peer-review to assess *soundness* and *sensibility*.

As these objectives are conflicting, the task is very challenging. While hiding author's style is a nontrivial task by itself, producing inconspicuous output is where most systems actually fail. At PAN'2016 [17], we proposed a method for author obfuscation that performed best in terms of *safety*, but lagged behind in *sensibility*. One of the other systems [14] generated text that scored high in terms of *sensibility* and *soundness*, but their system performed poorly in terms of *safety*. Below we describe and we evaluate both our original approach and a modification thereof that addresses the issue with poor results in *sensibility*. The evaluation results on the PAN'2016 dataset demonstrate the potential of our approach.

The remainder of this paper is organized as follows: Sect. 2 introduces related work. Section 3 describes our method: both the original one from PAN'2016 and the above-described modification thereof. Section 4 presents the evaluation setup and discusses the results. Finally, Sect. 5 concludes and points to some possible directions for future work.

2   Related Work

Author Identification is well-studied and has a long history [16]. The most common approach is to use variety of features [7,9,19,21] such as *punctuation* (e.g., relative frequency of commas), *lexical* (e.g., frequency of function words, average word and sentence length, etc.), *character-level* (e.g., n-gram frequencies), *syntactic* (e.g., frequencies of different parts of speech), *semantic* (e.g., frequency of various semantic relations), and *application-dependent* (e.g., style of writing email messages).

Author Identification has been a task at the PAN competition[1] since 2011. The most commonly-used features at PAN'2015 [22] include the lengths of words, sentences, and paragraphs, type-token ratios, the use of *hapax legomena* (i.e., words occurring only once), character n-grams (including unigrams), words, punctuation marks, stopwords, and part of speech (POS) n-grams. Other features that some participants used analyze the text more deeply by checking style and grammar.

Another line of research for author identification uses neural networks to induce features automatically. For example, Bagnal [2] used a recurrent neural network based on character unigrams, thus building a character-level language model. The idea is that a language model trained on texts by a particular author will assign higher probability to texts by the same author compared to texts written by other authors.

Author Obfuscation. Research in author obfuscation has explored manual, computer-aided, and automated obfuscation [19]. For manual obfuscation, people have tried to mask their own writing style as somebody else's, which was shown to work well [1,3,4]. Computer-aided obfuscation uses tools that identify and suggest parts of text and text features that should be obfuscated, but then the obfuscation is to be done manually [11,13,15].

Kacmarcik et al. [11] explored author masking by detecting the most commonly-used words by the target author and then trying to change them. They also mention the application of machine translation as a possible approach to author obfuscation. Other authors also used machine translation for author obfuscation [3,20], e.g., by translating passages of text from English to one or more other languages and then back to English. Brennan et al. [3] investigated three different approaches to adversarial stylometry: obfuscation (masking

---

[1] http://pan.webis.de.

author style), imitation (trying to copy another author's style), and machine translation. They further summarized the most common features people used to obfuscate their own writing style.

Juola et al. [9] developed a complex system for author obfuscation which consists of three main modules: canonization (unifying case, normalizing white spaces, spelling correction, etc.), event set determination (extraction of events significant for author detection, such as words, parts of speech bi- or tri-grams, etc.), and statistical inference (measures that determine the results and confidence in the final report). The authors used this same approach [8] to detect deliberate style obfuscation.

Kabbara et al. [10] used long short-term memory recurrent neural networks to transform the text in a similar fashion as in machine translation, but essentially 'translating' from one author's style to the style of other authors. While the approach looks promising, there was no proper evaluation in terms of safety and soundness.

3   Our Mediocrity Approach to Style Masking and Author Obfuscation

The main idea behind our approach is to measure the most significant features of the text used for author identification as mentioned in the work of Brennan et al. [3]. Then, we apply transformations of the text, so that the values of these metrics are pushed towards average.

Our approach consists of three main steps. First, we calculate "average" metrics based on the training corpus provided for the PAN-2016 Author Obfuscation task [19] and a corpus of several public-domain books from Project Gutenberg.[2] We will call these average values the *calculated averages*, and we will try to push the document metrics towards them. Second, we calculate the corresponding metrics for each target document. Then we apply ad hoc transformations on the text aiming to average those metric for the document. Third, we apply additional transformations aiming to randomly change the average metrics of the text. At this step, we apply very harmless transformations, so that we can preserve the meaning as much as possible. For example, we use dictionaries to transform abbreviations, equations, and short forms to listed alternatives.

3.1   Calculating Text Metrics

We calculate the following text metrics:

1. Average number of word tokens per sentence;
2. Punctuation to word token count ratio;

---

[2] We used the following books: *The Adventures of Sherlock Holmes* by Sir Arthur Conan Doyle, *History of the United States* by Charles A. Beard and Mary R. Beard, *Manual of Surgery Volume First: General Surgery* by Alexis Thomson and Alexander Miles. Sixth Edition., and *War and Peace*, by Leo Tolstoy.

3. Stop words to word token count ratio;
4. Word type to token ratio;
5. POS to word count ratio: measured for four part-of-speech groups: nouns, verbs, adjectives, and adverbs;
6. Uppercase word tokens to all word tokens count ratio;
7. Number of mentions of each word type in the text.

Given a document to be obfuscated, we calculate the above measures for it. Then, we split the document into parts, and for each part, we compare the measures for this part to the document-level averages. Finally, we apply transformations to push these values towards the corpus-level average.

3.2 Modulizing the Text

We used the PAN-2016 Author Obfuscation task setup, i.e., each text was to be split into parts of up to 50 words each. To do this, we first segmented the text into sentences using the NLTK sentence splitter. Then, we merged some of these sentences to get bigger segments, while keeping the segment lengths under 50 words. We ignored paragraph boundaries in this process.

3.3 Text Transformations

1. **Splitting or merging sentences**
   If the average sentence length of the whole document is below the *calculated average* of this metric, we perform merging of the sentences for each text part. We merge all the sentences for a given text part into one sentence. Merging is done by adding a random connecting word (*and, as, yet*) and randomly inserting punctuation - comma (,) or semicolon (;). When the average sentence length of the entire document is above the average, we split the sentence into shorter ones. We use a simple sentence splitting algorithm: we go through all POS-tagged words in the text, we count the nouns and the verbs, and when we reach a conjunction *and*, if the sentence so far contains a noun and a verb, we replace the *and* with a full stop (.) and we capitalize the next word as it will now start a new sentence.

2. **Stop Words**
   Stop words can be strong indicators for author identification as people have the tendency to use and overuse specific stop words. Thus, we perform two kinds of transformations regarding stop words:
   - removing stop words that carry little to no information;
   - replacing stop words with alternatives or with a phrase with the same meaning.

3. **Spelling**
   The spelling score of a document is high if there are no spelling mistakes, and low when there are some.
   - If we need to increase the spelling score, we apply automatic spelling correction. Our spell-checker uses a probabilistic model and a previously mentioned set of publicly-available books.

– If we need to decrease the score, we use a dictionary to insert common mistakes in the text. The dictionary was manually created using data from various sources.

4. **Punctuation**

If the punctuation use is above average, we remove some punctuation within the sentence. This is limited to the symbols comma (,), semicolon (;), and colon (:) If the punctuation use is below average, we apply the following two techniques to improve that score:

– We randomly insert comma or semicolon before prepositions. We insert comma with a higher probability compared to semicolon.
– We insert redundant symbols using the following schema:

```
! can be replaced with !, !!, or !!!
? can be replaced with ?, ??, ???, ?!?, or !?!
```

5. **Word Substitution**

In order to change the frequency of word types, we replace the most or the least common words. We use synonyms, hypernyms or word descriptions from WordNet [5, 18]. In particular, if the document type-token ratio is above average, we replace the most frequently used words in the document by random synonyms or hypernyms. In contrast, if the ratio is below average, we randomly replace the least frequently used words with their definitions.

6. **Paraphrase Corpus**

We randomly replace phrases from the text with variants from a paraphrase corpus. In particular, we use the short version of the phrasal corpus of PPDB, the Paraphrase Database [6]. As a result, the meaning of the text is still preserved, but there is improvement for the metrics for individual word frequencies and parts of speech.

7. **Uppercase Words**

If we need to decrease the proportion of uppercase words, we lowercase words that are all in uppercase and contain at least four letters. We assume that if a word is in uppercase and is less than four letters long, it is likely to be an acronym, and thus we should keep it in uppercase.

3.4   Noise

Having applied the above transformations, we then insert some random noise in the text, using the following two operations:

1. **Switching British and American English**

We randomly change words from British to American English and vice versa, using a lexicon.[3]

2. **Inserting random functional words**

We insert random functional words at the beginning of the sentence. The words are taken from a discourse marker lexicon.

---

[3] We have released our code, including all our lexicons, in the following repository: https://bitbucket.org/pan2016authorobfuscation/authorobfuscation/.

3.5   General Transformations

We also apply some general transformations that preserve the meaning of the text while helping mask the author style.

1. **Replacing short forms**
   We replace short forms such as *I've, I'd, I'm, I'll, don't, etc.* with their full forms.
2. **Replacing numbers with words**
   We replace tokens that are POS-tagged as numbers with their word representation in English.
3. **Replacing equations**
   As there were some examples of scientific text in the training corpus, if the text contains equations, the operations in them are being replaced with words. An equation is recognized if the text contains both comparison and inner equation symbols:

   ```
   ".[<>=]+." and ".[\+\-\*\/]+."
   ```

   We replace the following symbols if we find an equation: $+$ *(plus)*, $-$ *(minus)*, $*$ *(multiplied by)*, $/$ *(divided by)*, $=$ *(equals)*, $>$ *(greater than)*, $<$ *(less than)*, $<=$ *(less than or equal to)*, $>=$ *(greater than or equal to)*.
4. **Replace symbols and abbreviations with words**
   We further replace the following symbols and abbreviations with their word representations: currency symbols, *% (percent)*, *@ (at)*, abbreviations of person titles (such as *Prof., Mr., Dr.,* etc.).
5. **Simple transformations with regular expressions**
   Possessions (genitive markers) are replaced by a shorter form (e.g., *book of John* becomes *John's book*); here is the corresponding regular expression:

   ```
   "(\w+) of (\w+)"  is replaced with  "\2's \1"
   ```

   This will slightly change the stop words rate and could also obfuscate writing in general as the former way of expressing possession is generally less common.

3.6   Other Methods

1. **Machine translation**
   We also experimented with machine translation as described in [3]. We used the *Microsoft Translator API* to translate from English to Croatian and Estonian, and then back to English. The general assumption is that by applying machine translation using different languages, we will naturally paraphrase the text, while preserving the meaning. However, our manual evaluation has shown that this often yielded text whose meaning differed from that of the original text. This is consistent with the observations of Keswani et al. [12], who achieved poor sensibility and soundness by cyclic translations through several languages and then back to English for their PAN-2016 system [19].

2. **Simple word substitution**
   Another way to approach the task is to perform simple word substitution, e.g., using WordNet. Unfortunately, substituting a word does not always yield fluent text due to grammatical (e.g., wrong word inflection or wrong part of speech) or semantic mismatch (i.e., even though the substituted word may be a good paraphrase in general, it is not a good fit in the particular context). The problem can be alleviated to some extent by using multi-word paraphrases as longer phrases are less ambiguous and thus less context-dependent. In any case, the result of word/phrase substitution is not perfect; yet, it was found to perform much better than using machine translation in terms of sensibility and soundness. In particular, the two approaches were compared in the PAN-2016 Author Obfuscation task [19], where a substitution system outperformed a system based on round-trip machine translation [12] in terms of both soundness and sensibility; however, both systems scored low on safety. Mansoorizadeh et al. [14] performed substitution in a different manner, achieving high sensibility and soundness scores; however, their system was the worst in terms of safety in the PAN-2016 Author Obfuscation task. We tried their approach, and after some initial experiments, we concluded that word/phrase substitutions should be used not in isolation but rather together with our above-described transformations, which limits the use of such substitutions to some specific cases [17].

3.7 Transformation Magnitude

After reviewing the results for the system we submitted to PAN-2016 [17], we noticed that in some cases our transformations were too aggressive. In particular, if the value of some metric was below the average, we applied transformations to increase it, but we did not have a mechanism to control by how much we were boosting it. This sometimes resulted in undesired behaviour, e.g., when the value of a metric was close to the average, we could over-push it significantly over/below the average. Effectively, this goes against our aim to push it towards the average. As a side effect, in some cases, the text readability was affected negatively as well. In order to address the issue, we introduced an additional parameter, which tracks the magnitude of the desired change, i.e., the difference between the current value and the average value. We further modified the above transformations to keep track of and to update the value of this parameter accordingly.

4 Evaluation

The evaluation results in terms of *safety* are shown in Table 1. The table compares how well each of the three systems that participated in the PAN-2016 Author Obfuscation task can fool various author identification systems. A total of 44 authorship verification systems were used, which were submitted to the previous three shared tasks on Authorship Identification at PAN-2013, PAN-2014,

Table 1. Evaluation results in terms of *safety*. We compare the three obfuscation systems that participated in the PAN-2016 Author Obfuscation task. Shown is the average drop in performance for the 44 authorship verification systems submitted to the PAN-2013, PAN-2014, and PAN-2015 Author Identification tasks when running them on the obfuscated vs. the original versions of the test datasets.

Participant	PAN-2013	PAN-2014 EE	PAN-2014 EN	PAN-2015
Mihaylova *et al.* [17] (our)	−0.10	−0.13	−0.16	−0.11
Keswani *et al.* [12]	−0.09	−0.11	−0.12	−0.06
Mansoorizadeh *et al.* [14]	−0.05	−0.04	−0.03	−0.04

Table 2. Impact of the obfuscation on some text metrics. The first column shows the name of a text metric. The second column shows the value of the metrics for the input, i.e., *before* the obfuscation. Then follow the values after obfuscation, when using our *PAN-2016* and our *new* method, respectively. Finally, the values in the *average* column are calculated on the training dataset and on some texts from Project Gutenberg; these are the target values we want to push the metrics towards.

Text metric	Before	After obfuscation		Average
	(Input)	PAN-2016	New	(Target)
Punctuation to word token count ratio	0.14	0.14	0.15	0.15
Uppercase word tokens to all word tokens count ratio	0.03	0.01	0.02	0.02
Stop words to word token count ratio	0.52	0.45	0.50	0.50
Word type to token ratio	0.44	0.47	0.45	0.44
Number of nouns	0.23	0.24	0.24	0.24
Number of adjectives	0.08	0.09	0.08	0.06
Number of adverbs	0.07	0.09	0.08	0.07
Number of verbs	0.20	0.21	0.21	0.19

and PAN-2015. We can see that the output of our PAN-2016 system caused the performance for these 44 systems to drop the most for each of the three years; this means that it performs best in terms of *safety*.

While our PAN-2016 method is effective in terms of *safety*, it could not always produce text that is grammatically correct and contextually inconspicuous. That is why we introduced the use of transformation magnitude, which we evaluate below. For the evaluation, we use the data provided in the PAN-2016 Author Obfuscation task [19]. The data consists of 205 documents that have to be obfuscated.[4]

Table 2 shows the impact of the obfuscation, using both our PAN-2016 obfuscator and the new one that pays attention to transformation magnitude, on some text metrics. The first column shows the names of the text metrics. The second

---

[4] http://pan.webis.de/clef16/pan16-web/author-obfuscation.html.

column shows the value of the metrics for the input, i.e., *before* the obfuscation. Then follow the values after obfuscation, when using our *PAN-2016* and our *new* method, respectively. Finally, the values in the *average* column are calculated on the training dataset and on some texts from Project Gutenberg; these are the target values we want to push the metrics towards. We can see that, overall, the obfuscation methods do push the input metrics towards the target. We further see that the PAN-2016 method often overshoots and can push the metric even further away from the target compared to the input value. For example, for *Stop words to word token count ratio*, the input is 0.52, and it is pushed down to 0.45, which is further away from the target of 0.50 than the input was. In several other cases, the push went in the wrong direction, which can be due to the

Table 3. Obfuscation examples. Shown are examples of how the different systems that participated in the PAN-2016 Author Obfuscation task transform the original text; the last column shows the output of our new obfuscation method, which we introduced in this paper.

Original text	Machine translation [12]	Word substitution [14]	Our PAN-2016 obfuscation [17]	Our New Obfuscation
I am proud. Though I carry my love with me to the tomb, he shall never, never know it.	I believe expensive Though continue to never, tomb, it. ever be learned	Though I carry my love with me to the tomb, he shall never, never know it.	myself 'm proud in them, and though myself carry my beloved with me to the tomb he shall ever ever know it.	I 'm proud of them; and though I carry my beloved with me to the tomb he shall ever ever know it.
4) Religion discriminates. Sure, it unifies (...). On the other hand	4) religion discriminates. some people Sure, unifies (...) second, it condems	4) Religion discriminates. Sure, it unifies (...) On the other hand	Four) Religion discriminate; as sure, it unified (..), and on the other hand	Four) Religion discriminate, sure, it unified (...); on the other side
Consequently, they see a connection between development in spiritual life and professional economic development	they Consequently, a link between the spiritual development and economic professional development.	Consequently, they see a connection between growth in spiritual life and professional economic development.	Definitely, Consequently, they see a connection between development inside spiritual life also professional economic development;	Consequently, they see a link between development in spiritual life and professional economic development,

multi-objective optimization – when changing text to modify one text metric, we change some words, which could affect a number of other metrics.

We can further see in Table 2 that the new method, which takes transformation magnitude into account, gets much closer to the target compared to our PAN-2016 obfuscation: it matches exactly the values of four out of the eight metrics (compared to just two for the PAN-2016 version), and gets very close to the target for the other four metrics.

Finally, Table 3 looks into *sensibility*. In particular, it illustrates how the original texts are changed by the different systems (our initial PAN-2016 system, our new system that pays attention to transformation magnitude, and the remaining two systems that participated in PAN-2016). We can see from the examples that our methods perform better than the rivaling PAN-2016 systems, and that *sensibility* improves when using transformation magnitude. We can further see that the PAN-2016 system that uses machine translation [14] can alter the semantics of the input text, which often makes it meaningless. Moreover, the PAN-2016 system that uses word substitution [12] is too conservative and often does not change the text at all.

Overall, we can conclude that our method with transformation magnitude is promising and performs well (better than the three systems that participated in the PAN-2016 Author Obfuscation task) in terms of *sensibility* and *soundness*. In future work, we need to study how it performs in terms of *safety*.

5   Conclusion and Future Work

We have described our mediocrity approach to style masking and author obfuscation, which changes the text, so that it is pushed towards average values for some general stylometric characteristics, thus making these characteristics less discriminative.

In future work, we plan experiments with a richer set of features as well as with deep learning. We further aim to design an evaluation measure targeting soundness, which would be enabling for this kind of research. Finally, we plan to use the techniques used in this paper for author imitation. One key difference will be that the goal for the transformations should not be the average metrics, but the metrics for the target author who should be imitated.

Acknowledgments. We thank the anonymous reviewers for their constructive comments, which have helped us improve the quality of the present paper.

This research was performed by a team of students from MSc programs in Computer Science in the Sofia University "St. Kliment Ohridski". The work is supported by the NSF of Bulgaria under Grant No.: DN 02/11/2016 - ITDGate.

References

1. Almishari, M., Oguz, E., Tsudik, G.: Fighting authorship linkability with crowd-sourcing. In: Proceedings of the Second ACM Conference on Online Social Networks (COSN 2014), pp. 69–82. ACM, Dublin (2014)

2. Bagnall, D.: Author identification using multi-headed recurrent neural networks. In: Working Notes of CLEF 2015 - Conference and Labs of the Evaluation Forum (CLEF 2015), Toulouse (2015)

3. Brennan, M., Afroz, S., Greenstadt, R.: Adversarial stylometry: circumventing authorship recognition to preserve privacy and anonymity. ACM Trans. Inf. Syst. Secur. 15(3), 12:1–12:22 (2012)

4. Brennan, M.R., Greenstadt, R.: Practical attacks against authorship recognition techniques. In: Proceedings of the Twenty-First Innovative Applications of Artificial Intelligence Conference (IAAI 2009), Pasadena (2009)

5. Fellbaum, C.: WordNet: An Electronic Lexical Database. Bradford Books, Cambridge (1998)

6. Ganitkevitch, J., Van Durme, B., Callison-Burch, C.: PPDB: the paraphrase database. In: Proceedings of the 2013 Conference of the North American Chapter of the Association for Computational Linguistics: Human Language Technologies (NAACL-HLT 2013), Atlanta, pp. 758–764 (2013)

7. Holmes, D.I.: The evolution of stylometry in humanities scholarship. Lit. Linguist. Comput. 13(3), 111–117 (1998)

8. Juola, P.: Detecting stylistic deception. In: Proceedings of the Workshop on Computational Approaches to Deception Detection, Avignon, pp. 91–96 (2012)

9. Juola, P., Vescovi, D.: Analyzing stylometric approaches to author obfuscation. In: Peterson, G., Shenoi, S. (eds.) DigitalForensics 2011. IAICT, vol. 361, pp. 115–125. Springer, Heidelberg (2011). doi:10.1007/978-3-642-24212-0_9

10. Kabbara, J., Cheung, J.C.K.: Stylistic transfer in natural language generation systems using recurrent neural networks. In: Proceedings of the Workshop on Uphill Battles in Language Processing: Scaling Early Achievements to Robust Methods, Austin, pp. 43–47 (2016)

11. Kacmarcik, G., Gamon, M.: Obfuscating document stylometry to preserve author anonymity. In: Proceedings of the 21st International Conference on Computational Linguistics and 44th Annual Meeting of the Association for Computational Linguistics (COLING-ACL 2006), Sydney, pp. 444–451 (2006)

12. Keswani, Y., Trivedi, H., Mehta, P., Majumder, P.: Author masking through translation. In: Working Notes of CLEF 2016 - Conference and Labs of the Evaluation Forum (CLEF 2016), Évora, pp. 890–894 (2016)

13. Le, H., Safavi-Naini, R., Galib, A.: Secure obfuscation of authoring style. In: Akram, R.N., Jajodia, S. (eds.) WISTP 2015. LNCS, vol. 9311, pp. 88–103. Springer, Cham (2015). doi:10.1007/978-3-319-24018-3_6

14. Mansoorizadeh, M., Rahgooy, T., Aminiyan, M., Eskandari, M.: Author obfuscation using WordNet and language models. In: Working Notes of CLEF 2016 - Conference and Labs of the Evaluation Forum (CLEF 2016), Évora, pp. 932–938 (2016)

15. McDonald, A.W.E., Afroz, S., Caliskan, A., Stolerman, A., Greenstadt, R.: Use fewer instances of the letter "i": toward writing style anonymization. In: Fischer-Hübner, S., Wright, M. (eds.) PETS 2012. LNCS, vol. 7384, pp. 299–318. Springer, Heidelberg (2012). doi:10.1007/978-3-642-31680-7_16

16. Mendenhall, T.C.: The characteristic curves of composition. Science 9(214), 237–249 (1887)

17. Mihaylova, T., Karadjov, G., Nakov, P., Kiprov, Y., Georgiev, G., Koychev, I.: SU@PAN'2016: author obfuscation-notebook for PAN at CLEF 2016. In: Working Notes of CLEF 2016 - Conference and Labs of the Evaluation Forum (CLEF 2016), Évora (2016)

18. Miller, G.A.: WordNet: a lexical database for English. Commun. ACM **38**(11), 39–41 (1995)
19. Potthast, M., Hagen, M., Stein, B.: Author obfuscation: Attacking the state of the art in authorship verification. In: Working Notes of CLEF 2016 - Conference and Labs of the Evaluation Forum (CLEF 2016), Évora, pp. 716–749 (2016)
20. Quirk, C., Brockett, C., Dolan, W.: Monolingual machine translation for paraphrase generation. In: Proceedings of the 2004 Conference on Empirical Methods in Natural Language Processing (EMNLP 2004), Barcelona, pp. 142–149 (2004)
21. Stamatatos, E.: A survey of modern authorship attribution methods. J. Am. Soc. Inf. Sci. Technol. **60**(3), 538–556 (2009)
22. Stamatatos, E., Daelemans, W., Verhoeven, B., Juola, P., López-López, A., Potthast, M., Stein, B.: Overview of the author identification task at PAN 2015. In: Working Notes of CLEF 2015 - Conference and Labs of the Evaluation Forum (CLEF 2015), Toulouse (2015)

Author Clustering with an Adaptive Threshold

Mirco Kocher[✉] [iD] and Jacques Savoy

Computer Science Department, University of Neuchâtel, Neuchâtel, Switzerland
{Mirco.Kocher,Jacques.Savoy}@unine.ch

Abstract. This paper describes and evaluates an unsupervised author clustering model called SPATIUM. The proposed strategy can be adapted without any difficulty to different natural languages (such as Dutch, English, and Greek) and it can be applied to different text genres (newspaper articles, reviews, excerpts of novels, etc.). As features, we suggest using the m most frequent terms of each text (isolated words and punctuation symbols with m set to at most 200). Applying a distance measure, we define whether there is enough evidence that two texts were written by the same author. The evaluations are based on six test collections (PAN AUTHOR CLUSTERING task at CLEF 2016). A more detailed analysis shows the strengths of our approach but also indicates the problems and provides reasons for some of the potential failures of the SPATIUM model.

Keywords: Author clustering · Threshold · Author identification · PAN

1 Introduction

With the increased communication facilities and the ubiquity of social media, we encounter an enlarged number of authorship problems. With the believed anonymity offered by the Web, the number of anonymous and pseudonymous texts or threats is increasing. To be able to automatically determine the real author of a text presents a clear interest for criminal investigations as well as for historical or literature studies (e.g., who really is the novelist Elena Ferrante?).

In this perspective, the classical question is to determine the real author of a given text, usually based on a set of documents with known authorship. But the *author clustering* task is more demanding. This problem can be formulated as follows: given a corpus of n texts, regroup all documents written by the same author such that each of the k clusters corresponds to a distinct author. For example, based on a set of n passages extracted from a collaborative work, we should first determine the number of authors k and then regroup the texts into k clusters according to their real author.

This paper is organized as follows. The next section presents the related work while Sect. 3 briefly describes the test collections and the evaluation methodology used in our experiments. Section 4 describes our proposed algorithm based on the SPATIUM model. Section 5 evaluates the proposed scheme and compares it to the best performing schemes using six different test collections extracted from CLEF PAN 2016. Then, Sect. 6 provides an analysis to assess the variability of the performance measures. Finally, Sect. 7 exposes our adaptive threshold system that can extract some correct assignments

© Springer International Publishing AG 2017
G.J.F. Jones et al. (Eds.): CLEF 2017, LNCS 10456, pp. 186–198, 2017.
DOI: 10.1007/978-3-319-65813-1_19

even when the information available is rather limited. A conclusion draws the main findings of this study.

2 Related Work

The author clustering problem was introduced as a new task in the PAN CLEF 2016 track. In this view, Stamatatos *et al.* [16] provide a good overview of the proposed methods. Overall, the first main component for solving this issue is to define an effective distance measure between two text representations. Such a function returns a small value when the two documents are written by the same author, and a larger one otherwise. Of course, instead of defining a distance measure, one can propose a similarity measure and accept that two texts were written by the same person when the similarity value is high enough. The second problem consists of developing or applying a clustering procedure capable of establishing links between texts written by the same author. In this case, after assuming that Text A and B have the same author, as well as Text A and C, one can infer that Text B and C have been written by the same source as well (single link strategy).

An answer to the first question is related to classical authorship attribution, but in an unsupervised perspective. A first set of methods suggests defining an invariant stylistic measure [5] that must reflect the particular style of a given author and should vary from one person to another. Furthermore, we can assume that an author's writing style is stable over period of time (e.g., one decade) before showing measurable differences [4]. A multivariate method can be applied to project each document representation into a reduced dimensional space under the assumption that texts written by the same author will appear close together. Some of the main approaches applicable here are principal component analysis (PCA) [3], clustering [10], or discriminant analysis [6]. As stylistic features, these approaches tend to employ the top 50 to 200 most frequent word types (MFW), as well as some part-of-speech (POS) information. In a related vein, Layton *et al.* [11] also propose a clustering approach based on their iterative Silhouette method to determine the number of authors in a set of documents.

Based on the differences in word distribution between two texts, several distance-based measures have been proposed [9]. As well-known functions defined more specifically for solving the authorship attribution question, one can mention Burrows' Delta [2] using the top m MFW (with $m = 40$ to $1,000$), the Kullback-Leibler divergence [18] using a predefined set of 363 English words, or Labbé's method [10] using the whole vocabulary.

Finally, as a clustering algorithm, the complete link seems the more conservative strategy, requiring that all members in a cluster share a high similarity between them. As an alternative, the k-means procedure [17] can be applied. Based on PAN CLEF 2016 results [16], this approach tends to produce lower effectiveness levels than approaches based on distance measures.

3   Test Collections and Evaluation Methodology

To promote research and to evaluate author clustering algorithms, the CLEF PAN 2016 generated a benchmark composed of six test collections covering three languages (English, Dutch, and Greek) and two text genres (newspaper articles and customer reviews). For each of the six language/text genre combinations, one can find three "collections" denoted "problems" in the PAN parlance. Thus, for each language, one can find three problems composed of newspaper articles and three others containing reviews.

During the PAN CLEF 2016, there were 3×6 problems available for training with their main statistics as reported in Table 1. In this table, the number of texts belonging to each language/genre combination is indicated under the label "Texts". For example, with the EA (English Articles), one can find three problems, each containing 50 articles. The number of distinct authors per problem is indicated in the column "Authors", and the number of authors with a single document under the label "Single". Thus, the first problem in the EA test collection has 35 authors, from which 27 have written only one article. In the last column, the mean number of words per text is depicted.

Table 1. PAN CLEF 2016 *training* corpora statistics

Corpus	Texts	Training problems		
		Authors	Single	Words
English Articles (EA)	50	35; 25; 43	27; 17; 37	741; 745; 734
English Reviews (ER)	80	55; 70; 40	39; 62; 17	969; 1080; 1020
Dutch Articles (DA)	57	51; 28; 40	46; 20; 32	1086; 1334; 1026
Dutch Reviews (DR)	100	54; 67; 91	31; 44; 83	128; 135; 126
Greek Articles (GA)	55	28; 38; 48	10; 26; 42	756; 750; 735
Greek Reviews (GR)	55	50; 28; 40	46; 13; 29	534; 646; 756

During the PAN CLEF 2016 evaluation campaign, 18 additional problems were built (test phase) with the same distribution over the languages and text genres as the training collections (shown in Table 1). As the correct statistics for those corpora are still undisclosed, our study will focus mainly on the training corpora.

When inspecting the training problems, we note that the number of words available in DR is rather small (in mean, 130 words for each document). Moreover, there are many authors who only wrote a single text, so the number of authors per problem is rather large (as well as the number of expected clusters). This means that we should only regroup two documents if there is enough evidence for a single authorship.

As proposed in the PAN CLEF 2016 track, an author clustering algorithm is evaluated with two distinct metrics. First, the purity of the generated clusters is evaluated. In this perspective, a perfect system must create only k clusters, each containing all the documents written by the same person. The evaluation measures are the precision, the recall, and the harmonic mean between the two values (denoted BCubed F_1) [1]. Moreover, each document must belong to exactly one cluster. To achieve a perfect precision, the solution is to generate one cluster per document. Therefore, the purity of each cluster is maximal and the resulting precision is 1.0. On the other hand, to achieve a recall of 1.0, all documents can be regrouped into a single cluster. Thus, the two measurements are in opposition. The F_1 value will serve as an effectiveness measure of the resulting clusters, with a higher value meaning a better distribution.

As a second measure, one can ask the clustering algorithm to return a list of links between text pairs, ordered by an estimated probability of having the same author for the two cited documents. To evaluate such an ordered list, one can apply the mean average precision (MAP) [16]. As complementary measures, the precision after 10 ranks (P@10) or the RPrec can be computed. MAP is a classical evaluation measure in the IR domain [12]. It is known that this measure is sensitive to the first rank(s), and providing an incorrect answer in the top ranks intensively hurts the MAP value. On the other hand, MAP does not punish verbosity, i.e., every true link counts even when appearing near the end of the ranked list. Therefore, by providing all possible authorship links, one can attempt to maximize MAP, without penalizing the P@10.

4   Simple Clustering Algorithm

To solve the clustering problem, we propose an adapted approach based on a simple feature extraction and distance metric called SPATIUM [7]. The selected stylistic features correspond to the top m most frequent terms (isolated words without stemming, but with the punctuation symbols) from the query text. For determining the value of m, previous studies have shown that a value between 200 and 300 tends to provide the best perform- ance in the authorship attribution domain [2, 13]. Moreover, we will exclude the words appearing only once (*hapax legomenon*) in the text for the feature selection. This filtering decision was taken to prevent overfitting to single occurrences.

As shown in Table 1, some documents were rather short. Therefore, the real number of terms m was set to at most 200 terms but, in most cases, was well below. With this reduced number, the justification of the decision will be simpler to understand because it will be based on words instead of letters, bigrams of letters, or combinations of several representation schemes or distance measures.

To measure the distance between a Text A and another Text B, the SPATIUM model uses a weighted variant of the L^1-norm which was already found to be useful in a related task [9]. The Canberra distance suggests that the absolute differences of the individual terms are normalized based on the sum of them as indicated in Eq. 1.

$$\Delta_{AB} = \Delta(A, B) = \sum_{i=1}^{m} \frac{\left| P_A[t_i] - P_B[t_i] \right|}{P_A[t_i] + P_B[t_i]} \tag{1}$$

where m indicates the number of terms (words or punctuation symbols) occurring in A more than once, $P_A[t_i]$ and $P_B[t_i]$ represent the estimated occurrence probability of the term t_i in Text A and Text B respectively. To estimate these probabilities, we divide the term occurrence frequency (tf_i) by the length in tokens of the text (n), $Prob[t_i] = tf_i/n$, without smoothing and an estimation of 0.0 may occur in Text B.

For example, assume that Text A corresponds to "The fox, the moose, and the deer jump over a wolf." Based on the term frequency, the resulting vector is [the (3), (2)] after ignoring the letter case. The other words occurring once are ignored. The final representation is: [the (3/5), (2/5)]. Assuming Text B contains the following sentence: "The quick fox and the brown deer jump over the lazy dog and a cat." When computing the distance Δ_{AB}, the following terms are used {the ,} because they are extracted from the representation of Text A. The representation of Text B is therefore [the (3/3), (0/3)]. Applying Eq. 1 with these two terms gives us $\Delta_{AB} = 1.25$. On the other hand, when estimating the distance Δ_{BA}, only terms belonging to B's representation are considered, namely {the and}, giving us the representation [the (3/5) and (2/5)] for Text B and [the (3/4) and (1/4)] for Text A, resulting in a distance $\Delta_{BA} = 0.34$. This distance measure is not symmetric due to the choice of the terms.

Observing a small value for Δ_{AB} provides evidence that both documents are written by the same author. On the other hand, a large value suggests the opposite assuming the text length is long enough to support this finding. The real problem consists in defining precisely what a "small distance value" is. To verify whether the resulting Δ_{AB} value is small, a comparison basis must be determined.

To achieve this with a specific collection, the distance *from* A to all other texts is computed (or $\Delta(A, j)$). From this distribution, the mean (denoted $m(A, .)$) and standard deviation ($std(A, .)$) are estimated. Moreover, the distribution of distance values *to* Text B (or $\Delta(j, B)$) can be computed to provide the mean $m(., B)$ and the standard deviation $std(., B)$ of the intertextual distances *to* Text B.

As a first definition of a "small" distance, we can assume that a small distance value *from* Text A must respect Eq. 2. In this formulation, δ is a parameter to be fixed. Assuming a Gaussian distribution, setting $\delta = 1.645$ means that 5% of the observations are smaller than the *mean* $- 1.645 * std$.

$$\text{Hint 1: } \Delta(A, j) \leq \phi(A, .) = m(A, .) - \delta * std(A, .) \tag{2}$$

Similarly, a small distance *to* Text B can be defined as:

$$\text{Hint 2: } \Delta(j, B) \leq \phi(., B) = m(., B) - \delta * std(., B) \tag{3}$$

With these two decision rules, one can verify if a distance *from* Text A (Eq. 2) or *to* Text B (Eq. 3) is small or not. We propose to be more cautious, mainly because proposing an incorrect assignment must be viewed as more problematic than missing a link between two documents written by the same author.

To follow this idea, having a distance value Δ_{AB}, we can verify the magnitude of its value according to Eq. 2 (*from* A) and Eq. 3 (*to* B). In the same way, one can verify whether the resulting Δ_{BA} value is small or rather large. Therefore, we propose to create two additional decision rules with Eq. 4 (based on the distribution of distance values *from* Text B) and Eq. 5 (for distance *to* Text A) as follows:

$$\text{Hint 3: } \Delta(B, j) \leq \phi(B, .) = m(B, .) - \delta * std(B, .) \tag{4}$$

$$\text{Hint 4: } \Delta(j, A) \leq \phi(., A) = m(., A) - \delta * std(., A) \tag{5}$$

To ground our attribution decision on a solid foundation, we compute both the distance Δ_{AB} and Δ_{BA} and check all four hints. An authorship between Text A and B is expected if at least two of the four hints are satisfied.

The choice of the parameter value δ, and the number of limits to be respected (two in our case) indicate the willingness of having more or less strict assignments. A smaller value for δ generates more potential links between texts and thus increases the risk of observing incorrect assignments. If a corpus is composed of many authors with each cluster contains only a few items, the parameter δ can be fixed at a higher level (e.g., $\delta = 1.96$, corresponding to 2.5% of the values of a Gaussian distribution).

5   Evaluation

Based on the gold standard provided by the CLEF PAN 2016 dataset, the SPATIUM model with the threshold value $\delta = 2$ can be evaluated as shown in Table 2. This table reports the performance measures applied during the PAN CLEF campaign, namely the BCubed F_1 and the MAP presented in Sect. 3. These measures are not provided for each problem but only the average over the three problems included in each test collection. Under the term "Score" we report the mean between the F_1 and MAP value.

Table 2. Evaluation for the six *training* collections

Corpus	Score	F_1	MAP
English Article (EA)	0.4601	0.7972	0.1229
English Review (ER)	*0.4242*	*0.7656*	*0.0828*
Dutch Article (DA)	0.5184	0.8387	0.1981
Dutch Review (DR)	*0.4192*	*0.7895*	*0.0488*
Greek Article (GA)	0.5649	0.8294	0.3004
Greek Review (GR)	**0.6878**	**0.8588**	**0.5168**
Average	0.5124	0.8124	0.2116

The best performance values are depicted in bold. As one can see, the Spatium returns the best results for the GR collection with a final score of 0.6878 followed by the GA and DA test collection. The worst result is achieved with the ER and DR collections (values depicted in italics). Moreover, the BCubed F_1 is very similar over all collections but the variability of the MAP is remarkable. The achieved MAP with the GR corpus is almost ten times higher than in the DR or ER corpus.

The evaluation performed on the test set is depicted in Table 3. The differences between the training and test corpus are relatively small. Similar clustering performances can be achieved using either the training or test set, indicating a strong correlation between the two samples. Since our model is unsupervised, there is no influence of one collection on the other, and no resources have been used to fix any parameter values or to build a learning structure.

Compared to the other participants of the PAN 2016 author clustering task, we achieve the second best overall score with one of the fastest systems. Some texts were wrongly grouped up, which decreases the document precision part of the BCubed F-Score a bit. Overall, we cluster many documents correctly together (which increases document recall part) and assign them a high score for their authorship link (which increases MAP).

Table 3. Evaluation for the six *test* collections

Corpus	Score	F_1	MAP
English Article (EA)	0.4348	0.7518	0.1178
English Review (ER)	0.4320	0.7869	*0.0772*
Dutch Article (DA)	0.4742	0.8183	0.1301
Dutch Review (DR)	*0.4106*	0.7702	*0.0510*
Greek Article (GA)	0.4891	0.8005	0.1778
Greek Review (GR)	**0.5660**	**0.8326**	**0.2995**
Average	0.4678	0.7934	0.1422

6   Sensibility Assessment

To provide a fair evaluation methodology, we cannot simply compare the performance values (MAP, F_1, or Score) directly between two approaches. A leaving-one-out or cross-fold evaluation is not possible in this task. We need to estimate the underlying variability of each performance using, for instance, the bootstrap approach. In this approach, for each problem, the system must generate S new random bootstrap samples. More precisely, for each text, we will create $S = 200$ new copies having the same length. For each copy the probability of choosing one given term (word or punctuation symbol) depends on its relative frequency in the original text. This drawing is done with replacement; thus, the underlying probabilities are fixed.

Each resulting text must be viewed as a bag-of-words. As the syntax is not respected, each bootstrap text is not really readable but reflects the stylistic aspects as analyzed by the SPATIUM approach.

For each of the 200 generated collections of bootstrap samples, we have applied our approach and obtained the MAP and the BCubed F_1 values reported in Tables 4 and 5. In Table 4, the column F_1 (or MAP in Table 5) indicates the performance achieved with the original data (as presented in Table 2). Then the column labeled "$\bar{x}$" reports the mean of the F_1 (or MAP respectively) achieved with the 200 new collections, together with the limit of ± 2 standard deviations σ (last two columns) corresponding to a confidence interval of 95.4%.

As depicted in Table 4, the reported performance for the EA collection is 0.7972. With the bootstrap methodology, the 95.4% confidence interval is [0.7551; 0.8085] for this value. As one can see in Table 4 (F_1 values), the mean of the bootstrap sample is usually lower (around 2%) than the original performance values but the original performance is always within the confidence interval of the bootstrap sample. In Table 5, the difference between the original MAP performances and the mean of the bootstrap sample is larger. For the DR corpus (Table 5), the difference is rather small (around 4%) while in the EA collection a drop of over 50% can be observed. It is known that the MAP measure is more sensitive to variations because a misclassification in the highest ranks is strongly penalized leading to a higher standard deviation.

Table 4. Results for the BCubed F_1 after applying the bootstrap estimation

Corpus	F_1	$\bar{x}$	$\bar{x} - 2\sigma$	$\bar{x} + 2\sigma$
English Article (EA)	0.7972	0.7818	0.7551	0.8085
English Review (ER)	0.7656	0.7448	0.7091	0.7805
Dutch Article (DA)	0.8387	0.8210	0.7970	0.8450
Dutch Review (DR)	0.7895	0.7699	0.7394	0.8005
Greek Article (GA)	0.8294	0.8088	0.7777	0.8399
Greek Review (GR)	0.8588	0.8452	0.8173	0.8732
Average	0.8124	0.7952	0.7659	0.8246

Table 5. Results for the MAP after applying the bootstrap estimation

Corpus	MAP	$\bar{x}$	$\bar{x} - 2\sigma$	$\bar{x} + 2\sigma$
English Article (EA)	0.1229	0.0578	0.0179	0.0978
English Review (ER)	0.0828	0.0490	0.0138	0.0842
Dutch Article (DA)	0.1981	0.1159	0.0579	0.1740
Dutch Review (DR)	0.0488	0.0466	0.0317	0.0616
Greek Article (GA)	0.3004	0.2015	0.1013	0.3017
Greek Review (GR)	0.5168	0.4279	0.3271	0.5288
Average	0.2116	0.1498	0.0916	0.2080

7 Adaptive Thresholding

To improve our knowledge, it is important to understand why and when an automatic text categorization scheme fails to provide the correct answer. Such an analysis will reveal more precisely the advantages and drawbacks of a suggested scheme. In the current context, the important question is related to the definition of a pertinent threshold in defining our limits (see Eqs. 2 to 5).

In a previous study [8], we had to classify, under the same condition, 52 excerpts of English novels containing in mean 10,000 tokens [10]. In this corpus, nine authors had written multiple texts (specifically Hardy wrote 12 texts, Conrad wrote 8, Stevenson (7), Morris (6), Orczy (6), Butler (4), Chesterton (3), Forster (3), and Tressel (3)).

When analyzing the Canberra distance between all possible pairs of texts, the global distribution is a mixture of two distributions. The first one corresponds to the distance values obtained when the two texts are written by the same author (shown in blue or white on the left part of Fig. 1). The second one results from pairs composed of two texts written by two persons (depicted in red or gray on the right part in Fig. 1). In Fig. 1, one can see these two distributions in which the three means are indicated with the vertical lines. On the left part, one can observe the mean of the correct links (denoted by

"Mean(Blue)"), the mean of the mixed distribution ("Mean(MixDist)") and on the right, the mean of the incorrect links ("Mean(Red)"). In this figure, the limit proposed in our four hints in Eqs. 2 to 5 and corresponding to the mean − 1.64 * std ("Mean − 1.64 * SD") appears with a vertical line on the left. As we can see in this figure, all distances below this limit correspond to correct pairings.

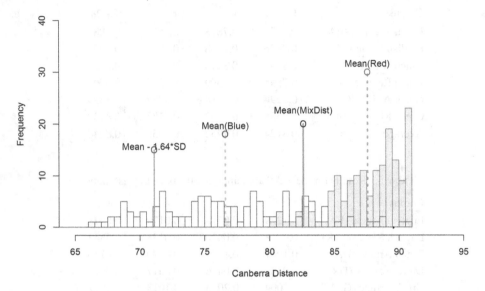

Fig. 1. Distribution of distances for a literary corpus. In white (blue) the correct links, and in grey (red) the wrong links. (Color figure online)

With the PAN data, we do not observe such a clear distinction between the two distance distributions. As an example, Fig. 2 visualizes the observed distributions of distances (on a logarithmic scale for the y-axis) in the Dutch Article corpus. This collection contains 57 documents out of which 20 have a single and unique author. From the remaining 37 documents, we should create one cluster of size two, three, four, six, and seven plus three clusters each containing five texts. Therefore, a total of 152 links (that is, 76 bidirectional links) must be created, out of the possible 3,192 links (57 * 57 − 57 in total, or 1,596 bidirectional links). Figure 2 is obtained when considering all link distance values. As we can see, there is an interleaving of the correct and incorrect links and the two means are almost identical.

Some texts are generally very close to many other texts, but they don't have sets of texts which are especially close to them. This results in a series of links that should be ignored. Then, there are texts that may be very far from some texts and very close to some other texts, meaning the link distance distribution has a large variance (or standard deviation). Again, those links should not be considered for a shared authorship due to the wide spread range of values. A correct authorship link could be detected if there are texts with a few link distance values that are substantially lower with respect to the text's general link distance distribution.

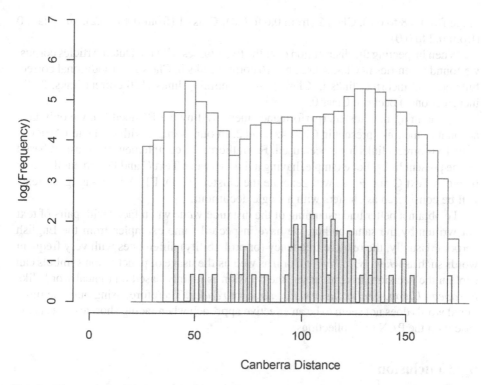

Fig. 2. Observed distribution of distance values (y-axis on a logarithmic scale) in the Dutch Article corpus (in dark (blue) the correct pairs, in white (red) incorrect pairs). (Color figure online)

As described in Sect. 4, we have two inequalities to determine when a distance value Δ_{AB} (or Δ_{BA}) can be viewed as "small" and thus hopefully reflecting a correct attribution. We use these limits to filter the distance values and to extract the more pertinent ones that are in the lower tail of a Gaussian distribution.

To generate the ranked list of links between two texts, the final attribution works as follows. After computing the distance values between all pairs of texts, we sort them from the smallest to the highest. Starting with the smallest (let's say Δ_{AB}), we also consider the opposite (Δ_{BA}). The link between the two texts is assigned in our Class 4 if the two distance values are smaller than the four limits (see Eqs. 2 to 5). If not, the link can be assigned to Class 3 (the two distances respect three limits), Class 2 (the two distances satisfy two limits), Class 1 (a single hint is available from the two distances), or Class 0 (the two distances are larger than the four limits).

To generate the final ranked list of links, we first consider Class 4. All links appearing in this group will obtain a probability of being correct between 1.0 and 0.8. For a given link, its probability depends on its position inside the Class 4. To define this position, we sort the links according to the sum of the two distance values (e.g., $\Delta_{AB} + \Delta_{BA}$) from the smallest to the largest. The smallest pair of distances will obtain the probability value of 1.0, the largest 0.8. The same sorting process is then applied for Class 3 (probability range from 0.8 to 0.6), Class 2 (from 0.6 to 0.4), Class 1 (from 0.4 to 0.2), and Class 0 (from 0.2 to 0.0).

When inspecting the distribution over the five classes with the Dutch Articles corpus, we found no entries in Class 4 or 3, but 16 correct links in Class 2, 16 additional correct links and 32 incorrect links in Class 1. All remaining links (120 correct links, 3,008 incorrect ones) occur in Class 0.

For defining the clusters (performance measured by the BCubed F_1), we only take account of the links present in Class 4, 3 and 2. In our example with the Dutch Articles, only 16 (correct) links have been used. From them, we complement the clusters based on the present links. For example, having a link between Text C and D, and another link between Text C and F, we will generate the cluster {C, D, F}. All non-assigned texts will be considered as clusters with a single document.

To obtain a better understanding of the distance value when faced with pairs of text not written by the same author, we have inspected some examples from the English corpora. Usually, the relative frequency (or probability) differences with very frequent words such as *when, is, in, that, to,* or *it* as well as the usage of punctuation symbols can explain the decision. In other cases, the decision is mainly based on topical words like *European Union, wealth, history, language,* or *reader.* Therefore, using only the functional words does not seem to be an effective approach when facing short texts, as is the case with the PAN test collections.

8   Conclusion

This paper evaluates a simple unsupervised technique to solve the author clustering problem. As features to discriminate between the proposed author and different candidates, we propose using the top 200 most frequent terms (isolated words and punctuation symbols). This choice was found effective for other related tasks, such as in authorship attribution [2]. Moreover, compared to various feature selection strategies used in text categorization [15], the most frequent terms tend to select the most discriminative features when applied to stylistic studies [14]. To make the author linking decision, we propose using a simple distance measure based on the SPATIUM model using a variant of the L^1 norm (Canberra). This choice seems a good one compared to other possible distance functions (such as Euclidean, Cosine, or Dice) [9].

When using the CLEF PAN test collections, several parameters having a clear impact on the text style have been fixed, such as the time period, the text genre, or length of the data. This strategy tends to minimize the possible sources of variation in the corpus. The most challenging aspect of those test collections are the rather short lengths of the texts. In this context, our main objective is to present a simple and unsupervised approach without many predefined arguments.

With an adapted version of the SPATIUM algorithm [7], the proposed clustering system could be explained because it is based on a reduced set of features on the one hand, and, on the other, those features are words or punctuation symbols. Thus, the interpretation for the final user could be clearer than when working with many features, dealing with numerous *n*-grams of letters or when combing several similarity measures. The SPATIUM decision can be explained by major differences in relative frequencies of frequent words, usually corresponding to functional terms.

To improve the current version of our classifier, we need to analyze in more detail the distance measurement. The current version ignores the terms appearing once and replaces all uppercase letters with their corresponding lowercase ones. It could be checked if such decisions are pertinent when facing short texts. Moreover, we think that replacing the single link agglomerative clustering by the complete or average link will provide a more robust solution. Furthermore, such strategies will reduce the risk of the chaining effect present in the single link approach.

Acknowledgments. The authors want to thank the task coordinators for their valuable effort to promote test collections in authorship attribution. This research was supported, in part, by the NSF under Grant #200021_149665/1.

References

1. Amigo, E., Gonzalo, J., Artiles, J., Verdejo, F.: A comparison of extrinsic clustering evaluation metrics based on formal constraints. Inf. Retr. **12**(4), 461–486 (2009)
2. Burrows, J.F.: Delta: a measure of stylistic difference and a guide to likely authorship. Lit. Linguist. Comput. **17**(3), 267–287 (2002)
3. Craig, H., Kinney, A.F.: Shakespeare, Computers, and the Mystery of Authorship. Cambridge University Press, Cambridge (2009)
4. Hernández, D.M., Bécue-Bertaut, M., Barahona, I.: How scientific literature has been evolving over the time? A novel statistical approach using tracking verbal-based methods. In: JSM Proceedings, Section on Statistical Learning and Data Mining, Alexandria, pp. 1121–1131. American Statistical Association (2014)
5. Holmes, D.I.: The evolution of stylometry in humanities scholarship. Lit. Linguist. Comput. **13**(3), 111–117 (1998)
6. Jockers, M.L., Witten, D.M.: A comparative study of machine learning methods for authorship attribution. Lit. Linguist. Comput. **25**(2), 215–223 (2010)
7. Kocher, M., Savoy, J.: A simple and efficient algorithm for authorship verification. J. Am. Soc. Inf. Sci. Technol. **68**(1), 259–269 (2017)
8. Kocher, M., Savoy, J.: Author clustering using spatium. In: Proceedings of ACM/IEEE Joint Conference on Digital Libraries (2017, to appear)
9. Kocher, M., Savoy, J.: Distance measures in author profiling. Inf. Process. Manag. **53**(5), 1103–1119 (2017)
10. Labbé, D.: Experiments on authorship attribution by intertextual distance in English. J. Quant. Linguist. **14**(1), 33–80 (2007)
11. Layton, R., Watters, P., Dazeley, R.: Evaluating authorship distance methods using the positive silhouette coefficient. Nat. Lang. Eng. **19**, 517–535 (2013)
12. Manning, C.D., Raghaven, P., Schütze, H.: Introduction to Information Retrieval. Cambridge University Press, Cambridge (2008)
13. Savoy, J.: Estimating the probability of an authorship attribution. J. Am. Soc. Inf. Sci. Technol. **67**(6), 1462–1472 (2016)
14. Savoy, J.: Comparative evaluation of term selection functions for authorship attribution. Digit. Scholarsh. Hum. **30**(2), 246–261 (2015)

15. Sebastiani, F.: Machine learning in automatic text categorization. ACM Comput. Surv. **34**(1), 1–27 (2002)
16. Stamatatos, E., Tschuggnall, M., Verhoeven, B., Daelemans, W., Specht, G., Stein, B., Potthast, M.: Clustering by authorship within and across documents. In: Working Notes of the CLEF 2016 Evaluation Labs, CEUR Workshop Proceedings, CEUR-WS.org (2016)
17. Witten, I.H., Frank, E., Hall, M.A.: Data Mining. Practical Machine Learning Tools and Techniques. Morgan Kaufmann, Burlington (2011)
18. Zhao, Y., Zobel, J.: Searching with style: authorship attribution in classic literature. In: Proceedings of the Thirtieth Australasian Computer Science Conference, Ballarat, pp. 59–68 (2007)

Segmenting Compound Biomedical Figures into Their Constituent Panels

Pengyuan Li$^{(\boxtimes)}$, Xiangying Jiang, Chandra Kambhamettu,
and Hagit Shatkay$^{(\boxtimes)}$

Department of Computer and Information Sciences,
University of Delaware, Newark, DE, USA
{pengyuan,shatkay}@udel.edu

Abstract. Many of the figures in biomedical publications are compound figures consisting of multiple panels. Segmenting such figures into constituent panels is an essential first step for harvesting the visual information within the biomedical documents. Current figure separation methods are based primarily on gap-detection and suffer from over- and under-segmentation. In this paper, we propose a new compound figure segmentation scheme based on Connected Component Analysis. To overcome shortcomings typically manifested by existing methods, we develop a quality assessment step for evaluating and modifying segmentations. Two methods are proposed to re-segment the images if the initial segmentations are inaccurate. Experiments and results comparing the performance of our method to that of other top methods demonstrate the effectiveness of our approach.

Keywords: Compound image separation · Biomedical image · Connected Component Analysis

1 Introduction

A fundamental task in biomedical informatics is to make information within documents available to researchers. Images convey essential information in biomedical publications. A few recent efforts started exploring the use of image information within biomedical documents [1,2]. However, many of figures within biomedical documents are compound images consisting of multiple panels, where each panel potentially carries a different type of information. To obtain the information embedded within each part of the image, it is essential to first segment each compound image into its constituent panels.

Current compound image segmentation methods are primarily based on finding gaps between panels [2–7]. The gaps, which are solid (typically white or black) bands in compound images, are commonly detected and used as panel separators. However, due to inconsistency in image quality gaps can be hard to detect, which leads to *under-segmentation*, that is, parts of the image may not be correctly segmented into individual panels. To overcome this issue, the image can be

© Springer International Publishing AG 2017
G.J.F. Jones et al. (Eds.): CLEF 2017, LNCS 10456, pp. 199–210, 2017.
DOI: 10.1007/978-3-319-65813-1_20

transformed, for instance via edge-detection [5,7], so that gaps are more readily detected. Notably, some white/black bands occurring in images are not necessarily panel separators. Still, gap-based segmentation methods tend to interpret all solid bands as gaps, and as a result, erroneously split images into too many panels, to which we refer as *over-segmentation*. To address under- and over-segmentation, captions and image labels have been used to estimate the number of panels in compound images and to identify true gaps of separation [2,4,5]. However, such methods are not always effective, and may not even be applicable, when captions and labels are not available. Additionally, extracting labels from images requires optical character recognition – a time consuming operation. An alternative approach [6], applies several rules, eliminating gaps that are not panel separators aiming to avoid over-segmentation. While this method does not require processing image captions or labels, it is still time consuming. Furthermore, its separation accuracy leaves much room for improvement.

Unlike the above methods that segment images through gap detection, Shatkay *et al.* [1] proposed a method based on first identifying connected contents within individual panels. They used Connected Components Analysis (CCA) to detect individual panels in images. Lopez *et al.* [9] and Kim *et al.* [8] also used the same method for panel separation. Similar to the gap-based approach discussed earlier, CCA can also suffer from over-segmentation; unconnected small objects may be detected as individual panels and segmented off the main image-panel. Aiming to address a different task, namely the *identification of multi-paneled images*, Wang *et al.* [10] used a post-processing step by setting a threshold on panel-size to avoid fragmentation into very small panels. However, their work was not applied to the image-segmentation task, but rather aimed only to identify whether an image is compound or not. Notably, none of the above methods can segment *stitched compound images* whose panels are not separated by visible gaps. Santosh *et al.* [11] first proposed a method to separate stitched compound images based on straight lines detected in the images. Their method is applicable only to stitched compound images and as such relies on a manual selection step in which such images are identified within the dataset.

In this paper, we present a new CCA-based scheme for separating compound figures, including stitched compound images. To do this, we first introduce a preprocessing step to broaden and un-blur gaps in images. We then present the CCA method for segmenting images into panels. To avoid over- and under-segmentation, we extend our method by adding an assessment step to detect, evaluate and modify segmentation errors, and re-separate some of the images accordingly. The rest of the paper is organized as follows: Sect. 2 describes the complete framework of our method; in Sect. 3 we discuss experiments used to assess performance and present related results; Sect. 4 concludes and outlines directions for future work.

2   Methods

Our goal is to segment compound images appearing in biomedical documents. As noted above, compound images consist of several panels, typically separated by gaps, which appear as vertical or horizontal light/dark bands; such gaps may be blurry or too thin to recognize. We first preprocess compound images by resizing, adjusting, and cropping them to make the gaps in the images clearer and broader. We then apply Connected Component Analysis (CCA) to segment compound images into individual panels. This approach eliminates small objects and keeps only the main components as individual panels. We assess separation quality of the extracted panels, and modify them if the image segmentation quality appears to be low.

We note that CCA may not correctly segment panels whose contents are not well-connected, highly blurred images, and stitched compound images. We thus introduce specific methods for handling blurry and fragmented images as well as stitched images. We assess the segmentation quality of the panels obtained, and modify the segmentation if needed. The complete framework is shown in Fig. 1. The rest of this section introduces these methods.

Image Preprocessing. Gaps in compound images typically separate panels into distinct individual components. However, some panels may be positioned too close to one another, or a thin gap may be noisy or blurred, making separation hard. To address this issue we apply the bicubic interpolation [12] to the image I, of size $m \times n$; this scales up the image ($2m \times 2n$) and enhances contrast between image regions and gaps. The gaps in the scaled image, $I_{resized}$, thus become broader and clearer.

Notably, the separating gaps are not always white or black, that is, the intensity of pixels in gaps can be non-binary. To improve gap clarity and detectability we adjust the intensity of images by mapping pixel intensities whose values are in the interval $[T_{low}, T_{high}]$ to the entire intensity interval $[0, 1]$ using *linear mapping*. This mapping enhances contrast within the image so that gaps, which are the lightest or the darkest bands in compound images, become clearer. In the experiments described here, we set T_{low} to 0.05 and T_{high} to 0.95.

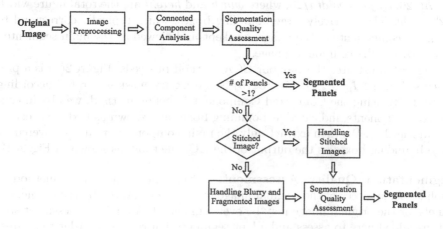

Fig. 1. Our framework for compound image segmentation.

We also note that it is hard to distinguish between the external boundary of the image as a whole and the boundaries of individual panels. To disambiguate image-boundaries, we crop the image borders by removing rows and columns of pixels whose maximum gradient value is 0. We denote the image obtained by applying all these preprocessing steps by $I_{processed}$.

Connected Component Analysis (CCA). To segment a preprocessed image, we first detect connected components within it. We assume that gaps among image-panels are white (which can be reversed later by inverting pixel values). To identify gaps among panels, a binary mask M is generated as:

$$M_{(x,y)} = \begin{cases} 1 & \text{if } I_{processed}(x,y) \leq t\,; \\ 0 & \text{if } I_{processed}(x,y) > t\,, \end{cases} \tag{1}$$

where $I_{processed}(x,y)$ denotes the pixel at row x and column y in the preprocessed image $I_{processed}$. By setting the threshold t, each pixel $I_{processed}(x,y)$ in the preprocessed image is labeled as *background* $M(x,y) = 0$ if $I_{processed}(x,y) > t$ and as *foreground* $M(x,y) = 1$ otherwise. In our experiments the threshold t is set to 0.95. Based on the mask M we detect connected components by applying the *Connected Component Labeling* method [13]. This method works by scanning the mask M and assigning labels to pixels. Adjacent pixels sharing the same pixel intensity are assigned the same label. A connected component is a set of pixels that have the same label value. In this paper we set the connectivity to 4, which means we count pixels above and below the central pixel, as well as those to the left and right of the central pixel as the adjacent pixels.

Using CCA may give rise to many small connected components due to small and unconnected objects in the image, such as text. A panel bounding box is set around the smallest rectangle that contains all pixels in each connected component. To initially eliminate connected components covered by bounding boxes of very small box-height or box-width, we thus set two thresholds: $t_{height} = height/20$, $t_{width} = width/20$, where *width* and *height* are the total figure width and height. The relatively large bounding boxes, which typically correspond to the main components of the image, are kept and viewed as the main segmented panels within the compound image.

Figure 2 illustrates the way our CCA method proceeds. Figure 2(a) is a pre-processed image $I_{processed}$. Figure 2(b) is the binary mask generated according to Eq. 1. By using the Connected Component Labeling method, we obtain connected components, indicated as bounding boxes and shown as textured rectangles in Fig. 2(c). We then extract only the main components that are covered by large bounding boxes as the output of the CCA method, as shown in Fig. 2(d).

Segmentation Quality Assessment. After the segmentation method is applied, some pieces of the original image may not be covered by the segmented panels, or the original image may be over-segmented. A quality assessment step is thus added here to assess and adapt segmentation results in order to address these shortcomings. We assess segmentation quality by employing the five steps

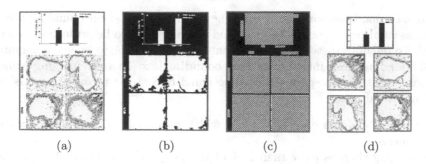

(a)   (b)   (c)   (d)

Fig. 2. Steps in Connected Components Analysis. The original image is Fig. 2, in Publication PMID: 21040544. (a) The preprocessed image. (b) The binary mask generated according to Eq. 1. (c) The Connected Component Labeling result. (d) The segmented image resulting from CCA.

described below. Steps 1–3 are used to evaluate and modify individual panels obtained by the segmentation methods discussed here, while steps 4 and 5 are used to assess and adapt the overall segmentation result.

1. Merge overlapping panels: Components within a panel may be erroneously detected by CCA as individual panels. As the largest connected component within a panel is typically indicative of the panel's boundary, the bounding boxes of smaller components within the same panel will typically overlap with the bounding box of the largest component. For example, the bounding box of legends may overlap the bounding box of corresponding line graph. We thus compute the ratio between the intersection area and the area of the minimum intersecting bounding box, and merge two bounding boxes when their overlap ratio exceeds 0.1.

2. Temporarily eliminate small components: Similar to the elimination step in CCA, we eliminate bounding boxes that are small (less than 1/5 in height or width) compared to the largest bounding box, thus reducing noise.

3. Recover missing panels: Due to blurred or disconnected contents in compound figures, some panels may be omitted in the initial segmentation process. We thus introduce a recovery step, in which missing panels are detected and recovered. We assume that the missing panel is similar in size and symmetric in position to present panels. We thus check for each panel whether there is enough space for another bounding box to its left, right, as well as above or below it. The space available for a bounding box next to a present panel indicates the position of a candidate panel. The candidate panels are expected to have similar content area, calculated as the number of non-white pixels within it, as that of the present panel and the same intensity values of all boundary pixels.

4. Check segmentation area: To detect incorrect segmentation, we compute the ratio between the sum of the areas of segmented panels and the area of the original image. If this ratio is below 0.5, we consider the segmentation to be incorrect. Incorrect segmentations are discarded leaving the image contents unsegmented.

5. Recover small components: During the elimination of small bounding boxes, some essential parts, such as the text and legend may also be erroneously eliminated. To re-adopt these small components into the panels, we merge eliminated small bounding boxes into their nearest bounding box. To avoid merging bounding boxes that are not part of the same panel during the recovery process, we employ several rules:

- If merging changes both height and width of a qualified bounding box - do not merge.
- If merging changes more than 20% of the height or the width of a qualified bounding box - do not merge.
- If the change of height or width for a qualified bounding box is more than 20% through the small components recovery step, this qualified bounding box keeps its original size.
- An eliminated small bounding box is merged at most once.

Handling Blurry and Fragmented Images. Through the steps above, several cases may not correctly be segmented, namely: very blurry images, fragmented images that have components with very low internal connectivity, and stitched images. Notably, stitched compound images are different from the other two kinds, which consist of panels that are separated by gaps. To segment these images, we employ a classifier to distinguish stitched images from the other two kinds of images. We define a *gap* as a row or a column whose minimum gray value is above 0.95. If a gap is found in a compound image, the image is classified as a compound image with gaps; otherwise, it is labeled as stitched.

To handle blurry images and fragmented images, we apply an edge detector, which sharpens blurry components in the $I_{processed}$. The corresponding edge image, denoted I_{edge}, may still have poor connectivity. To enhance connectivity of components in I_{edge}, we dilate the connected regions within the edge image using the minimum gap-width in the image as the dilation factor. After dilation, the connectivity within the dilated edge image is increased. We then apply the CCA method on the dilated edge image again to obtain the segmentation.

Figure 3 illustrates the handling of blurry and fragmented images. Figure 3(a) is a blurry compound image containing many small pieces. By applying an edge detector, we unblur the blurry components, as shown in Fig. 3(b). We then find the gaps in the edge image along the horizontal and the vertical directions and use the width of the thinest gap as the dilation factor. Figure 3(c) is the dilated edge image, while Fig. 3(d) shows the segmentation result obtained by using CCA on the dilated edge image. Three panels are detected and highlighted by bounding boxes in Fig. 3(d). The augmented method thus correctly handles this blurry and fragmented image and identifies the segments within it.

Handling Stitched Images. Stitched compound images do not contain any gap between panels, and as such, cannot be directly segmented by the CCA method. Identifying panel boundaries in such images is thus the main challenge.

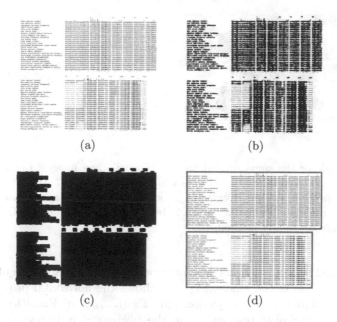

(a) (b)

(c) (d)

Fig. 3. Example in which we handle blurry and fragmented images. (a) The compound figure is Fig. 2, in Publication PMID: 20649995. (b) The edge image of the figure shown in (a). (c) The dilated edge image. (d) The result of CCA applied to dilated image.

Edge detection is applied to identify pixels whose neighbors' intensity sharply changes. Using edge detection, we can create an edge image that clearly shows the boundaries between panels. The edge detector is applied to the preprocessed image to generate a binary edge image in which the boundaries between panels are intensified (see e.g. Fig. 4(b)). The objective thus becomes that of detecting boundaries in the resulting edge image I_{edge}. Given an edge image I_{edge}, if pixel (x, y) is detected as a pixel along an edge we set $I_{edge}(x, y) = 1$. Summing the pixel value along the horizontal and the vertical directions gives rise to two projections: $Proj_{horizontal}$ and $Proj_{vertical}$, which are calculated as:

$$Proj_{horizontal} = I_{edge}(x, y),\ y \in 1...2n\,;$$
$$Proj_{vertical} = I_{edge}(x, y),\ x \in 1...2m\,. \tag{2}$$

The values $2n$ and $2m$ are the width and height of the edge image, respectively. The panel segmentation takes place along the horizontal or the vertical line that goes through the highest projection position. For images with complex layout, the boundary between panels may not cross the whole image; in such cases we recursively segment the image along one direction at a time, where the projection peak value is at least 0.7 of the height or the width of the region currently considered for segmentation.

Figure 4 shows an example of the steps applied for handling stitched images. Figure 4(a) is the original stitched compound image; Fig. 4(b) shows the edge

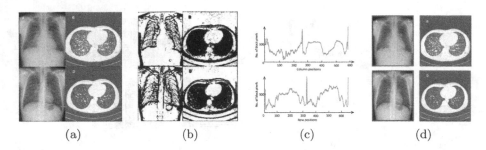

(a)   (b)   (c)   (d)

Fig. 4. Example of the steps applied for handling stitched images. (a) The stitched compound image taken from Fig. 1, Publication PMID: 16480497. (b) The edge image of the figure shown in (a). (c) Horizontal projection (top plot) and vertical projection (bottom plot) calculated according to Eq. 2, based on the edge image. (d) A complete segmentation result by our method.

image obtained by applying an edge detector. Panel boundaries are observed as straight black lines in the image; Fig. 4(c) shows the horizontal projection plot $Proj_{horizontal}$ and vertical projection plot $Proj_{vertical}$ of Fig. 4(b). By recursively choosing the peak position along the horizontal projection and vertical projection as panel separators, we segment the image into individual panels shown in Fig. 4(d).

3   Experiments and Results

3.1   Experiments

To evaluate our method we conducted two sets of experiments using datasets from the Figure Separation task in the ImageCLEF Medical tasks. In the first experiment, we assess the separation accuracies obtained by the different steps of our segmentation method. We use the training and test datasets of Image-CLEF'16 [16] to train our system and test its performance.

In the second experiment, we compare the separation accuracy of our comprehensive method against that of state-of-the-art systems using test datasets from ImageCLEF'13, '15 and '16 [14–16]. Additionally, to demonstrate the general applicability of our method, we test our method, trained over ImageCLEF'15 dataset, on the ImageCLEF'13 test dataset. For selecting an edge-detector, we experimented with several methods, and decided to use the SUSAN edge detector [17] as it has demonstrated the best performance in this context.

3.2   Datasets and Evaluation

We used five imageCLEF datasets in this study, two for training (ImageCLEF'15 and '16) and three for testing (ImageCLEF'13, '15 and '16). The images in the datasets are first extracted from the biomedical publications stored in PubMed Central and then identified as compound images through manual classification.

The ground truth tagging pertaining to the five datasets used in our experiments was provided by ImageCLEF organizers. To evaluate our image separation performance, we use the tool provided by ImageCLEF Medical [14]. This tool computes the accuracy of the separation result for a compound image I_i as:

$$Accuracy_i = \frac{C}{max(N_G, N_D)},$$

where C is the number of detected panels that overlap with at least 2/3 of the area of the ground-truth panel, N_G is the true number of panels in the image, and N_D is the number of panels we detected. The overall accuracy for the dataset as a whole is then calculated by averaging the accuracies of all separations.

3.3   Results

Table 1 shows the separation accuracies obtained in our first set of experiments using different combination of steps within our method over the imageCLEF'16 test dataset. The dataset contains 1615 compound figures, which include 8528 individual panels. The CCA method alone achieves 73.57% accuracy, where 162 images remain unsegmented. Proceeding the CCA method by a preprocessing step leads to an increase of 0.73% in accuracy. Table 1 also shows that 16 additional images are segmented when the preprocessing step is added. By combining the segmentation-quality-assessment step and the CCA method, 40 fewer images are separated compared to CCA-alone, but the separation accuracy increases by

Table 1. Segmentation accuracies obtained and numbers of images that remain unsegmented by employing different combinations of steps within our method.

Methods used	Separation accuracy	# of unsegmented images
CCA-alone	73.57%	162
Preprocessing + CCA	74.30%	146
CCA + Segmentation quality assessment	75.27%	202
Preprocessing + CCA + Segmentation quality assessment	74.38%	243
Preprocessing + CCA + Segmentation quality assessment + Handling blurry and fragmented images	81.23%	85
Preprocessing + CCA + Segmentation quality assessment + Handling stitched images	84.03%	13
The combination of all methods	**84.43%**	9

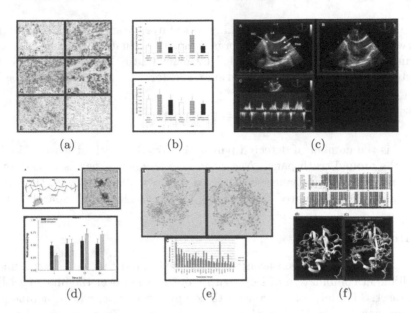

(a)   (b)   (c)

(d)   (e)   (f)

Fig. 5. Examples of successful segmentation obtained by our method. The original images of (a)–(f) are taken from: Publication PMID: 18282279, Fig. 6; PMID: 20955558, Fig. 1; PMID: 19439081, Fig. 14; PMID: 16930490, Fig. 1; PMID: 21073692, Fig. 6; PMID: 21129218, Fig. 1, respectively.

1.7% (compared to the first row in the table). Thus, the segmentation-quality-assessment step improves the correctness of separation result. Combining the image preprocessing step and the segmentation-quality-assessment step with the CCA method (Row 4 in the table), the overall accuracy reaches 74.38%, but 243 images remain unsegmented.

To reduce the number of images that remain unsegmented, we utilize additional steps, as described in Sect. 2. Applying the step for handling blurry and fragmented images, the accuracy reaches 81.23% and the number of compound images that remain unsegmented decreases to 85. Similarly, applying the step for handling stitched images, the accuracy over the whole dataset reaches 84.03% and the number of compound images that remain unsegmented decreases to 13. Combining all the steps leads to the highest accuracy of 84.03% on the Image-CLEF'16 test dataset while only 9 figures remain unsegmented.

Figure 5 shows several examples of successful compound figure separation results. Our method not only correctly segments figures containing a single type of image type such as microscopy, graphs, or medical images, but also images containing multiple types of panels.

In the second set of experiments, we compare the results obtained by our comprehensive method with those of other systems submitted to ImageCLEF'15 Medical, using the 2015 test dataset. Santosh et al.'s method [18] (based on their previous work [2,11]) achieves an accuracy of 84.64%, while Taschwer et al. [7]

report an accuracy of 84.90%. Our method performs significantly better than all other systems with an accuracy of 90.65%.

To demonstrate the general applicability of our method, we used parameters obtained by training over one of the datasets (ImageCLEF'15) to segment images provided in another dataset (ImageCLEF'13); Our result shows 84.47% accuracy. The other three top performers reported accuracy of 68.59% [21], 69.27% [20], and 84.64% [19]. We note that while the performance of our method is slightly lower than that reported by de Herrera et al. [19] using method proposed by Chhatkuli et al. [6], the average time required to process one image by our system is 0.74 s (Wall-clock), which is much faster than that reported by de Herrera et al. [19], i.e. 2.4 s.

For ImageCLEF2016 [16], as the only team participating in the Figure Separation task, we achieved 84.43% accuracy on the Figure Separation task test dataset. The segmentation accuracy is similar to the best result obtained in ImageCLEF'15. This result is particularly noteworthy, given that the difficulty of the Figure Separation task was increased in 2016 by adding more stitched compound images and compound images containing multiple types of panels, as indicated in the task description [16].

4 Conclusion

We have presented a new scheme for segmenting compound figures, including stitched compound images. We first proposed a preprocessing step to make gaps clearer so that more images are segmented. We then introduced a method based on Connected Components Analysis to segment images into panels. Segmentation errors were addressed through a step of segmentation quality assessment. Notably, this step evaluates separation quality, back-tracks separation errors, and ensures that only panels that are likely to be correct are extracted from images. Error stemming from over- and under-segmentation in very blurry images, fragmented images, and stitched images are more difficult to address. As such, we proposed two advanced methods to directly handle blurry and fragmented images, and stitched images accordingly. The results demonstrate that our comprehensive method improves upon the panel-segmentation performance of state-of-the-art methods.

While our method achieves a high accuracy in segmenting compound images, there are still challenging cases that are not perfectly addressed. For example, compound images in which both panels and gaps vary in size are hard to segment accurately. We plan to develop methods to identify such cases and address them.

Acknowledgments. This work was partially supported by NIH grant R56LM011354A.

References

1. Shatkay, H., Chen, N., Blostein, D.: Integrating image data into biomedical text categorization. Bioinformatics **22**(14), e446–e453 (2006)

2. Apostolova, E., You, D., Xue, Z., Antani, S., Demner-Fushman, D., Thoma, G.R.: Image retrieval from scientific publications: text and image content processing to separate multipanel figures. J. Am. Soc. Inf. Sci. Technol. **64**(5), 893–908 (2013)

3. Murphy, R.F., Velliste, M., Yao, J., Porreca, G.: Searching online journals for fluorescence microscope images depicting protein subcellular location patterns. In: Proceedings of the IEEE International Symposium on on Bioinformatics and Bioengineering, pp. 119–128 (2001)

4. Antani, S., Demner-Fushman, D., Li, J., Srinivasan, B.V., Thoma, G.R.: Exploring use of images in clinical articles for decision support in evidence-based medicine. In: Proceedings of SPIE Document Recognition and Retrieval XV, 68150Q (2008)

5. Cheng, B., Antani, S., Stanley, R.J., Thoma, G.R.: Automatic segmentation of subfigure image panels for multimodal biomedical document retrieval. In: Proceedings of SPIE Document Recognition and Retrieval XVIII, 78740Z (2011)

6. Chhatkuli, A., Foncubierta-Rodrguez, A., Markonis, D., Meriaudeau, F., Müller, H.: Separating compound figures in journal articles to allow for subfigure classification. In: Proceedings of SPIE Medical Imaging, 86740J (2013)

7. Taschwer, M., Marques, O.: Compound figure separation combining edge and band separator detection. In: Tian, Q., Sebe, N., Qi, G.J., Huet, B., Hong, R., Liu, X. (eds.) MultiMedia Modeling. LNCS, vol. 9516, pp. 162–173. Springer, Cham (2015). doi:10.1007/978-3-319-27671-7_14

8. Kim, D., Ramesh, B.P., Yu, H.: Automatic figure classification in bioscience literature. J. Biomed. Inform. **44**(5), 848–858 (2011)

9. Lopez, L.D., Yu, J., Arighi, C., Tudor, C.O., Torii, M., Huang, H., Vijay-Shanker, K., Wu, C.: A framework for biomedical figure segmentation towards image-based document retrieval. BMC Syst. Biol. **7**(4), 1 (2013)

10. Wang, X., Jiang, X., Kolagunda, A., Shatkay, H., Kambhamettu, C.: CIS UDEL Working Notes on Image-CLEF 2015. CLEF Working Notes (2015)

11. Santosh, K.C., Antani, S., Thoma, G.: Stitched multipanel biomedical figure separation. In: Proceedings of IEEE International Symposium on Computer-Based Medical Systems, pp. 54–59 (2015)

12. Keys, R.: Cubic Convolution Interpolation for Digital Image Processing. IEEE Trans. Acoust. Speech Signal Process. **29**(6), 1153–1160 (1981)

13. Gonzalez, R.C., Woods, R.E.: Digital Image Processing. Prentice Hall, Upper Saddle River (2002)

14. de Herrera, A.G.S., Kalpathy-Cramer, J., Demner-Fushman, D., Antani, S., Müller, H.: Overview of the ImageCLEF 2013 Medical Tasks. CLEF Working Notes (2013)

15. de Herrera, A.G.S., Mller, H., Bromuri, S.: Overview of the ImageCLEF 2015 Medical Classification Task. CLEF Working Notes (2015)

16. de Herrera, A.G.S., Schaer, R., Bromuri, S., Müller, H.: Overview of the ImageCLEF 2016 Medical Tasks. CLEF Working Notes (2016)

17. Smith, S.M., Brady, J.M.: SUSAN-a new approach to low level image processing. Int. J. Comput. Vis. **23**(1), 45–78 (1997)

18. Santosh, K.C., Xue, Z., Antani, S., Thoma, G.: NLM at ImageCLEF 2015: Biomedical Multipanel Figure Separation. CLEF Working Notes (2015)

19. de Herrera, S., Garcia, A., Markonis, D., Schaer, R., Eggel, I., Müller, H.: The medGIFT group in ImageCLEFmed 2013. CLEF Working Notes (2013)

20. Simpson, M.S., You, D., Rahman, M.M., Demner-Fushman, D., Antani, S., Thoma, G.R.: ITI's Participation in the 2013 Medical Track of ImageCLEF. CLEF Working Notes (2013)

21. Kitanovski, I., Dimitrovski, I., Loskovska, S.: FCSE at Medical Tasks of ImageCLEF 2013. CLEF Working Notes (2013)

An Analysis of Cross-Genre and In-Genre Performance for Author Profiling in Social Media

Maria Medvedeva[(✉)], Hessel Haagsma, and Malvina Nissim

University of Groningen, Groningen, The Netherlands
m.medvedeva@student.rug.nl, {hessel.haagsma,m.nissim}@rug.nl

Abstract. User profiling on social media data is normally done within a supervised setting. A typical feature of supervised models that are trained on data from a specific genre, is their limited portability to other genres. Cross-genre models were developed in the context of PAN 2016, where systems were trained on tweets, and tested on other non-tweet social media data. Did the model that achieved best results at this task got lucky or was it truly designed in a cross-genre manner, with features general enough to capture demographics beyond Twitter? We explore this question via a series of in-genre and cross-genre experiments on English and Spanish using the best performing system at PAN 2016, and discover that portability is successful to a certain extent, provided that the sub-genres involved are close enough. In such cases, it is also more beneficial to do cross-genre than in-genre modelling if the cross-genre setting can benefit from larger amounts of training data than those available in-genre.

Keywords: Author profiling · Cross-genre · Twitter · Blog · Social media

1 Introduction and Background

Promoting and evaluating research on user profiling, especially in social media, has been a key objective of the PAN evaluation Labs[1] in the past years [7–9]. The tasks have regularly attracted substantial numbers of participants, and by providing a variety of annotated datasets, have basically established the benchmark for author profiling in social media.

Social media, however, is a rather broad concept. It has been observed that at least in the context of user profiling, even within the realm of *social media* in general, not all data is the same. At the 2014 edition of the PAN Labs, the organisers provided four different types of widely defined social media data, namely Twitter, blogs, reviews, and an underspecified "social media" set to develop author profiling systems [8]. Training and testing was done *in-genre*, and average results in English are highest for Twitter, followed by blogs, reviews, and Social Media,

---

[1] pan.webis.de.

© Springer International Publishing AG 2017
G.J.F. Jones et al. (Eds.): CLEF 2017, LNCS 10456, pp. 211–223, 2017.
DOI: 10.1007/978-3-319-65813-1_21

with performance differences up to 10% points. While the organisers suggested that the highest accuracies on Twitter might be due to the larger number of documents in that dataset [8] compared to blogs and reviews, the results might also be an indication that surface cues for predicting demographics are easier to grasp on Twitter than blogs, and that reviews are harder than either of them.

How much of such results depends on size and how much depends on genre is still an unanswered question, though it would be useful to have an answer. Assuming that we have very many tweets and not too many blogs, can we train our model on tweets and successfully test it on blogs, as both pertain to the *social media* wide genre? It is known that supervised models are pretty much bound to the training data they are exposed to. So, if social media data were more or less interchangeable in their sub-genres (e.g. Twitter, blogs), at least for profiling, one sub-genre dataset could serve for all and we could save a lot of annotation effort.

This kind of reflections, we believe, prompted the settings for the author profiling competition at PAN 2016, which was indeed organised in a *cross-genre* fashion [10]: while training data was provided in the form of user generated tweets, the test data was only known to be some kind of non-Twitter social media texts, with no other information available at system development time regarding its nature. After submission, the test set was eventually revealed to be blogs.

In building a system for this cross-genre task, we had aimed at exploiting features that might be typical for social media texts in general and focusing on language characteristics rather than metadata or genre-specifics (e.g. mentions, hashtags, etc.) [4]. This strategy proved successful, yielding indeed the best overall performance at the PAN 2016 cross-genre profiling task [10], training on tweets, and (unknowingly) testing on blogs. But beside the specifics of the PAN 2016 competition, have we really succeeded in developing a robust cross-genre model for social media? Operationally, in the context of this paper, we translate this question into the following two research questions:

Q1: is the Twitter-trained model that won the PAN 2016 cross-genre task on author profiling truly cross-genre so that good results can be observed in datasets other than the PAN 2016 test set?

Q2: if the features truly capture some general aspects of demographics, can the model be trained on datasets other than Twitter and still yield a good performance?

In order to provide an answer to Q1, we test our models on existing non-Twitter datasets from PAN 2014, and compare its performance to *in-genre* results in two ways, by cross-validation and by comparison to the official results of the PAN 2014 competition. To answer Q2, we train our model on existing non-Twitter datasets and test it on the same sets we test the Twitter-model on, still in a *cross-genre* setting. All these experiments are described in Sect. 3, results are provided in Sect. 4, and discussed in depth in Sect. 5. The GronUP system, which we use to run all experiments, is briefly summarised in Sect. 2.

2   GronUP

GronUP is a Support Vector Machine classifier that performs age and gender classification. It was developed in the context of the PAN 2016 author profiling competition, where it achieved best results. We offer here a short overview of its main features, while a full system description, together with the details of its performance at PAN 2016, can be found in [4].

2.1   Preprocessing

The first module in our system preprocesses the data. In the case of tweets, we use the tweets as is. For blogs, we split the document into sentences, in order to convert them into a similar structure as the tweets. That is, we treat tweets as being rough equivalents of sentences in other data types. The items (tweets/sentences) are then tokenised, using the Natural Language Toolkit (NLTK, [2]) TweetTokenizer.[2] Before tokenisation, some tokens are converted to placeholders: all URLs are replaced with the string 'URL', and all numbers are replaced by 'NUMBER'. Additionally, any HTML markup is deleted.

2.2   Classifier

For developing our system we used the Python Scikit-learn machine learning library [5]. Building on previous work and insights from participating systems in previous editions of this task, we opted to train a Support Vector Machine (SVM) with a linear kernel. We shortly evaluated the effect of different parameter values on performance using cross-validation on training data, but as a general approach we did not want to tune the system on tweets too closely. Indeed, while we observed that increasing the value of the cost parameter C led to an improvement in performance, we suspected that allowing for fewer incorrect classifications in the training data would lead to overfitting and worse performance on data from a different genre. We therefore focused on feature engineering rather than SVM settings, and used default parameters ($C = 1.0$) for all our models.

2.3   Features

In coming up with features to model user gender and age, we relied both on previous work and our own intuitions.[3] From a language perspective, we have aimed at creating general models, without language-specific features. In this section, we describe all of the features of GronUP.

N-grams. N-gram-based features have proven to be highly useful indicators of various linguistic differences between authors [1,8]. We use 1- to 3-grams for

---

[2] Contrary to what was stated in [4], TweetTokenizer is used for blogs, not word_tokenizer.

[3] In this paper we do not investigate the contribution of individual features, but insights on this can be found in the detailed description of GronUP [4].

tokens, and 2- to 5-grams for characters. We also add 1- to 3-grams for part-of-speech, since previous work indicates that female writers on social media tend to use more pronouns while male writers use more articles and prepositions, independently of their age [11]. We assign PoS-tags using the TnT tagger [3] trained on the Penn Treebank for English and on the CoNLL 2002 data for Spanish.

Capitalisation. We incorporate three capitalisation-related phenomena, which we believe to be indicative of user age: we expect younger users to be less keen on using sentence-initial capitals, we expect younger users to be more lenient when it comes to the capitalisation of specific tokens, such as proper nouns, and we expect them to be more likely to either use only lower-case or only upper-case letters. The sentence-initial capitalisation feature is implemented as the percentage of sentences (≈tweets) which start with a capital letter. For capitalisation of tokens, implementing a check for the correct application of language-specific capitalisation rules is too time-consuming and to language-specific for our purposes. Hence, we add this feature as the user-level average proportion of capitalised words in a sentence w.r.t. the total number of words in that sentence. The final capitalization feature is represented by the proportion of capital letters in a sentence as the total number of characters in that sentence. The resulting feature is the average of this proportion for a user, across documents.

Punctuation. The second group of orthographic features captures punctuation. We expect younger users to use 'proper' sentence-final punctuation less often, and to be more likely to use either very long sequences of punctuation (e.g. 'Nice!!!!!!!!!') or no punctuation at all. The first feature is represented as the percentage of items which have ".", "!" or "?" as their final token, the latter as the average proportion of punctuation characters w.r.t. all characters, across documents, for each user.

Average Word Length and Average Sentence Length. The use of longer words and sentences can be an indicator of a more advanced writing style, and it has been observed that older users tend to use longer words and sentences. We represent this characteristic as the average word length in characters and the average sentence/tweet length in tokens, per user.

Out of Dictionary Words. We hypothesise that the number of typos and slang or non-standard words can be a useful feature for distinguishing different age groups. For this feature we calculate the percentage of misspelled and out-of-dictionary words out of the total number of words per user. To detect out of dictionary words in English and Spanish we use the Python Enchant Library[4] [6] with `aspell` dictionaries for all available dialects.

Vocabulary Richness. The level of variety of a person's language use, represented here by the richness of their vocabulary, is a salient characteristic of their writing style. A way to measure the vocabulary richness of a person is to count how many words used by the user were only used once. The more unique words

---

[4] http://abisource.com/projects/enchant.

written by a user, the richer their vocabulary. We represent vocabulary richness as the percentage of words used only once w.r.t. to the total number of words used by an author.

High-Frequency Words. Highly frequent have been shown to be a key feature for determining the gender and age of an author [11]. Highly frequent, here, means those words which are used more frequently by users in one category than by users in other categories. Examples include sports-related words for male users versus female users, or popular textisms for the youngest users versus older users.

$$r f_{tc} = \frac{t f_{tc}}{t f_{t \neg c}} \tag{1}$$

The relative frequency (rf) of a term was calculated as in Eq. 1, where t denotes the term and c the class. We then used a ranked list of the top 2500 most frequent words for each category. The feature itself is a vector containing the occurrences of function words for each category in a document.

Emoticons. It has been found that, on American Twitter, younger people are more likely to use emoticons without noses (e.g. ':)') than older people, who prefer emoticons with noses (e.g. ':-)') [12]. Also, there is a correlation between the overall frequency of emoticon use, usage of non-nose emoticons, and age. As such, we implement features capturing the proportion of emoticons with noses out of all emoticons, and the proportion of emoticon tokens out of all tokens. In addition, based on our own intuitions about emoticon use among different user groups, we add features capturing the percentage of reverse emoticons (e.g. '(:') out of total emoticons, and the proportion of happy emoticons out of total emoticons.

Second-Order Attributes. Given that the goal of our model is to be able to generalise author profiling from Twitter to a different social media domain, we want to have feature representations that are not too closely tailored to the training data. For example, sentence length should ideally be represented as a relative value, indicating whether users from a certain class write longer sentences, on average, than users from another class. This is especially the case here, since tweets are not usually written as full sentences but other social media can be, and absolute values thus might not be have any meaning outside of the Twitter domain.

We implement second-order attribute representation of features to deal with this problem, similar to the approach in [1]. For the real-valued features (i.e. all except n-grams, function words), the mean is calculated for the training data, and the relative distance of the classes in each category is determined. Applying this to test data, the scores for a user are compared to the mean for the test data, and the difference is then compared to those found for the training data. If, in training, it is found that female users use five percent less capitals than the mean, and an unseen author gets the same relative score compared to the mean for the test data, the feature vector would indicate that the unseen author is more likely to be female than male.

3   Experiments

3.1   Datasets

We use several datasets to test the cross-genre robustness and portability of our system. Although GronUP at PAN 2016 was run on English, Spanish, and Dutch, the other datasets we use do not have Dutch data (and in one case only English data), so that we run experiments on English and Spanish only.[5] An overview of genres, source, and size in terms of users and total tokens is provided for each set in Table 1.

Table 1. Size of training sets (number of tokens and users) used in this paper. (*) We only use a portion of the original dataset of Social Media from PAN 2014 as it is far too large to use as such for training to carry out proper comparisons.

Genre	Source	ID	English tokens	Spanish tokens	English users	Spanish users
Twitter	PAN 16 training set	T2016	4306 K	1876 K	436	250
Blogs	PAN 14 training set	B2014	653 K	770 K	147	88
Social Media	PAN 14 training set[(*)]	S2014	23551 K	8494 K	1381	1271
Reviews	PAN 14 training set	R2014	1128 K	–	4160	–

All of the sets that we use for training and for testing in our experiments are *training sets* that come from the PAN competitions, since *test sets* are usually not available outside of the live shared task. Note that we want to profile authors based on their textual content, rather than using metadata, which is not included in the datasets we use.

Twitter. The portion of the training data that we use consists of tweets of users in English (436 users) and Spanish (250 users). We do not include Dutch. Also, we do not use the PAN 2014 tweets distribution because that dataset is almost entirely (96%) subsumed in the PAN 2016 Tweet training set.

Blogs. We use PAN 2014 blogs dataset, consisting of 147 manually annotated blogs in English and 88 in Spanish with up to 25 posts per blog. For the PAN 2016 evaluation the combination of training and test blog data of 2014 has been used (B2016-test).

Reviews. For the 'reviews' genre we have used hotel reviews dataset that have also been a part of PAN 2014, containing 4160 users and their reviews on TripAdvisor.[6]

---

[5] The choice for these languages is due to the availability of data, but the model could be trained on any language for which preprocessing tools (tokenizer, PoS-tagging, dictionary) and training data are available.

[6] http://www.tripadvisor.com.

Social Media. We have also used the 'social media' dataset distributed at PAN 2014 (non-Twitter, non-blogs, but not otherwise defined), which is a portion of the PAN 2013 corpus. In order to control for the size of the training data in the experiments, we took a smaller random sample of the Social Media dataset and reduced it to the size of Twitter PAN 2016, which still resulted in higher number of users and tokens (see. Table 2); the data in Spanish was of comparable size. The distribution of gender and age groups remained very similar to the original.

While gender is completely balanced in all datasets that we use for training with a 50/50 distribution of male and female users in all sets, the age groups are not so evenly distributed, with dominant classes being 35–49 and 25–34 in both tweets and blogs.

3.2   Experimental Settings

The experiments we run are aimed at targeting and answering the two research questions put forward in Sect. 1. We briefly report them again, and explain which experiments will answer which question, and how. For the sake of clarity, we refer to the specific lines in Tables 2 and 3 that provide experimental evidence towards each answer.

Q1: Is the Twitter-trained GronUP model truly cross-genre so that good results can be observed in datasets other than the PAN 2016 test set?

To answer this question, we need a measure of what "good performance" means on test sets other than the PAN 2016 blogs. We compare our cross-genre performance on other datasets (lines #6–#8) to new in-genre results, obtained in two different ways:

– Cross-validation on the same datasets the Twitter-trained model is tested on (lines #2–#4).
– The scores obtained by the systems officially submitted to the PAN 2014 competition, on the same datasets (lines #13–#15)

Two of the datasets that we consider (reviews and blogs, see Table 1) are smaller in size than the Twitter dataset, so cross-validation should yield, at least due to training data size, lower results than the cross-genre setting, unless the benefit of training and testing in-genre contributes highly. It should also be remembered that the Twitter model we submitted could not have been tuned to the test set, as nothing about test data was known to us, but the good performance could have still been chance.

Q2: If the features truly capture some general aspects of demographics, can the model be trained on datasets other than Twitter and still yield a good performance on different test sets?

Table 2. Cross-genre and within-genre results on profiling age and gender in Spanish (es) and English (en), for the various settings described in Sect. 3.2. For dataset abbreviations, please refer to Table 1.

#	Training set	Test set	Age (en)	Gender (en)	Age (es)	Gender (es)	Average
1	T2016	(cross-val)	0.4573	0.7067	0.4899	0.7085	0.5906
2	B2014	(cross-val)	0.3810	0.7143	0.4091	0.6590	0.5409
3	R2014	(cross-val)	0.3236	0.6526	–	–	*0.4881*
4	S2014	(cross-val)	0.3277	0.4946	0.3855	0.5951	0.4507
5	T2016	B2016-test	0.5897	0.6410	0.5179	0.7143	0.6157
6	T2016	B2014	0.5374	0.7347	0.4205	0.6818	0.5936
7	T2016	R2014	0.2377	0.5000	–	–	*0.3689*
8	T2016	S2014	0.3273	0.4960	0.3097	0.5628	0.4240
9	S2014	B2014	0.4354	0.5714	0.4318	0.5795	0.5045
10	S2014	R2014	0.2413	0.5115	–	–	*0.3764*
11	S2014	T2016	0.3601	0.5367	0.3985	0.5188	0.4535

Table 3. Best results from PAN 2014, per sub-task (i.e. not necessarily by the same participant). Scores are obtained from the official report [8].

#	Training set	Test set	Age (en)	Gender (en)	Age (es)	Gender (es)	Average
12	T2014	T2014-test	0.5065	0.7338	0.6111	0.6556	0.6268
13	B2014	B2014-test	0.4615	0.6795	0.4821	0.5893	0.5531
14	R2014	R2014-test	0.3502	0.7259	–	–	*0.5381*
15	S2014	S2014-test	0.3652	0.5421	0.4894	0.6837	0.5201

To answer Q2, we train our model on the Social Media dataset rather than Twitter and test it on the same sets we test the Twitter-model on, still in a *cross-genre* setting. As mentioned in Sect. 3.1, we subsampled the Social Media dataset which was originally much larger than the Twitter dataset to control for size of training data. We do not repeat this experiment on reviews/blogs as these datasets are too small in terms of tokens. Evidence towards answering this question will be provided by comparing lines #6–#8 with lines #9–#11 in Table 2.

Both age and gender predictions are structured as classification tasks, where gender is a two class problem, and age values are binned into five separate classes. In all experiments we use GronUP for classification.

4   Results

The results of the experiments described in Sect. 3 are shown in Table 2. In what follows, we refer to each setting by means of its line number in the table.

We distinguish three types of experiments: in-genre using cross-validation (#1–#4), cross-genre, training on tweets (#6–#8), and cross-genre, training on Social Media (#9–#11). Each result is represented by the accuracy score on each subtask (English/Spanish & age/gender), and the average over subtasks. The exception here are the results on review data, which do not include Spanish (#3, # 7, and #10). Generally, there is little difference between the performance on each subtask and in the overall average. That is, when the system does better in one setting than another, it also tends to do better on each of the subtasks in the one setting. This makes the results easier to interpret, since we can focus mainly on the average accuracy score, when comparing performances between settings.

Note that we also report GronUP's results for the official setting of PAN 2016 (#5, [4,10]). This data is not available to use in any of the other experiments, so we cannot compare these results directly to the new results reported in this work. However, the test set of PAN 2016 is a superset of the PAN 2014 blogs training data, and we can use the latter as a proxy for the PAN 2016 test set. This is corroborated by the fact that the performance on the PAN 2016 test set (#5, 0.6157) is very similar to the performance on the PAN 2014 blogs training set (#6, 0.5936), using the same training data.

By comparing the system's performance in different settings, we hope to answer the research questions posited in Sect. 3.2. First, we compare the results in cross-validated in-genre settings to the cross-genre settings where we train on Twitter data. We see that results are mixed: on blog data, cross-genre performance (#6, 0.5936) is surprisingly higher than in-genre performance (#2, 0.5409). On review data, on the other hand, the system performs clearly worse in the cross-genre setting (#7, 0.3689) than in the cross-validation setting (#3, 0.4881). On Social Media, cross-genre performance (#8, 0.4240) is also lower than in-genre (#4, 0.4507), but the difference is much smaller, with a drop of only 2.5% points, compared to 12% points for reviews.

As a second point of comparison, we use the best performances in the PAN 2014 shared tasks. The highest accuracy score on each subtask for PAN 2014 is displayed in Table 3. Note that these results are on the test sets of PAN 2014, whereas the results in Table 2 are on the cross-validated training sets of PAN 2014, since we do not have access to the test data. Still, we can assume that these sets are drawn from the same underlying pool of data, and thus can make valid comparisons. We can compare these state-of-the-art results to both the cross-validated in-genre performance and the cross-genre Twitter-trained performance of GronUP. For blog data, we see that our system performs very similarly to the best systems of 2014 when cross-validating (#2, 0.5409 vs. #13, 0.5531) and does even better in the cross-genre setting (#6, 0.5936). On review data, the picture is the complete opposite. In-genre, we see lower results (#3, 0.4881 vs. #14, 0.5381), and cross-genre, they are a lot worse (#7, 0.3689). Social Media shows a similar effect, with an in-genre difference of 7% points (#4, 0.4507 vs. #15, 0.5201), and a cross-genre difference of 10% poings (#8, 0.4240 vs. #14, 0.5201).

Lastly, we assess whether GronUP is dependent on having tweets as training data, or whether it can generalise to having different social media text as training data. For review data, this seems to be the case, as performance is low, but similar, in both the Twitter-trained setting (#7, 0.3689) and the social-media-trained setting (#10, 0.3764). On blog data, there is a clear drop, from 0.5936 (#6) when trained on Twitter, to 0.5045 (#9) when trained on Social Media. However, given that the within-genre state-of-the-art on blogs is 0.5531 (#13), results in both settings are far from poor.

Additional insight can be gained by looking at the same cross-genre setting, with different directionality, as in #8 and #11. Here, accuracy is somewhat lower when going from Twitter to Social Media (#8, 0.4240) than when going from Social Media to Twitter (#11, 0.4535), but the difference between cross-validation on tweets (#1, 0.5906) and cross-validation on Social Media (#4, 0.4507) is much larger.

5   Analysis

Results appear to provide a rather mixed picture in terms of performance, but we believe they can be explained according to three aspects, namely *size of training data*, *gap in genre*, and *quality of data*.

The fact that a Twitter-trained model performs better on blogs than cross-validating on blogs (#6 vs. #2) can be explained with a substantial difference in size in training sets, and the cross-genre potential loss is limited because of a narrow genre-gap. That is, we hypothesise that tweets and blogs are relatively similar, especially topic-wise. Conversely, we expect the difference between tweets and reviews to be much larger; the main source of difference, in addition to writing style, being that the content of tweets (and blogs) mainly concerns the author themselves, whereas reviews are mostly about the thing that is being reviewed.

This, indeed, explains why training on tweets and testing on reviews (#7) yields a significant drop in performance, when compared to cross-validated results on reviews (#3), official results from PAN 2014 (#14), and also with respect to testing on blogs (#6), which are a domain closer to training data than reviews. Therefore, we can speculate that, in a cross-genre setting, the main influencing factor is the difference in genre, and that, if genres are sufficiently similar, an increase in the amount of training data can further boost performance. This is corroborated by the performance of the social media-trained model (#9–#10), which is also clearly better on blogs than on reviews (assuming the genre-gaps are very similar to those for Twitter).

Such a hypothesis makes sense theoretically and is supported by the data, but leaves us with one additional question: why does GronUP perform about 10% points worse when going from Social Media to blogs (#9) than when going from Twitter to the same blogs (#6)? Similarly, cross-validation on Social Media (#4) is a lot less accurate than on Twitter (#1). The genre-gap we discussed above cannot explain this, since we assess tweets, blogs, and social media to be all quite similar to each other, and distinct from reviews. One possibility is that GronUP has limited portability: it manages to create a model of age and gender from Twitter data, for which it was developed, but this does not extend to other types of training data, even similar ones like social media.

This is possible, but another explanation seems more likely: based on manual inspection of the social media data, we believe that the performance difference is due to the relatively lower quality of the Social Media data. This hypothesis is supported by the fact that Social Media results are the lowest in the PAN 2014 official results (#15) in spite of this dataset being the largest of all in that competition. At PAN 2014, moreover, models were developed in-genre, and therefore should have been more capable of utilising the social media data for training.

We would like to conclude the analysis of the results with an observation regarding GronUP's in-genre performance overall. In [4], it was suggested that "*We believe that our lower scores [w.r.t. the in-genre state-of-the-art] are likely due to the conscious choice to avoid the use of potentially strong age/gender indicators which would work on Twitter only.*" The additional results seem to confirm this idea, although the difference between GronUP's in-genre performance (#1–#4) and the state-of-the-art (#12–#15) is not excessively big. On tweets, the difference is 4% points, on blogs 1, on reviews 5, and 7 on Social Media. In making such comparison, it should also be noted that our scores are on the cross-validated training sets of PAN 2014, while the official results reported in Table 3 are based on training on the whole set (thus slightly larger than what we used) and testing on separate data.

6   Conclusions

We set out to investigate the robustness of our cross-genre profiling system, beyond the good performance obtained at the PAN 2016 author profiling task. We did so via a series of experiments that revolved around two main research questions, and involved training and testing models on datasets pertaining to different sub-genres under a more general *social media* label. We ran in- and cross-genre evaluations, compared results across the various settings, and also against official results that had been obtained on the same datasets at the PAN 2014 evaluation campaign. How then do the obtained results and our analysis answer our research questions?

In Q1, we were asking if the GronUP Twitter-model would perform well beyond the PAN 2016 dataset. The answer is yes, it does, as long as the genre-gap is not too broad. Good performance on blogs, and poorer performance on reviews, provide support for this claim. Additionally, a general advice that we glean from the experiments from a practical perspective, is that, provided the genres are similar enough, size matters, and a large cross-genre training set can be more useful than a small in-genre one.

In Q2, we were asking whether GronUP's features do indeed capture demographics beyond Twitter, and could therefore be used to build satisfactory models on training sets of different genres. We trained a profiling models on the

PAN 2014 Social Media data, and observed that results are completely comparable to those obtained when training on Twitter when testing on reviews, but are quite almost 10 point lower when testing on blogs. The answer is not definite, but the picture is particularly blurred by the quality of the Social Media data, which makes this set rather unreliable to draw any safe conclusions.

As a byproduct of our investigation, we believe we can also provide a preliminary answer to the question the organisers of PAN 2014 put forward when observing the discrepancy in in-genre performance between tweets (higher), blogs (lower), and reviews (even lower) [8]. As mentioned in Sect. 1 of this paper, they had wondered whether the gap was due to training size (larger for tweets and smaller for the other datasets) or some intrinsic complexity of the data itself, leaving this issue to further investigation. Our analysis seems to point towards the size explanation, as we have seen that cross-validating on blogs yields a worse performance than training on tweets and testing on blogs. Some more specific data characteristic appears to play a role in cross-genre settings.

This last point is important with respect to the direction the development of manually annotated training data should take (if the genres are similar enough, can we just concentrate on one and build a single very large training set?), and definitely deserves further investigation. Indeed, in future work, we would very much like to better understand and model the trade-off between these two crucial aspects: size and genre-gap. While the first one is straightforward to quantify, the second one will require the development of some more complex, dedicated measure. Moreover, we plan to explore the contribution of additional datasets, especially to provide better evidence in answering Q2, potentially also in different experimental settings, so as to test more robustly the benefits and drawbacks on cross-genre profiling.

References

1. Álvarez-Carmona, M.A., López-Monroy, A.P., Montes-y Gómez, M., Villaseñor-Pineda, L., Jair-Escalante, H.: INAOE's participation at PAN'15: author profiling task. In: Proceedings of CLEF (2015)
2. Bird, S., Klein, E.: Natural Language Processing with Python. O'Reilly Media Inc., Sebastopol (2009)
3. Brants, T.: TnT: a statistical part-of-speech tagger. In: Proceedings of the Sixth Conference on Applied Natural Language Processing, pp. 224–231. Association for Computational Linguistics (2000)
4. Busger op Vollenbroek, M., Carlotto, T., Kreutz, T., Medvedeva, M., Pool, C., Bjerva, J., Haagsma, H., Nissim, M.: GronUP: Groningen user profiling. In: Working Notes of CLEF, pp. 846–857. CEUR Workshop Proceedings. CEUR-WS.org (2016)
5. Pedregosa, F., Varoquaux, G., Gramfort, A., Michel, V., Thirion, B., Grisel, O., Blondel, M., Prettenhofer, P., Weiss, R., Dubourg, V., Vanderplas, J., Passos, A., Cournapeau, D., Brucher, M., Perrot, M., Duchesnay, E.: Scikit-learn: machine learning in Python. J. Mach. Learn. Res. **12**, 2825–2830 (2011)
6. Perkins, J.: Python 3 Text Processing with NLTK 3 Cookbook. Packt Publishing Ltd. (2014)

7. Rangel, F., Rosso, P., Potthast, M., Stein, B., Daelemans, W.: Overview of the 3rd author profiling task at PAN 2015. In: CLEF (2015)
8. Rangel, F., Rosso, P., Chugur, I., Potthast, M., Trenkmann, M., Stein, B., Verhoeven, B., Daelemans, W.: Overview of the author profiling task at PAN 2014. In: Cappellato, L., Ferro, N., Halvey, M., Kraaij, W. (eds.) Working Notes for CLEF 2014 Conference, Sheffield, 15–18 September 2014. CEUR Workshop Proceedings, vol. 1180, pp. 898–927. CEUR-WS.org (2014)
9. Rangel, F., Rosso, P., Moshe Koppel, M., Stamatatos, E., Inches, G.: Overview of the author profiling task at PAN 2013. In: CLEF Conference on Multilingual and Multimodal Information Access Evaluation, pp. 352–365. CELCT (2013)
10. Rangel, F., Rosso, P., Verhoeven, B., Daelemans, W., Potthast, M., Stein, B.: Overview of the 4th author profiling task at PAN 2016: cross-genre evaluations. In: Working Notes Papers of the CLEF 2016 Evaluation Labs. CEUR Workshop Proceedings. CLEF and CEUR-WS.org, September 2016
11. Schler, J., Koppel, M., Argamon, S., Pennebaker, J.: Effects of age and gender on blogging, vol. SS-06-03, pp. 191–197. AAAI Press (2006)
12. Schnoebelen, T.: Do you smile with your nose? Stylistic variation in Twitter emoticons. Univ. Pa. Work. Pap. Linguist. 18(2), 117–125 (2012)

TimeLine Illustration Based on Microblogs: When Diversification Meets Metadata Re-ranking

Philippe Mulhem, Lorraine Goeuriot$^{(\boxtimes)}$, Nayanika Dogra,
and Nawal Ould Amer

Univ. Grenoble Alpes, CNRS, Grenoble INP, LIG, 38000 Grenoble, France
{philippe.mulhem,lorraine.goeuriot,nayanika.dogra,nawal.amer}@imag.fr

Abstract. This paper presents one approach used for the participation of the task 3 (TimeLine illustration based on Microblogs) for the CLEF Cultural Microblog Contextualization track in 2016. This task deals with the retrieval of tweets related to cultural events (music festivals). The idea is mainly to be able to get tweets that describe what happened during the shows of one festival. For the content-based aspects of the retrieval, we used the classical BM25 model [12]. Our concern was to study the impact of duplicate removal and several ways to re-ranks tweets. The obtained recall/precision evaluation results are biased by the limited number of runs considered in the pooling set for manual assessment, but the evaluation of results according to several informativeness measures show that adequate filtering increases such measure. We also describe the lessons learned from the first edition of this task and present how this impacts 2017's edition of the task.

Keywords: Tweet retrieval · Diversification · Re-ranking · Evaluation

1 Introduction

Retrieval of microblogs is a hard task due to the shortness of these documents. Such documents have however other data on which we may rely on, like their timestamps or authors for instance. In a general case, it is difficult to study retrieval on localized (in topic, time, and space) flow of documents. That is why the problem of illustrating what happened during one festival, provided a large set of microblogs that are related to such an event, is interesting. Among the possible directions to study, the goal of the Timeline illustration based on microblogs subtask[1] is to provide, for each event of a cultural festival, the most interesting tweets.

The goal of our participation to this retrieval task was to investigate the use of an information retrieval (IR) document index as a basis for near-duplicate

---

[1] https://mc2.talne.eu/~cmc/spip/Tasks/task-3-timeline-illustration-based-on-microblogs.html.

© Springer International Publishing AG 2017
G.J.F. Jones et al. (Eds.): CLEF 2017, LNCS 10456, pp. 224–235, 2017.
DOI: 10.1007/978-3-319-65813-1_22

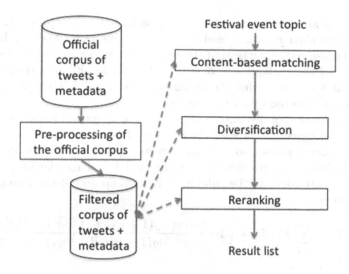

Fig. 1. Overview of the query processing

removal. Using such index allows to avoid complex partial string inclusion processes and to use simpler overlap measures.

In Sect. 2 we describe the evaluation task, and give an overview of the related work in Sect. 3. Section 4's organization follows the structure of our overall approach described in Fig. 1. From the initial tweet set provided for the task, we filter (pre-process) the potentially relevant tweets as described in Sect. 4.1. Then Sect. 4.2 presents the content-based retrieval achieved. In a second step, a diversification process is achieved through a simple instance-based duplicate removal, as presented in Sect. 4.3. The re-ranking of the diversified tweets, in Sect. 4.4, is then performed in three different ways: timeline, tweet author activity, and tweet author popularity. Experimental results are presented in Sect. 5, evaluated in Sect. 6 and discussed in Sect. 7. We finally give our conclusions in Sect. 8.

2   Description of the Task

The goal of this task is to link the events (mostly shows) of a given festival program to related microblog posts from a collection. Such information is useful for attendees of festivals, for people that are interested in knowing what happens in a festival, and for organizers to get feedback [10]. So, accessing information is important, but allowing diversity in results are also important.

Microblogs are provided with their timestamps, which are crucial as a basis for the requested linking. However, such timestamps must be use with care: they do not necessarily give accurate enough information (for instance in the case of parallel sessions), or might even generate noise (microblogs about one event may be posted before, during, or after the actual event).

Participants were required to provide, for each event of the program, the best tweets based on their relevance and diversity. In this task, diversity has to be evaluated because retrieving several times the same post is not really beneficial. The first part of the evaluation is based on classical recall/precision values: the relevance is only characterizing the topicality of the results. The diversity of the result set is not reflected by such a metric.

For the evaluation of diversity, namely Div, of a run, we make use of informativeness divergence [3] between the content of tweets in a run and the content of the manually-assessed relevant tweets (ground truth). Compared to the original divergence, we define 4 divergences based on the first N words ($N \in 100, 250, 500, 1000$) in the submitted text composed of the concatenation of retrieved tweets:

$$Div(R, S, N) = \sum_{t \in R} P(t|R) \times \left(1 - \frac{min(log(1 + P(t|R)), log(1 + P(t|S_N)))}{max(log(1 + P(t|R)), log(1 + P(t|S_N)))} \right)$$
(1)

where:

- R: a set of terms present in the first reference text composed of relevant tweets;
- S_N: a set of terms present in the first N words of the submitted text;
- conditional probabilities $P(.|.)$ are a computed using classical maximum likelihood estimation of the occurrences of one term according to the reference text length.

As the *Informativeness* is a dissimilarity, a low value of informativeness indicates that the texts are highly similar.

Three different distributions are considered:

- Unigrams (Uni) consider single lemmas (after removing stop-words).
- Bigrams (Bi) are made of pairs of consecutive lemmas (in the same sentence, assuming that two successive tweets are in different sentences).
- Bigrams with 1-gap ($Skip$) also consider pairs of consecutive lemmas but allowing the insertion between them of one lemma.

These distributions have varied flexibility: Unigrams only consider simple overlap, where Bigrams take into account successive words. Classically between two texts the Uni informativeness divergence value is lower than the Bi informativeness divergence value, and the Bi informativeness divergence value is lower than the $Skip$ divergence value.

3   Related Work

The goal of this paper is to study the impact of removal of duplicate tweets on the results and the impact of several re-ranking techniques on the results.

A large study of near-duplication detection for tweets has been presented in [13]. Several levels of duplication detections have been studied and a study of the impact of duplicate detection on the performances of IR systems is proposed. We choose, among the proposed features of [13], the one that is compatible with information retrieval indexes, i.e., compatible with sparse vector representation of term frequencies: the overlap of terms. Another advantage of the computation of this feature, according to [13], is that it is highly correlated with the actual overlap of such near-duplicates. Many overlap measures exist, as described in [8]. In our case we focus on very classical overlap coefficients, namely Jaccard, Szymkiewicz-Simpson, and Sørensen-Dice.

Classical ranking of tweets according to queries depends of course on their content, but also on many other features like the timestamps of the tweets [6, 15] or the authors of the tweet as in [4]. What we propose here is to mix content-based retrieval and these additional features using a classical pipeline: the content-based retrieval is performed in the first step, and then the additional features are used to re-rank the results of the first step. With such an approach, we are able to smoothly take into account these elements, which are different in nature, and to study experimentally their impact on the overall results. Re-ranking has been classically used in images [5] and videos [14] retrieval, as well as for personalization purposes [2]. In the approach described here, we re-rank the content-based results according to simple computed features: the timestamps of the retrieved tweets and two features that characterize some aspects of the popularity of tweet's authors.

We do not claim that the theoretical proposal made here is new, but in the context of microblogs related to cultural events, it is quite difficult to guess a priori the impact of such re-ranking when applied after duplication removal.

4   Description of the Approach

4.1   Pre-processing of the Official Tweet Corpus

The official corpus contains tweets crawled during the months of July and December 2015, containing the word `festival` (see Sect. 5.1).

Before indexing it, we filtered the provided dataset to work on a subset of the official set of tweets provided. The filtering is based on the tweets timestamp, corresponding to the dates of the festivals, and text matching patterns (location or festival name for instance). The obtained subset consists of 243,643 tweets.

We chose to keep the entire text of the initial tweets: we remove the '@' and '#' characters, and use a classical stoplisting process and Porter stemmer.

4.2   Content-Based Matching

The content-based retrieval is a simple process that uses the topic as the query, and matches each against the documents of the filtered corpus described in Sect. 4.1. The content based retrieval uses BM25 [12] model.

4.3 Diversification

The second step of the query processing aims at diversifying the results. In the state of the art, several ways to diversify the results are proposed [1]. The authors of [9] mention that most of the state-of-the-art diversification processes re-rank a set of firstly retrieved documents: that is also our approach here. In the case of tweets, i.e. very short documents, we chose to tackle this problem by removing duplicate tweets that correspond to *retweets*[2]. In fact, our proposal does not limit the process to retweets but to very similar tweets. Here we propose:

- to keep the original tweet t when t and its retweets are in the result list;
- to keep the most relevant retweet (with the highest rank) of t, when several retweets are in the result list, but t is not retrieved.

Unlike we may think, this approach is not similar to achieving a flat clustering on the tweets, as we defined an iterative process that goes from the top results to the last ones.

Similarity-based filtering is performed on the tweets as they are indexed (i.e. on stemmed terms), and not on the initial tweet. This allows to avoid storing the original tweets. We use then an overlap function over the index of compared tweets, and a threshold above which the tweets are considered similar.

The results with duplicates removed is considered to be a diversified result list.

4.4 Re-ranking

Once tweets are filtered, we explore several re-ranking methods:

1. No re-ranking (NO): The result of step Sect. 4.3 is directly given as an answer;
2. Time-based re-ranking (TIM): The result is re-ranked according to the creation date of the tweets. Having time-ranked tweets allows event organizers to pinpoint when something happened. Such re-ranking may however emphasize redundancy when many people report the same element at the same time;
3. Social-based re-ranking: we defined two social based re-ranking functions as follows:
   - ACT: this re-ranking function is related to the activity of a tweet author and is defined as the number of tweets he wrote. We assume that, the more active an author is, the more interesting his tweets are. In our case, the activity is festival related, as we want to integrate the activity of an author in a specific context. Hence, we compute author's activity on the subset of tweets representing the festival;
   - POP: this re-ranking function is based on the popularity of the tweet author. The underlying assumption being that the more the author is mentioned in tweets of the corpus, the more interesting his tweets are. The popularity of a user is the number of times he is mentioned. Similarly, the popularity is computed only on the subset of tweets representing the festival.

---

[2] One feature of Twitter is to allow users to "forward" (with or without alteration), or retweet, received tweets.

5 Experimental Results

5.1 Dataset

The Timeline Illustration Subtask focuses on two French music festivals (the "festival des vieilles charrues" and the "Transmusicales").

The full microblog collection available for the CMC campaign contained more than 50 million tweets. This full collection gathers tweets containing the term "festival" (and other specific names related to the considered festivals). Participants would use a subset of the microblog collection, matching the months the targeted festivals were organized at (July and December 2015). The data contain a total of 6,296,611 tweets (4,197,324 for July 2015, and 2,099,287 for December 2015).

Overall, there are 53 topics evaluated for this subtask, and these topics are all the live-events of one full day for each festival. One example of topic depicts the show of *Khun Narin's Electric* that took place at the *Transmusicales* on the 04/12/15:

```
<topic>
  <id>1</id>
  <title>Khun Narin's Electric</title>
  <festival>Transmusicales</festival>
  <begindate>04/12/15-14:00</begindate>
  <enddate>04/12/15-16:30</enddate>
</topic>
```

Manual assessments were conducted on the entire filtered collection. The assessments were conducted on both initial and retweeted tweets. For the diversity-based evaluation using the informativeness divergence, only the text of manually selected tweets is assessed, in a way to ensure that no duplicated tweets are considered.

5.2 Parameters Settings

We only indexed the filtered corpus as detailed in Sect. 5.1. The content-based retrieval uses the Terrier system [11], that implements BM25, using the default parameters (stoplist, Porter stemming, $b = 0.75$).

To remove the duplicate tweets, we compute overlap values between tweet terms as detailed in Sect. 4.3. We tested three overlap measures, considering two tweets t_i and t_j, so that the vector corresponding to t_i (resp. t_j) is Vi (resp. Vj) and a vocabulary size of k:

1. The Jaccard overlap coefficient, defined as:

$$Overlap_{JACC}(V_i, V_j) = \sum_k \frac{min(V_i[k], V_j[k])}{max(V_i[k], V_j[k])} \qquad (2)$$

2. The Szymkiewicz-Simpson coefficient, defined as:

$$Overlap_{SZY}(V_i, V_j) = \frac{\sum_k min(V_i[k], V_j[k])}{min(|V_i|_1, |V_j|_1)} \tag{3}$$

with $|V|_1$ the L_1 norm of a vector V.

3. The Sørensen-Dice coefficient defined as:

$$Overlap_{DICE}(V_i, V_j) = \frac{2 * \sum_k min(V_i[k], V_j[k])}{|V_i|_1 + |V_j|_1} \tag{4}$$

These three overlap coefficients are not so different, they mainly differ on the definition of the denominator, by considering at the vector level one (Formula 3) or both vectors (Formula 4), or considering the dimension vectors one by one as in formula 2.

After some preliminary tests on tweet sets, the overlap threshold value is fixed to 0.75 for the Jaccard and Szymkiewicz-Simpson coefficients, and for the Sørensen-Dice coefficient the overlap is fixed to 0.8.

5.3   Runs Submitted

We submitted 7 runs:

- BM25: The baseline, based only on the content, after the step 1 of the query processing described in Sect. 4.2. On average, each topic obtain a result list of 67 tweets;
- BM25+JACC: Jaccard coefficient diversified-only run, obtained as the result of the step 2 of the query processing described in Sect. 4.3. On average, each topic obtained a 36 tweets long result list, so the diversity removes 45% results from the RUN BM25.
  Because the runs BM25+JACC+{TIM, ACT, POP} only reorder the results, they have the same result sizes;
- BM25+SZY: Szymkiewicz-Simpson coefficient diversified-only run, obtained as the result of the step 2 of the query processing described in Sect. 4.3. On average, each topic obtained a 28 tweets long result list, so the diversity removes 59% results from the RUN BM25;
- BM25+DICE: Sørensen-Dice Coefficient diversified-only run, obtained as the result of the step 2 of the query processing described in Sect. 4.3. On average, each topic obtained a 42 tweets long result list, so the diversity removes 38% results from the RUN BM25;
- BM25+JACC+TIM: As described in Sect. 4.4, the results corresponding to the timeline re-ranking of RUN BM25+JACC, TIM;
- BM25+JACC+ACT: The results corresponding to the social activity-based re-ranking of RUN BM25+JACC, ACT, as described in Sect. 4.4;
- BM25+JACC+POP: The results corresponding to the social popularity-based re-ranking of RUN BM25+JACC, POP, as described in Sect. 4.4.

6 Evaluation Results

6.1 Classical IR Metrics

Table 1 shows the results obtained according to classical IR evaluation measures. Our experiment is closely related to focused information retrieval (as proposed by INEX), in the sense that we are retrieving very short documents, that are much closer to text passages than full web documents. Hence, we are focusing on similar metrics: precision at 5, 10 and 30, interpolated precision at recall 0 and MAP [7].

The best results are obtained with our run BM25+JACC+ACT. This run performs diversification with Jaccard coefficient and reorders results according to the tweet author's social activity. In general, runs including re-ranking give better results than the BM25 baseline, while diversification gives similar results.

The depth of the pool an impact on the MAP value, it is indeed quite low on all the runs and does not help discriminating efficiently the runs.

Table 1. Precision at 5, 10 and 30 documents, MAP, and iprec at recall 0 for the runs. (Best results per measure in bold.)

Run	P@5	P@10	P@30	MAP	IP0
BM25	0.5667	0.5310	0.4794	0.0109	0.6913
BM25+JACC	0.5571	0.5262	0.4754	0.0079	0.7026
BM25+SZY	0.5571	0.5238	0.4659	0.0077	0.7096
BM25+DICE	0.5571	0.5262	0.4754	0.0079	0.7026
BM25+JACC+TIM	0.5524	0.4857	0.4405	0.0038	0.6680
BM25+JACC+ACT	**0.6238**	**0.5508**	**0.4960**	0.0069	**0.7258**
BM25+JACC+POP	0.6000	0.5643	0.4817	0.0063	0.7181

6.2 Informativeness

In this section, we focus on the results obtained according to the informativeness divergence [3] between the content of tweets in a run and the content of the manually-assessed relevant tweets (ground truth). This evaluation measure takes into account unigrams (*Uni*), bigrams (*Bi*) or skip grams with a gap of 1 (*Skip*). Table 2 presents the results for the first 100, 250, 500 and 1000 words in the results, ranked by Unigram for the first 100 words retrieved. Because this value is a divergence measure, lower values are better.

Looking at the first 100 words in the table, we find that the run BM25+SZY, corresponding to Zymkiewicz-Simpson coefficient diversified run, largely outperform the other runs. This shows that accurate filtering of redundant tweets play a important role in the diversification. The run BM25+JACC+POP re-ranks run BM25+JACC, using the author's popularity. This run BM25+JACC+POP slightly outperforms run BM25+JACC, meaning that integrating the popularity of the authors

Table 2. Averaged Diversity evaluations for first 100, 250, 500 and 1 000 words of the result lists. (Best results per measure in bold.)

Run	Uni	Bi	Skip	Uni	Bi	Skip
	First 100 words			First 250 words		
BM25	0.5977	0.6670	0.6797	0.5663	0.6394	0.6537
BM25+JACC	0.5689	0.6331	0.6534	0.5288	0.5965	0.6170
BM25+SZY	**0.5541**	**0.6201**	**0.6392**	0.5305	**0.5962**	**0.6149**
BM25+DICE	0.5899	0.6553	0.6730	0.5458	0.6126	0.6315
BM25+JACC+TIM	0.6223	0.6968	0.7065	0.5721	0.6442	0.6573
BM25+JACC+ACT	0.6021	0.6784	0.6902	0.5499	0.6229	0.6373
BM25+JACC+POP	0.5681	0.6424	0.6567	**0.5281**	0.6023	0.6190
	First 500 words			First 1 000 words		
BM25	0.4740	0.5478	0.5637	**0.3510**	**0.4290**	**0.4470**
BM25+JACC	0.4588	0.5307	0.5533	0.4355	0.5099	0.5343
BM25+SZY	0.4814	0.5513	0.5710	0.4794	0.5515	0.5711
BM25+DICE	**0.4436**	**0.5154**	**0.5362**	0.4089	0.4787	0.5012
BM25+JACC+TIM	0.4628	0.5365	0.5582	0.4355	0.5099	0.5343
BM25+JACC+ACT	0.4634	0.5357	0.5564	0.4355	0.5099	0.5343
BM25+JACC+POP	0.4542	0.5267	0.5493	0.4355	0.5099	0.5343

according to the festival is interesting. One explanation for such results is that the same author does probably not send duplicates.

We also observe that runs BM25+JACC+TIM and BM25+JACC+ACT, are outperformed by run BM25+JACC which they are based on. This indicates that they might not diversify enough the results.

Results on the first 250 words, the best unigram informativeness is achieved by the run BM25+JACC+POP. For the two other evaluation measures, the best results are achieved by BM25+SZY (as with the first 100 words). This means again that the filtering plays an important role in gaining informativeness.

The results obtained using the first 500 words, give a very different ranking of runs: the top run is BM25+DICE. This lighter duplicate removal (compared to runs BM25+JACC+POP and BM25+JACC for instance), leads to larger results (in terms of words) and more diverse.

For the first 1000 words, the runs BM25+JACC, and BM25+JACC+{TIM, ACT, POP} obtain the same value: this is explained by the fact that these runs never return more than 1000 words, so the evaluation measures are the same. For such evaluation measures, we see that the baseline RUN1 (i.e. classical BM25) outperforms all the other runs, these other runs have duplicate tweets removed. This means that the ranking of BM25, according to the first 1000 words, covers most of the relevant topics. One interesting finding here is that, for the first 1 000 words evaluation measures, that the ranking is strictly inversely proportional to the number of results obtained by the runs (see above).

7 Discussion

7.1 Evaluation of Tweet Retrieval

From Sect. 6.1, we see that the classical IR measures might not well fitted to our search task. Firstly, they do not seem to be accurate enough: having a larger pool set may help to solve part of this problem. Moreover, traditional pooling and relevance assessment methods are not adapted to tweet search: as documents are shorter, perception of relevance very. The informativeness measures seem to be much more interesting, especially since our purpose is to give an overview of a cultural event. When focusing on a small number of words, these measures can help discriminating between the runs submitted.

7.2 Building a Tweet Retrieval Benchmark

Defining clearly what to evaluate as relevant was a challenge in our case. Documents were assessed by several assessors, who had different perception of the relevance (i.e. assessors from social sciences considered as relevant only informative tweets, while IT assessors considered as relevant informative as well as on topic tweets). We also observed that assessors would sometimes contradict themselves in the case of retweets. In the future, we will need to anticipate this issue. Cossu et al. for instance proposed a propagation approach to avoid contradictions in the assessment set: once retweets were detected, if one had been assessed as relevant, they would propagate its score over all the retweets.

At the beginning we were considering that integrating retweets in the assessment pool was interesting, as a retweet may add important information to an initial tweet (the *strong duplicates* of [13]). However, keeping all similar tweets without any control may lead to uninteresting results according to a user that looks for a survey about what happened during one show. So, providing a second set of evaluations measures that are able to tackle with diversity was also required. Retweets certainly raised various challenges throughout the entire process: firstly for the retrieval itself and pooling, but also for the evaluation and the assessment.

To concentrate on one aspect of the quality of a tweet retrieval system, the timeline illustration campaign in 2017 is specifically dedicated to evaluate the "gathering" of tweets related to one event. The idea is to study the behaviour of retrieval systems regarding their ability to retrieve **all** the relevant tweets related to one event: we are then more focusing on recall than precision. In 2017, we still consider all the tweets (with retweets) in the pool of assessed documents, but we also need to assess tweets that may not match any topic term.

8   Conclusion

The participation to the subtask TimeLine illustration based on Microblogs of the Cultural Microblog Contextualization Workshop allowed us to define a comprehensive process for the retrieval of tweets. The pre-processing allows us to focus on a subset of the whole official set of tweets provided for the task. The content-based retrieval is a classical one. We used three variations of duplicate removal (diversification) methods that take into account the specificity of the tweets. We applied 3 ways to re-rank the results in a third step of the query processing.

The impact of the pre-processing of the original corpus should be measured in the future, because it impacts the content-based matching, but also the activity and popularity values of tweet authors. Other variations of diversity algorithms also have to be studied, taking into account the specificity of tweets (especially their length, and their metadata), or even the choice of the kept tweet when we have duplicates.

References

1. Agrawal, R., Gollapudi, S., Halverson, A., Ieong, S.: Diversifying search results. In: Proceedings of the Second ACM International Conference on Web Search and Data Mining, WSDM 2009, pp. 5–14. ACM, New York (2009)
2. Amer, N.O., Mulhem, P., Géry, M.: Personalized parsimonious language models for user modeling in social bookmaking systems. In: Proceedings of the Advances in Information Retrieval - 39th European Conference on IR Research, ECIR 2017, Aberdeen, UK, 8–13 April, 2017, pp. 582–588 (2017)
3. Bellot, P., Moriceau, V., Mothe, J., SanJuan, E., Tannier, X.: INEX tweet contextualization task: evaluation, results and lesson learned. Inf. Process. Manag. **52**(5), 801–819 (2016)
4. Ben Jabeur, L., Damak, F., Tamine, L., Cabanac, G., Pinel-Sauvagnat, K., Boughanem, M.: IRIT at TREC Microblog Track 2013. In: Text REtrieval Conference - TREC 2013, Gaithersburg, United States, November 2013
5. Cai, J., Zha, Z.-J., Zhou, W., Tian, Q.: Attribute-assisted reranking for web image retrieval. In: Proceedings of the 20th ACM International Conference on Multimedia, MM 2012, pp. 873–876. ACM, New York (2012)
6. Efron, M., Lin, J., He, J., de Vries, A.: Temporal feedback for tweet search with non-parametric density estimation. In: Proceedings of the 37th International ACM SIGIR Conference on Research & #38; Development in Information Retrieval, SIGIR 2014, pp. 33–42. ACM, New York (2014)
7. Kamps, J., Pehcevski, J., Kazai, G., Lalmas, M., Robertson, S.: Inex 2007 evaluation measures. In: Focused Access to XML Documents: Sixth Workshop of the Initiative for the Evaluation of XML Retrieval (INEX 2007) (2008)
8. Kocher, M., Savoy, J.: Distance measures in author profiling. Inform. Process. Manag. **53**(5), 1103–1119 (2017)
9. Kuoman, C., Tollari, S., Detyniecki, M.: Using tree of concepts and hierarchical reordering for diversity in image retrieval. In: 2013 11th International Workshop on Content-Based Multimedia Indexing (CBMI), pp. 251–256, June 2013

10. Leskovec, J., Backstrom, L., Kleinberg, J.: Meme-tracking and the dynamics of the news cycle. In: Proceedings of the 15th ACM SIGKDD International Conference on Knowledge Discovery and Data Mining, KDD 2009, pp. 497–506. ACM, New York (2009)
11. Ounis, I., Amati, G., Plachouras, V., He, B., Macdonald, C., Lioma, C.: Terrier: a high performance and scalable information retrieval platform. In SIGIR 2006 Workshop on Open Source Information Retrieval (OSIR 2006) (2006)
12. Robertson, S.E., Walker, S., Jones, S., Hancock-Beaulieu, M.M., Gatford, M.: Okapi at trec3. In: Overview of the Third Text Retrieval Conference (TREC-3), pp. 109–126. NIST, Gaithersburg, January 1995
13. Tao, K., Abel, F., Hauff, C., Houben, G.-J., Gadiraju, U.: Groundhog day: near-duplicate detection on Twitter. In: Proceedings of the 22nd International Conference on World Wide Web, pp. 1273–1284. International World Wide Web Conferences Steering Committee (2013)
14. Tian, X., Yang, L., Wang, J., Yang, Y., Wu, X., Hua, X.-S.: Bayesian video search reranking. In: Proceedings of the 16th ACM International Conference on Multimedia, MM 2008, pp. 131–140. ACM, New York (2008)
15. Vosecky, J., Leung, K.W.-T., Ng, W.: Collaborative personalized twitter search with topic-language models. In: Proceedings of the 37th International ACM SIGIR Conference on Research & #38; Development in Information Retrieval, SIGIR 2014, pp. 53–62. ACM, New York (2014)

Labs Overviews

CLEF 2017 NewsREEL Overview:
A Stream-Based Recommender Task
for Evaluation and Education

Andreas Lommatzsch[1]([✉]), Benjamin Kille[1], Frank Hopfgartner[2],
Martha Larson[3,4], Torben Brodt[5], Jonas Seiler[5], and Özlem Özgöbek[6]

[1] TU Berlin, Berlin, Germany
{andreas.lommatzsch,benjamin.kille}@dai-labor.de
[2] University of Glasgow, Glasgow, UK
frank.hopfgartner@glasgow.ac.uk
[3] Radboud University, Nijmegen, Netherlands
[4] TU Delft, Delft, Netherlands
m.a.larson@tudelft.nl
[5] Plista GmbH, Berlin, Germany
{torben.brodt,jonas.seiler}@plista.com
[6] NTNU, Trondheim, Norway
ozlem.ozgobek@ntnu.no

Abstract. News recommender systems provide users with access to
news stories that they find interesting and relevant. As other online,
stream-based recommender systems, they face particular challenges,
including limited information on users' preferences and also rapidly fluctuating item collections. In addition, technical aspects, such as response
time and scalability, must be considered. Both algorithmic and technical
considerations shape working requirements for real-world recommender
systems in businesses. NewsREEL represents a unique opportunity to
evaluate recommendation algorithms and for students to experience realistic conditions and to enlarge their skill sets. The NewsREEL Challenge
requires participants to conduct data-driven experiments in NewsREEL
Replay as well as deploy their best models into NewsREEL Live's 'living lab'. This paper presents NewsREEL 2017 and also provides insights
into the effectiveness of NewsREEL to support the goals of instructors
teaching recommender systems to students. We discuss the experiences
of NewsREEL participants as well as those of instructors teaching recommender systems to students, and in this way, we showcase NewsREEL's
ability to support the education of future data scientists.

Keywords: Recommender systems · News · Evaluation · Living lab ·
Stream-based recommender

1 Introduction

Many recommender systems operate in large-scale, highly dynamic environments. Thousands of users must simultaneously receive suggestions fitting their

© Springer International Publishing AG 2017
G.J.F. Jones et al. (Eds.): CLEF 2017, LNCS 10456, pp. 239–254, 2017.
DOI: 10.1007/978-3-319-65813-1_23

individual preferences. In order to serve these users, system designs have to fulfill multiple requirements. These requirements include accurate predictions, reliability, responsiveness, and maintainability among others. Unfortunately, a majority of university curricula falls short of providing students with the necessary background to design systems that address these requirements. As a result, students have to pick up the skills necessary to design, maintain, and optimize recommender systems outside of the classroom, for example, on the job. Students complete their degrees with a lopsided profile. Some students may have proficient skills to accurately compute relevance estimates, but struggle to implement scalable systems. Conversely, some students may be excellent programmers, but lack the knowledge of statistics needed to develop better models.

Providing students with the comprehensive background needed to develop and deploy recommender systems is a challenging problem. Universities may be aware of the issue, but are still unable to address it effectively. For example, when teaching courses that focus on Recommender Systems, the lack of resources to conduct multi-criteria evaluation represents a major reason for universities to focus on simplified aspects only. The NewsREEL challenge is well-suited to allow educators to move beyond current restrictions. With NewsREEL 2017, we provide the required resources such that students can experience authentic conditions. They can deepen their skills in many ways and prepare for a career as system engineer, data scientist, or business analyst. All these roles command an understanding for a variety of aspects related to the performance of intelligent systems.

The world of news offers a use case that is widely familiar. Nearly everyone is a consumer of news in some form. Societies in the information age demand a continuous influx of information pieces, steadily provided by busy journalists. News recommender systems filter the information flow for news readers. They automatically select a subset of articles in pursuit of the goal of high user engagement with the recommendations.

Participants in the CLEF NewsREEL (News REcommendation Evaluation Lab) challenge face a complex environment. They have to define a strategy to produce accurate recommendations. At the same time, they have to deploy the strategy onto a server accessible to recommendation requests. They have to maintain the server and assure reliability even at times with plenty of simultaneous requests. In addition, the rapidly changing sets of users and items demand the recommendation algorithms to respond quickly to updates. This environment forces participants to develop systems that perform well with respect to real-world multi-criteria requirements.

In 2017, NewsREEL features a set of changes compared to the former editions. We have released a renewed data set, which covers more recent events from February 2016. We continue our cooperation with plista, a company offering personalization and targeted advertising services. Plista has revised the Open Recommendation Platform (ORP) in 2016. The revised platform operates with a RESTful API which facilitates automating administrative processes such as starting and stopping recommendation services. Finally, we have published a new evaluator for the offline evaluation along with a tutorial. Completing the tutorial took some time, which meant that it was not available until some time after the start of NewsREEL in October 2016.

The purpose of this paper is to introduce the NewsREEL 2017 challenge, and to discuss its ability to foster education and provide students with skills necessary to succeed in industry. We give a general overview of how NewsREEL provides an opportunity for multi-dimensional benchmarking in stream-based scenarios. We then turn to the discussion of the potential of NewsREEL in higher education. We report the results of a survey of past participants that gives us insight into how the challenge has contributed to the development of the participants' skills. We also discuss the contribution of NewsREEL to education from an instructor's point of view.

The remainder of this paper conveys the following parts. First, Sect. 2 reviews previous work on news recommender systems and resources used for teaching information access systems. Section 3 introduces the tasks defined for News-REEL 2017 and presents the main results. Section 4 looks at the challenge from different perspectives. We highlight results from a participant survey and discuss experiences of using NewsREEL as practical course work. Finally, Sect. 5 concludes the paper and anticipates future directions for research on news recommender systems.

2   Related Work

NewsREEL, as mentioned in the introduction, is now in its fourth year. While its principles and practices of benchmarking recommender systems have been covered in the overview papers of the previous years, in this paper we focus on the use of NewsREEL as a resource for teaching and learning. In this section, we first briefly introduce the importance of stream-based recommendation in an industry context in Sect. 2.1. In Sect. 2.2 we then discuss NewsREEL in the context of higher education.

2.1   Evaluation of News Recommendation Systems

News Recommender Systems, a type of information access system, facilitate finding relevant news articles (*cf.* [4]). The evaluation of recommender systems represents a challenging endeavor. Unlike for information retrieval systems, a consistent notion of relevance has not been established. Shani and Gunawardana [25] distinguish three evaluation methodologies: offline experiments, user studies, and online evaluation. In NewsREEL, our focus is on the static environments of offline experimentation and dynamic environments of online evaluation.

Benchmarking in Static Environments. A myriad of offline experiments emerged from academic research on recommender systems. Data sets facilitate repeating experiments under identical conditions. Initially, large-scale data sets focused on movie ratings (*cf.* [3,10]). In 2013, *plista* released a data set specifically for news recommender systems [14]. Subsequently, multiple updates of the data set have been released in scope of CLEF NewsREEL [13,15,18]. Li *et al.* [19] model news recommendation as contextual bandit problem. They define an evaluation procedure yielding valid results in offline settings. Similarly, Joachims *et al.* [12] apply counterfactual reasoning to logs of a news recommender system. They show how estimating propensity scores yields meaningful insights even if the ranking strategy had not been applied to collect the data.

Benchmarking in Dynamic Environments. Online evaluation has been established as the preferred mode for industrial applications. Das *et al.* [6] describe how Google's news aggregator presents news stories in a personalized fashion. The system combines MinHash clustering, probabilistic latent semantic indexing, and covisitation counts to estimate how relevant stories are to a given user. Garcin *et al.* [9] present the case of a Swiss news publisher. They devise a system using context trees to capture changing preferences. Finally, we mention the literature documenting CLEF NewsREEL, which offers participants the opportunity to evaluate their ideas on an industrial news recommender system (*cf.* [13,16,17]).

2.2 Education

In 2011, the Royal Academy of Engineering announced that teaching STEM is vital for the UK given the large numbers of industries that "depend on engineering knowledge and skills and [all] are signaling increasing demand and experiencing a scarcity of supply of suitable qualified young people" [1]. This suggests that students of STEM programs need to gain a technical skillset that will allow them to thrive in industry.

Addressing this, many computer science courses consist of lectures and accompanying lab sessions in which students are required to implement pieces of software to better understand the techniques taught in the course. This approach is based on the idea of deep learning outlined by Fry *et al.* [8] where students aim to "gain maximum meaning from their studying", *i.e.*, they are learning by doing. Similarly, Barr [2] highlights the importance of interactive learning environments for the development of valuable skills such as problem solving and the ability to communicate. Smart and Csapo [26] argue that such learning-by-doing activities can engage students, hence supporting active learning.

As outlined by Efthimiadis *et al.* [7], leading educators follow a very similar format when teaching information retrieval. For example, Mizzaro [21] first teaches the theoretical foundations of the subject in the lectures and then asks students to develop a search engine using open source software components. Lopez-Garcia and Cacheda [20] refer to this teaching methodology as "technical-oriented IR methodology" consisting of theoretical lectures and practical work.

Although this technical-oriented teaching method has been introduced in good faith to familiarize students with the theoretical foundations as well as appropriate technical skill sets, Hopfgartner *et al.* [11] argue that the technical challenges addressed in these courses are often too limited and therefore do not support the students in gaining the more advanced skill sets required to thrive in our technology-oriented economy. Hoping to address this shortcoming, they suggest incorporating realistic and complex challenges that model real-world problems faced in industrial settings. In this paper, we provide a preliminary analysis of the potentials of NewsREEL for student learning.

3   News Recommendation Scenario

Recommender systems reduce a large collection of items, news articles in our case, to a manageable subset. Early recommender systems addressed the reduction problem by optimizing a single criterion on a fixed data set. More recently, researchers have pointed out that multiple criteria affect recommendations' perception. Castells *et al.* [5] introduce novelty and diversity as additional criteria. Ribeiro *et al.* [23] emphasize that multiple criteria can be considered when learning suited reduction strategies. Said *et al.* [24] elaborate on the need to consider non-functional criteria. NewsREEL's scenario encompasses multiple criteria in two tasks.

3.1   NewsREEL Live

Participants deploy their recommendation algorithms into a *living lab* environment. The environment consists of three major parts: recommendation services, communication platform, and publishers' webservers. Participants contribute the recommendation services. They run these on their own systems connected to the communication platform via HTTP. Alternatively, plista offered virtual servers to participants who could not afford their own servers or were located far away. Increased network latency puts server located far off at a disadvantage. The communication platform orchestrates messages and monitors performances and issues. Publishers' webservers interface with visitors and initiate recommendation requests. We consider four basic types of messages. Recommendation requests arise from visitors reading news articles. The communication platform receives recommendation requests, forwards them to available recommendation services, and randomly selects a valid list of recommendations to return to the publisher. The publisher displays the recommendations to the reader. The second type of message serves to keep the article collection up to date. Whenever publishers add new articles or update existing ones, they inform the communication platform, which subsequently forwards the information to all connected recommendation services. The third type of message concerns actions of readers. Readers can access news articles, thus generating impressions. Alternatively, they can click on recommendations thus creating clicks. The communication platform recognizes these events and forwards the information to all connected recommendation services. Simultaneously, it keeps track of click events to measure to what degree individual recommendation services succeed. Finally, the fourth type of message represents errors, *i.e.* cases of failure. Errors occur if the communication

Table 1. Observations from the algorithms run in NewsREEL Live from 24 April–7 May, 2017 except 28 April, 2017

Recommender	Clicks	Impressions	CTR
baseline	726	62,052	0.0117
5	58	3,708	0.0156
9	879	77,723	0.0113
21	817	61,524	0.0133
33	166	23,023	0.0072
34	600	49,830	0.0120
35	810	68,768	0.0118
45	813	79,120	0.0103
55	2	349	0.0057
56	747	60,814	0.0123
59	764	75,535	0.0101
61	875	63,950	0.0137
62	813	59,227	0.0137
63	925	68,582	0.0135
64	1,139	72,601	0.0157
65	1,268	81,245	0.0156
66	896	42,786	0.0209
67	6	816	0.0074
70	12	443	0.0271

fails or an invalid list of recommendation is produced. Recommendation lists are invalid if they include invalid or too few articles. Table 1 summarizes the results of NewsREEL Live. Eighteen algorithms provided recommendations in addition to the baseline. The evaluation period lasted from 24 April to 7 May, 2017, excluding 28 April due to technical difficulties with the logging. Participants collected up to 1,268 clicks with a maximum of 81,245 impressions. We observe click through rates of up to 2.71%. Click through rates describe the proportion of supplied recommendations which users subsequently clicked.

3.2   NewsREEL Replay

Participants receive a large-scale data set comprising messages similar to those exchanged in the living lab environment. Each message has a timestamp assigned. Thus, participants can replay the sequence of messages creating conditions similar to the living lab. In contrast to NewsREEL Live, NewsREEL Replay allows participants to issue each request to multiple recommendation algorithms. This enables them to compare algorithms in a repeatable fashion. The spectrum of algorithms include relatively simplistic methods, for example based on popularity or freshness, content-based filtering, and collaborative filtering. Participants can measure predictive accuracy as well as scalability. For instance, they may increase the rate at which requests arrive and compare how many requests various algorithms process within a specified time. An evaluator has been made available to participants. It takes chronologically ordered messages and sends them to a recommendation service. Subsequently, the evaluator checks whether any of the recommended articles appear in the session's future impressions. In addition, the evaluator records the time elapsing until the recommendations arrive. The evaluator produces click rates and a response time distributions based on the records. The response time distribution enables us to assess how quickly and reliably an algorithm generates recommendations. Supplying recommendations quickly is necessary for publishers to include them as the webpage is loaded. The more comfortably the average response time ranges below the permitted limit, the less likely the recommender will fail to supply recommendations. Comparing algorithms' response time distributions, we expect to find differences with respect to all major characteristics describing distributions. These characteristics include the average, dispersion, and skewness. For instance, recommendation algorithms may exhibit slightly higher average response rates yet less dispersion.

3.3 Summary

NewsREEL allows participants to experience realistic conditions as they evaluate recommendation algorithms. They may use the data set to establish repeatable results. Conversely, the living lab setting highlights the necessity to pay attention to non-functional aspects such as response time limitations. NewsREEL is particularly interesting for researchers in academia and students. Researchers get access to actual users unavailable in the majority of evaluation initiatives. Students can develop many skills required for future careers in industry. These skills are difficult to obtain via toy examples on static data sets prevalent at university courses.

4   NewsREEL for Learning and Teaching

In [11], Hopfgartner *et al.* argue that the technical skills that are taught in STEM courses at higher education institutes often are limited in scope and therefore do not support students in learning the more advanced skill sets that are required by industry nowadays. They therefore suggest to incorporate realistic and complex challenges that model real-world problems faced in industrial settings in the teaching curriculum. More specifically, focusing on the recommender systems domain, they hypothesize that campaigns such as NewsREEL can be employed to teach students the skills required by modern data scientists. Following through on this line of thinking, this section addresses this hypothesis from two directions. In Sect. 4.1, we present the results of a survey that was sent out to anyone who had signed up for the lab since it became part of CLEF. Our main motivation for this study was to gain further insights on who is interested in the campaign, and to understand *if* and *how* they benefited from NewsREEL. Section 4.2 presents these challenges and opportunities from the perspective of a course instructor who embedded NewsREEL as a case study in a Data Science course. Section 4.3 concludes this part.

4.1   Participant's Perspective

In order to identify and assess the potentials of NewsREEL as an innovative tool to learn new technical skills that are in high demand in industry, we sent out an online survey to everyone who had registered for any of the NewsREEL tasks since CLEF 2015. The survey consisted of three main parts: With the first part, we aimed to better understand the demographics of our participants. The second part focused then on gathering information about the participants' technical skill set with an emphasis on recommender systems. In the last part, we focused on learning about the participants' motivation to register for NewsREEL and on gathering feedback about what they experienced while participating in the campaign. The participants' responses are summarized and discussed in the remainder of this section.

Demographics. The survey was sent by email to 160 people who had registered for NewsREEL in the past. It was completed by ten subjects, nine male and one female (ca. six per cent response rate). Although ten respondents were enough to provide us with valuable insight, ideally, we would have liked to have heard back from more of the registrants who we had contacted. Here, we comment briefly on why the expectation of a higher response rate was probably unrealistic. First of all, it is important to know that a significant number of participants decided to register for multiple, if not even all labs that were organized as part of CLEF. We realize that it is unlikely that these registrants really had the intention to participate in all labs. Moreover, various participants registered with an email address that suggests that they are students at a higher education institution. While these individuals might have participated in any of the tasks, *e.g.*, as part of their training or teaching, they may have graduated by now and are no longer interested in academic work. In fact, a few emails that were send out to the registrants bounced since the email addresses no longer existed. While we had alternative email addresses for a few of those students, there remained four registrants whom we could not reach.

Our ten respondents were a diverse group. While two participants stated that they are between 18–25 years old and two others indicated that they are between 26–30 years of age, six participants reported that they are in their thirties. Ninety per cent of participants stated they either already hold a postgraduate degree (*i.e.*, MSc or PhD) or that they currently study towards such a degree. Only one participant stated that he has no academic degree. When asked what they currently study, computer science and related degrees were named. Four participants stated that they are currently employed in a university teaching position, two described their job position as programmer or developer. We conclude from this that the group of our participants consists of students, academics, and IT professionals.

Technical Skills. In order to better understand the technical skill sets of our participants, we focused in the second part of our survey on aspects related to the implementation and operation of recommender systems.

In the first question, we asked whether our participants had any experience in setting up a recommender system (*e.g.*, as part of their studies or job) before registering for NewsREEL. While 50% confirmed that they did have prior experience, the other half did not have any experience. When asked to outline their experience further, participants reported that they had performed offline evaluation using publicly available datasets, that they had developed such system as part of their thesis, or that they had worked on recommender systems for study and research purposes. One participant indicated that he was involved in the implementation of a commercial enterprise search and recommender system.

In the next question, we asked participants to select from a list of applications and frameworks that they had used before. Multiple answers were possible. The most common answers were Mahout and Idomaar with 40% each, followed by Lenskit and MyMediaLite with 10% each. Three users indicated that they had used none of these.

Next, we wanted to learn more about the datasets that they have used in the past when implementing a recommender system by asking them to choose from a list of the most commonly used datasets. Seventy per cent of participants selected the plista dataset that is used in the NewsREEL challenge. Half of the participants have experimented with the MovieLens dataset. This is hardly surprising given the dominant usage of this dataset for research purposes in the past. Other options that were chosen (by 10% each) include the Million Song Dataset, the Netflix Prize, MovieTweetings dataset, datasets provided as part of the ACM RecSys Challenge, and datasets shared on Kaggle.

Finally, we wanted to learn more about the challenges that the participants faced while developing a recommender system. To this end, we asked them in an open question to outline their experiences. Answers included issues such as evaluation, scalability and responsiveness, data size, lack of information about user and historical data, and finding the right tradeoff between the quality of recommendations and the performance of the algorithms.

Feedback on NewsREEL. In the final part of the survey, we explicitly asked the students about their experience in participating in NewsREEL.

First, we asked the participants to indicate the year in which they registered for CLEF NewsREEL. Multiple answers were possible since participants could also register for more than one iteration of NewsREEL in the past few years. We were pleased to see that participants from all three iterations provided feedback in this survey. Two subjects had participated since 2015, six subjects had participated in NewsREEL'16, and five subjects had registered for the most recent iteration. One subject indicated that he did not remember the year in which he registered.

Next, we wanted to learn which task the participants were most interested in. Although the description of the individual tasks has evolved slightly throughout the past three years, all tasks can broadly be categorized as online or offline evaluation tasks. Again, we were happy to see that participants stated interest in both tasks: 80% expressed interest in the online evaluation task, referred to as Online Evaluation in a Living Lab or NewsREEL Live, and 60% were interested in the offline task using the plista dataset, also referred to as NewsREEL Replay. When asked whether they managed to participate in NewsREEL, only one participant stated that he did not participate. When asked for the reason, he stated that he "did not solve it on time".

One of the main motivations for us for developing this survey was to understand what motivated participants most to register for NewsREEL. Our main assumption is that participants participated because they would like to develop a new technical skill set. In order to gain more insights into this issue, we explicitly asked the participants to indicate on a five-point Likert scale whether they are "looking for experience both with software engineering and recommender systems algorithms". A vast majority of 70% either fully agreed or agreed with this statement. Moreover, we asked participants to select from a list the technical aspects that motivated them most to participate. The most commonly chosen answer, selected by 80% of participants, is the challenge of providing recommendations in real-time, followed by the possibility to benchmark recommenders in a large-scale setting. Half of the participants stated that they were attracted by the challenging scenario of news recommendation, and by the opportunity of getting access to a new dataset.

Further, we asked the participants to rate on a five-point Likert scale whether "NewsREEL allowed [them] to acquire new skills relevant for [their] career". An overwhelming majority of 80% either fully agreed or agreed with this statement. When asked to indicate which skills they have learned while participating, the most commonly chosen answers included stream processing, real-time processing, providing recommendations, software development, and data analysis. This supports the hypothesis put forward by Hopfgartner *et al.* [11] that NewsREEL provides the opportunity to acquire new skill sets that are in high demand by industry.

Different from traditional evaluation campaigns that follow the Cranfield evaluation paradigm, one of the challenges that participants of NewsREEL face is the increased complexity that comes with setting up and connecting with the Open Recommendation Platform. Given this complexity, we assume that it might be easier to work on NewsREEL as a team, thus splitting the workload. While the majority of 70% stated that they worked alone or mostly alone but with input from others, three participants stated that they worked as a team. When asked to elaborate on the advantages and disadvantages of this, they argued that "evaluating many approaches in a team is more efficient and usually yields better results" and that it is "a lot of work for a single person". Although these statements support our assumption we were surprised to learn that the majority of participants worked on their own to solve the NewsREEL challenge. We aim to address this issue by developing a communication channel among participants that would allow them to collaborate with each other further.

An obvious choice for such channel would be to use the CLEF conference as a venue that allows participants to engage and network with each other. Interestingly, only 30% stated that the possibility to present their work at an Academic conference motivated them to participate in NewsREEL. Addressing this issue further, we also asked participants to rate on a five-point Likert scale how much they agree with the statement that "the possibility to network with other researchers of [their] area at the CLEF conference is a motivating factor". Here, we received very mixed responses, suggesting that the academic networking component is of lesser interest to the participants than the opportunity to acquire new technical skills.

4.2 An Instructor's Perspective

After having reported on people's motivation to participate in NewsREEL, in this section we now provide insights and discuss lessons learned from the perspective of a university instructor who embedded NewsREEL in their teaching.

Course Details. In the 2016/17 semester, NewsREEL was used as a learning & teaching resource of the Web Intelligence course at Norwegian University of Science and Technology (NTNU). The course targets postgraduate students of the Master and PhD programs of the Computer Science Department. The course objective is to train students to use semantic technologies and open linked data to analyze unstructured content as well as teaching the theoretical foundations of building recommender systems. The course consists of theoretical classes, guest lectures from industrial partners, practical group project assignment and exercise classes.

The course's group project seeks to have students solve a practical problem. They ought to apply methods covered in the lectures to a problem they might encounter in an industrial setting. In previous years, students were assigned different tasks. In the Spring 2017 edition, all students have been tasked to take part in CLEF NewsREEL Replay. The forty students were assigned to eleven groups with three to five members each. All students were in the first year of their master studies except three PhD students. The allowed time to complete the assignment was set to 18 January to 7 April, 2017. Students had to conduct the experiment, present their findings, and write a delivery report in order to pass the assignment.

Why CLEF NewsREEL? News domain in recommender systems exhibits specific properties and challenges compared with other recommender systems domains such as movies or music [4,22]. Articles being textual objects support content-based filtering techniques. The data set provided for NewsREEL Replay includes articles' textual content along with interaction data. The latter can be used to apply collaborative filtering techniques. These characteristics facilitate applying the theoretical models taught in the course. These models include recommender systems, text analytics/natural language processing, open linked data, and further semantic technologies. The data set's scale lets students experience conditions in which processing all data on a single computer becomes infeasible. These conditions are expected to become increasingly prevalent as systems keep producing increasing amounts of data. Finally, NewsREEL Live offers students the opportunity to evaluate their ideas with real user feedback. Participation in NewsREEL Live was left as a voluntary exercise for students curious on how well their ideas would perform with real user feedback.

In short, assigning the CLEF NewsREEL challenge to students as a group project naturally fits to the content of the course and the intended learning outcomes of the group project.

Course Assessment. NTNU employs multiple tools to assert high quality education. At the beginning of the semester, a student group is voluntarily selected as "reference group" to represent students' interests during the course. The group's task is to give feedback to the teachers. Three meetings between the reference group and the teachers took place. Two meetings took place during the semester and another meeting at the end of the course. In addition, the course organizers held a review session in the course of which all students presented their results and gave feedback about their experiences.

Results and Lessons Learned. Two teaching assistants were available to help students with questions and problems throughout the semester. Exercise classes took place on a bi-weekly basis and provided a forum for discussion. Additionally, an online discussion group was added to the official NTNU e-learning platform. At the end of the semester, the reference group stated that more teaching assistants with more experience in CLEF NewsREEL challenge would be helpful. When we consider the high number of questions through e-mails, messages on the discussion platform and face to face meetings, previous hands on experience with CLEF is definitely recommended for the teaching assistants. A lot of the questions from students were about updated framework, documentation and data set inconsistencies and, practical problems about setting up and running the framework, which requires up to date experience on the challenge to answer the questions. This could be related to the fact that NewsREEL Replay does not require using a particular set of tools. Instead, participants are free to use whatever technology they like.

In NewsREEL challenge there are no restrictions for the choice of technology. But in order to make the challenge more efficient within the course context, some restrictions provided by the instructors could be useful and prevent students to get lost in the very many choices of technologies available.

As the representative of all the students in the classroom, the reference group stated at the end of the semester that the group project topic was exciting and up to date, and that they gained a lot of experience about recommender systems technology. They also stated that it was fun to work with a practical project where they can apply what they have learned in the class theoretically. As a downside of the group project, they stated that they have spent much more time than expected to set up and run the framework before they can start implementing their recommender system algorithm. However, with the experience and existing implementations from this year's students, this problem can be minimized in the coming years.

Naturally, there are no textbook clear, step by step instructions for the framework and challenge. And this is quite different from what most of the students get used to. So students should be clearly told what they will deal with during the CLEF challenge and what prior knowledge is expected. Even though we mentioned these, we have observed some confusion within students as a result of dealing with a real challenge.

One of the main goals of this course was to encourage the students to collaborate with each other, hence helping them to develop graduate attributes. Throughout the group project, we observed that students collaborated not only within their own group but also between the groups.

Even though only a few groups could come up with some evaluation results, getting the best results from the challenge was not the main goal in this course's group project assignment. As stated above, the main goal of this assignment was to teach the students about applying theoretical knowledge in practice as well as challenge them with real-world implementation problems and encourage them to work in teams. As a result, we believe that we have reached to the learning goals of this group project.

4.3 Summary

In this section, we have discussed the opportunities that NewsREEL brings for higher education. In particular, we studied the hypothesis presented in [11] that participants of NewsREEL can gain important skills that are of importance in the labor market. For this, we first approached all past and current registrants of NewsREEL to better understand their motivation for registering for the lab. The survey results suggest that the participants did indeed acquire new technical skills while participating. Moreover, we presented a reflection of an instructor who embedded NewsREEL in their teaching. We argue that these insights can serve as guidelines for Academics who might consider using NewsREEL as a tool for learning & teaching as well.

5 Conclusion and Outlook

This paper has discussed the NewsREEL 2017 news recommendation challenge with a particular emphasis on the contribution it makes to education in the area of recommender systems. Our point of departure has been the observation that universities often fall short of teaching students the full range of skills necessary in order to develop and deploy effective recommender system algorithms. In order to be effective, real-world recommender systems must not only maintain high prediction performance, but must also fulfill requirements of scalability, availability, and response time. The opportunities to evaluate recommender systems offered by NewsREEL allow students to gain first-hand experience in addressing the challenges faced in the types of stream-based scenarios typical for today's large online recommender systems.

The NewsREEL Live task offers a living lab environment in which participants can test their stream-based recommendation algorithms online. The News-REEL Replay task allows more detailed examination of algorithms by supporting the replay of the information (items, requests, interactions) in the stream. Taken together students have exercised a full spectrum of skills from algorithm design, to implementation, to performance analysis with respect to multiple criteria.

We reported information collected from a survey of past registrants of the NewsREEL task. We found that there was great interest in the opportunity for online recommendation, and also for access to real-world data. The survey revealed that past participants indeed acquired skills in multiple areas important for recommender systems.

In the future, we would like to invest explicit effort into bringing people interested in NewsREEL together in order to solve the problem as a team. This will help to lighten the load on any individual participants. For students, support of people with previous experience was identified as being important. Here, explicit attention to bringing the right people together to address NewsREEL together could be particularly important.

We close by noting that here we have focused on the education aspects of NewsREEL: what past participants have learned, and how NewsREEL supports teachers in the university setting. Moving forward, we are interested in gaining further insight into industry perspectives. In particular, we want to understand whether the skills learned by NewsREEL participants prove valuable in practice "on the job" in an industry setting.

References

1. Banks, F., Barlex, D.: Enabling the 'E' in STEM. In: Teaching STEM in the Secondary School. Routledge, London (2014)
2. Barr, M.: Video games can develop graduate skills in higher education students: a randomised trial. Computers & Education, (2017) (Accepted for publication)
3. Bennett, J., Lanning, S., et al.: The Netflix prize. In: Proceedings of KDD Cup and Workshop, New York, NY, USA, vol. 2007, p. 35 (2007)
4. Billsus, D., Pazzani, M.J.: Adaptive news access. In: Brusilovsky, P., Kobsa, A., Nejdl, W. (eds.) The Adaptive Web. LNCS, vol. 4321, pp. 550–570. Springer, Heidelberg (2007). doi:10.1007/978-3-540-72079-9_18
5. Castells, P., Hurley, N.J., Vargas, S.: Novelty and diversity in recommender systems. In: Ricci, F., Rokach, L., Shapira, B. (eds.) Recommender Systems Handbook, pp. 881–918. Springer, New York (2015)

6. Das, A., Datar, M., Garg, A., Rajaram, S.: Google news personalization - scalable online collaborative filtering. In: WWW, pp. 271–280. ACM, New York (2007)
7. Efthimis, E., Fernandez-Luna, J.M., Huete, J.F., MacFarlane, A. (eds.): Teaching and Learning in Information Retrieval. Springer, Heidelberg (2011)
8. Fry, H., Ketteridge, S., Marshall, S.: Understanding student learning. In: A Handbook for Teaching and Learning in Higher Education. Routledge, London (2003)
9. Garcin, F., Faltings, B., Donatsch, O., Alazzawi, A., Bruttin, C., Huber, A.: Offline and online evaluation of news recommender systems at swissinfo.ch. In: RecSys, pp. 169–176 (2014)
10. Harper, F.M., Konstan, J.A.: The Movielens datasets: history and context. ACM Trans. Interact. Intell. Syst. (TiiS) 5(4), 19 (2016)
11. Hopfgartner, F., Lommatzsch, A., Kille, B., Larson, M., Brodt, T., Cremonesi, P., Karatzoglou, A.: The potentials of recommender systems challenges for student learning. In: Proceedings of CiML 2016: Challenges in Machine Learning: Gaming and Education, October 2016
12. Joachims, T., Swaminathan, A., Schnabel, T.: Unbiased learning-to-rank with biased feedback. In: The Tenth ACM International Conference, pp. 781–789. ACM Press, New York (2017)
13. Kille, B., Brodt, T., Heintz, T., Hopfgartner, F., Lommatzsch, A., Seiler, J.: NEWSREEL 2014: summary of the news recommendation evaluation lab. In: Working Notes for CLEF 2014 Conference, Sheffield, UK, 15–18 September 2014, pp. 790–801 (2014)
14. Kille, B., Hopfgartner, F., Brodt, T., Heintz, T.: The plista dataset. In: International News Recommender Systems Workshop and Challenge, pp. 16–23. ACM, New York, October 2013
15. Kille, B., Lommatzsch, A., Gebremeskel, G.G., Hopfgartner, F., Larson, M., Seiler, J., Malagoli, D., Serény, A., Brodt, T., Vries, A.P.: Overview of NewsREEL'16: multi-dimensional evaluation of real-time stream-recommendation algorithms. In: Fuhr, N., Quaresma, P., Gonçalves, T., Larsen, B., Balog, K., Macdonald, C., Cappellato, L., Ferro, N. (eds.) CLEF 2016. LNCS, vol. 9822, pp. 311–331. Springer, Cham (2016). doi:10.1007/978-3-319-44564-9_27
16. Kille, B., Lommatzsch, A., Hopfgartner, F., Larson, M., Seiler, J., Malagoli, D., Serény, A., Brodt, T.: CLEF NewsREEL 2016: Comparing multi-dimensional offline and online evaluation of news recommender systems. In: Working Notes of CLEF 2016 - Conference and Labs of the Evaluation forum, Évora, Portugal, 5–8 September, 2016, pp. 593–605 (2016)
17. Kille, B., Lommatzsch, A., Turrin, R., Serény, A., Larson, M., Brodt, T., Seiler, J., Hopfgartner, F.: Overview of CLEF NewsREEL 2015: news recommendation evaluation lab. In: Working Notes of CLEF 2015 - Conference and Labs of the Evaluation forum, 8–11 September 2015, Toulouse, France (2015)
18. Kille, B., Lommatzsch, A., Turrin, R., Serény, A., Larson, M., Brodt, T., Seiler, J., Hopfgartner, F.: Stream-based recommendations: online and offline evaluation as a service. In: Mothe, J., Savoy, J., Kamps, J., Pinel-Sauvagnat, K., Jones, G.J.F., SanJuan, E., Cappellato, L., Ferro, N. (eds.) CLEF 2015. LNCS, vol. 9283, pp. 497–517. Springer, Cham (2015). doi:10.1007/978-3-319-24027-5_48
19. Li, L., Schapire, R.E., Chu, W., Langford, J., Langford, J.: A contextual-bandit approach to personalized news article recommendation. In: The 19th International Conference, pp. 661–670. ACM Press, New York (2010)
20. Lopez-Garcia, R., Cacheda, F.: A technical approach to information retrieval pedagogy. In: Teaching and Learning in Information Retrieval. Springer (2011)

21. Mizzaro, S.: Teaching web information retrieval to computer science students: concrete approach and its analysis. In: Teaching and Learning in Information Retrieval (2011)
22. Özgöbek, Ö., Gulla, J.A., Erdur, R.C.: A survey on challenges and methods in news recommendation. In: WEBIST (2014)
23. Ribeiro, M.T., Lacerda, A., Veloso, A., Ziviani, N.: Pareto-efficient hybridization for multi-objective recommender systems. In: Proceedings of the Sixth ACM Conference on Recommender Systems, pp. 19–26. ACM (2012)
24. Said, A., Tikk, D., Stumpf, K., Shi, Y., Larson, M., Cremonesi, P., Evaluation, R.S.: A 3D benchmark. In: RUE@ RecSys, pp. 21–23 (2012)
25. Shani, G., Gunawardana, A.: Evaluating recommendation systems. In: Recommender Systems Handbook, pp. 257–297. Springer, Boston, October 2010
26. Smart, K.L., Csapo, N.: Learning by doing: engaging students through learner-centred activities. Bus. Prof. Commun. Q. **40**, 451–457 (2007)

LifeCLEF 2017 Lab Overview: Multimedia Species Identification Challenges

Alexis Joly[1]([✉]), Hervé Goëau[2], Hervé Glotin[3], Concetto Spampinato[4],
Pierre Bonnet[2], Willem-Pier Vellinga[5], Jean-Christophe Lombardo[1],
Robert Planqué[5], Simone Palazzo[4], and Henning Müller[6]

[1] Inria, LIRMM, Montpellier, France
alexis.joly@inria.fr
[2] CIRAD, UMR AMAP, Montpellier, France
[3] AMU, CNRS LSIS, ENSAM, University of Toulon, IUF, Toulon, France
[4] University of Catania, Catania, Italy
[5] Xeno-canto Foundation, Meijendel, The Netherlands
[6] HES-SO, Sierre, Switzerland

Abstract. Automated multimedia identification tools are an emerging
solution towards building accurate knowledge of the identity, the geographic distribution and the evolution of living plants and animals. Large
and structured communities of nature observers as well as big monitoring equipment have actually started to produce outstanding collections
of multimedia records. Unfortunately, the performance of the state-of-
the-art analysis techniques on such data is still not well understood and
far from reaching real world requirements. The LifeCLEF lab proposes
to evaluate these challenges around 3 tasks related to multimedia information retrieval and fine-grained classification problems in 3 domains.
Each task is based on large volumes of real-world data and the measured
challenges are defined in collaboration with biologists and environmental stakeholders to reflect realistic usage scenarios. For each task, we
report the methodology, the data sets as well as the results and the main
outcomes.

1 LifeCLEF Lab Overview

Identifying organisms is a key for accessing information related to the uses and
ecology of species. This is an essential step in recording any specimen on earth
to be used in ecological studies. Unfortunately, this is difficult to achieve due
to the level of expertise necessary to correctly record and identify living organisms (for instance plants are one of the most difficult group to identify with an
estimated number of 400,000 species). This *taxonomic gap* has been recognized
since the Rio Conference of 1992, as one of the major obstacles to the global
implementation of the Convention on Biological Diversity. Among the diversity
of methods used for species identification, Gaston and O'Neill [11] discussed in
2004 the potential of automated approaches typically based on machine learning and multimedia data analysis methods. They suggested that, if the scientific

© Springer International Publishing AG 2017
G.J.F. Jones et al. (Eds.): CLEF 2017, LNCS 10456, pp. 255–274, 2017.
DOI: 10.1007/978-3-319-65813-1_24

community is able to (i) overcome the production of large training datasets, (ii) more precisely identify and evaluate the error rates, (iii) scale up automated approaches, and (iv) detect novel species, it will then be possible to initiate the development of a generic automated species identification system that could open up vistas of new opportunities for theoretical and applied work in biological and related fields.

Since the question raised in Gaston and O'Neill [11], *automated species identification: why not?*, a lot of work has been done on the topic (*e.g.* [5,24,33,45,46]) and it is still attracting much research today, in particular on deep learning techniques. In parallel to the emergence of automated identification tools, large social networks dedicated to the production, sharing and identification of multimedia biodiversity records have increased in recent years. Some of the most active ones like eBird[1] [41], iNaturalist[2], iSpot [38], Xeno-Canto[3] or Tela Botanica[4] (respectively initiated in the US for the two first ones and in Europe for the three last one), federate tens of thousands of active members, producing hundreds of thousands of observations each year. Noticeably, the Pl@ntNet initiative was the first one attempting to combine the force of social networks with that of automated identification tools [24] through the release of a mobile application and collaborative validation tools. As a proof of their increasing reliability, most of these networks have started to contribute to global initiatives on biodiversity, such as the Global Biodiversity Information Facility (GBIF[5]) which is the largest and most recognized one. Nevertheless, this explicitly shared and validated data is only the tip of the iceberg. The real potential lies in the automatic analysis of the millions of raw observations collected every year through a growing number of devices but for which there is no human validation at all.

The performance of state-of-the-art multimedia analysis and machine learning techniques on such raw data (e.g., mobile search logs, soundscape audio recordings, wild life webcams, etc.) is still not well understood and is far from reaching the requirements of an accurate generic biodiversity monitoring system. Most existing research before LifeCLEF has actually considered only a few dozen or up to hundreds of species, often acquired in well-controlled environments [14,31,36]. On the other hand, the total number of living species on earth is estimated to be around 10 K for birds, 30 K for fish, 400 K for flowering plants (cf. State of the World's Plants 2017[6]) and more than 1.2 M for invertebrates [3]. To bridge this gap, it is required to boost research on large-scale datasets and real-world scenarios.

In order to evaluate the performance of automated identification technologies in a sustainable and repeatable way, the LifeCLEF[7] research platform was cre-

---

[1] http://ebird.org/content/ebird/.
[2] http://www.inaturalist.org/.
[3] http://www.xeno-canto.org/.
[4] http://www.tela-botanica.org/.
[5] http://www.gbif.org/.
[6] https://stateoftheworldsplants.com/.
[7] http://www.lifeclef.org/.

ated in 2014 as a continuation of the plant identification task [25] that was run within the ImageCLEF lab[8] the three years before [13–15]. LifeCLEF enlarged the evaluated challenge by considering birds and marine animals in addition to plants, and audio and video contents in addition to images. In this way, it aims at pushing the boundaries of the state-of-the-art in several research directions at the frontier of information retrieval, machine learning and knowledge engineering including (i) large scale classification, (ii) scene understanding, (iii) weakly-supervised and open-set classification, (iv) transfer learning and fine-grained classification and (v), humanly-assisted or crowdsourcing-based classification. More concretely, the lab is organized around three tasks :

📷 **PlantCLEF**: an image-based plant identification task making use of Pl@ntNet collaborative data, Encyclopedia of Life' data, and Web data

🎵 **BirdCLEF**: an audio recordings-based bird identification task making use of Xeno-canto collaborative data

📠 **SeaCLEF**: a video and image-based identification task dedicated to sea organisms (making use of submarine videos and aerial pictures).

As described in more detail in the following sections, each task is based on big and real-world data and the measured challenges are defined in collaboration with biologists and environmental stakeholders so as to reflect realistic usage scenarios. The main novelties of the 2017th edition of LifeCLEF compared to the previous years are the following:

1. **Scalability:** To fully reach its objective, an evaluation campaign such as LifeCLEF requires a long term research effort so as to (i) encourage non incremental contributions, (ii) measure consistent performance gaps and (iii), progressively scale up the problem. Therefore, the number of species was increased considerably between the 2016-th and the 2017-th edition. The plant task, in particular, made a big jump with 10,000 species instead of 1,000 species in the training set. This makes it one of the largest image classification benchmark.

2. **Noisy + clean data:** The focus of the plant task this year was to study the impact of training identification systems on noisy Web data rather then clean data. Collecting clean data massively is actually prohibitive in terms of human cost whereas noisy Web data can be collected at a very cheap cost. Therefore, we built two large-scale datasets illustrating the same 10 K species: one with clean labels coming from the Web platform Encyclopedia Of Life[9] [47], and one with a high degree of noise - domain noise as well as category noise - crawled from the Web without any filtering.

---

[8] http://www.imageclef.org/.
[9] http://eol.org/.

3. **Time-coded soundscapes:** As the soundscapes data appeared to be very challenging in 2016 (with an accuracy below 15%), we introduced in 2017 new soundscape recordings containing time-coded bird species annotations thanks to the involvement of expert ornithologists. In total, 4,5 h of audio recordings were collected and annotated manually with more than 2000 identified segments.
4. **New organisms and identification scenarios:** The SeaCLEF task was extended with novel scenarios involving new organisms, *i.e* (i) salmons detection for the monitoring of water turbine, and (ii), marine animal species recognition using weakly-labeled images and relevance ranking.

Overall, 130 research groups from around the world registered to at least one task of the lab. Seventeen of them finally crossed the finish line by participating in the collaborative evaluation and by writing technical reports describing in details their evaluated system.

2 Task1: PlantCLEF

The 2017-th edition of PlantCLEF is an important milestone towards working at the scale of continental floras. Thanks to the long term efforts made by the biodiversity informatics, it is actually now possible to aggregate clean data about tens of thousands species world wide. The international initiative Encyclopedia of Life (EoL) in particular is one of the biggest resource of plant pictures. However, the majority of plant species are still very poorly illustrated in such expert databases (or often not illustrated at all). A much larger number of plant pictures are spread on the Web through botanist blogs, plant lovers web-pages, image hosting websites and on-line plant retailers. The LifeCLEF 2017 plant identification challenge proposes to study to what extent a huge but very noisy training set collected through the Web is competitive compared to a relatively smaller but trusted training set checked by experts. As a motivation, a previous study conducted by Krause et al. [29] concluded that training deep neural networks on noisy data was unreasonably effective for fine-grained recognition. The PlantCLEF challenge completes their work in several points:

1. it extends their result to the plant domain. The specificity of the plant domain is that it involves much more species than birds. As a consequence the degree of noise might be much higher due to scarcer available data and higher confusion risks.
2. it scales the comparison between clean and noisy training data to 10 K of species. The clean training sets used in their study were actually limited to few hundreds of species.
3. it uses a third-party test dataset that is not a subset of either the noisy dataset or the clean dataset. More precisely, it is composed of images submitted by the crowd of users of the mobile application Pl@ntNet [23]. Consequently, it exhibits different properties in terms of species distribution, pictures quality, etc.

In the following subsections, we synthesize the resources and assessments of the challenge, summarize the approaches and systems employed by the participating research groups, and provide an analysis of the main outcomes. A more detailed description of the challenge and a deeper analysis of the results can be found in the CEUR-WS proceedings of the task [12].

2.1   Dataset and Evaluation Protocol

To evaluate the above mentioned scenario at a large scale and in realistic conditions, we built and shared three datasets coming from different sources. As training data, in addition to the data of the previous years, we provided two new large data sets both based on the same list of 10,000 plant species (living mainly in Europe and North America):

Trusted Training Set *EoL10K:* a trusted training set based on the online collaborative Encyclopedia Of Life (EoL). The 10 K species were selected as the most populated species in EoL data after a curation pipeline (taxonomic alignment, duplicates removal, herbaria sheets removal, etc.). The training set has a massive class imbalance with a minimum of 1 picture for *Achillea filipendulina* and a maximum of 1245 pictures for *Taraxacum laeticolor.*

Noisy Training Set *Web10K:* a noisy training set built through Web crawlers (Google and Bing image search engines) and containing 1.1M images. This training set is also imbalanced with a minimum of 4 pcitures for *Plectranthus sanguineus* and a maximum of 1732 pictures for *Fagus grandifolia.*
The main idea of providing both datasets is to evaluate to what extent machine learning and computer vision techniques can learn from noisy data compared to trusted data (as usually done in supervised classification). Pictures of EoL are themselves coming from several public databases (such as Wikimedia, Flickr, iNaturalist) or from some institutions or less formal websites dedicated to botany. All the pictures can be potentially revised and rated on the EoL website. On the other side, the noisy training set will contain more images for a lot of species, but with several type and level of noises which are basically impossible to automatically filter: a picture can be associated to the wrong species but the correct genus or family, a picture can be a portrait of a botanist working on the species, the pictures can be associated to the correct species but be a drawing or an herbarium sheet of a dry specimen, etc.

Mobile search test set: the test data to be analyzed within the proposed challenge is a large sample of the query images submitted by the users of the mobile application Pl@ntNet (iPhone[10] & Androïd[11]). It contains covering a large number of wild plant species mostly coming from the Western Europe Flora and the North American Flora, but also plant species used all around the

---

[10] https://itunes.apple.com/fr/app/plantnet/id600547573?mt=8.
[11] https://play.google.com/store/apps/details?id=org.plantnet.

world as cultivated or ornamental plants, or even endangered species precisely because of their non-regulated commerce.

2.2 Participants and Results

80 research groups registered to LifeCLEF plant challenge 2017 and downloaded the dataset. Among this large raw audience, 8 research groups succeeded in submitting *runs*, i.e., files containing the predictions of the system(s) they ran. Details of the methods and systems used in the runs are synthesised in the overview working note of the task [12] and further developed in the individual working notes of the participants (CMP [40], FHDO BCSG [35], KDE TUT [18], Mario MNB [32], Sabanci Gebze [2], UM [34] and UPB HES SO [44]). We report in Fig. 1 the performance achieved by the 29 collected runs. The PlantNet team provides a baseline for the task with the system used in Pl@ntNet app, based on inception model Szegedy [43] and describe in Affouard [1].

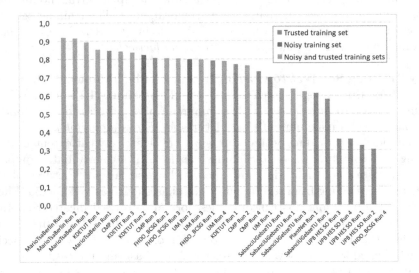

Fig. 1. Performance achieved by all systems evaluated within the plant identification task of LifeCLEF 2017.

Trusted or noisy? As a first noticeable remark, the measured performances are very high despite the difficulty of the task with a median Mean Reciprocal Rank (MRR) around 0.8, and a highest MRR of 0.92 for the best system Mario MNB Run 4. A second important remark is that the best results are obtained mostly by systems that learned on both the trusted and the noisy datasets. Only two runs (KDE TUT Run 2 and UM Run 2) used exclusively the noisy dataset but gave better results than most of the methods using only the trusted dataset. Several teams also tried to filter the noisy dataset, based on the prediction of a preliminary system trained only on the trusted dataset (*i.e.* by rejecting pictures whose label is contradictory with the prediction). However, this strategy did not improve the final predictor and even degraded the results. For instance Mario MNB Run 2 (using the raw Web dataset) performed better than Mario MNB Run 3 (using the filtered Web dataset).

Succeeding strategies with CNN models: Regarding the used methods, all submitted runs were based on Convolutional Neural Networks (CNN) confirming definitively the supremacy of this kind of approach over previous methods. A wide variety of popular architectures were trained from scratch or fine-tuned from pre-trained weights on the ImageNet dataset: GoogLeNet [43] and its improved inception v2 [21] and v4 [42] versions, inception-resnet-v2 [42], ResNet-50 and ResNet-152 [19], ResNeXT [19], VGGNet [39] and even the AlexNet [30]. One can notice that inception v3 was not experimented despite the fact it is a recent model giving state of art performances in other image classification benchmarks. It is important to note that the best results were obtained with ensemble classifiers as for the KDE TUT team who learned and combined predictions from the ResNet-50 and two declinations of this architecture, as the CMP and FHDO BCSG teams with the inception-resnet-v2. Bootstrap aggregating (bagging) was also very efficient to extend the number of classifiers by learning several models with the same architecture but on different training and validation subsets. This is the case of the best run Mario MNB Run 4 for instance where at the end a total of 12 CNNs were learned and combined (7 GoogLeNet, 2 ResNet-152, 3 ResNeXT). CMP team combined also numerous models, a total of 17 models for instance for the CMP Run 1 with various sub-training datasets and bagging strategies, but all with the same inception-resnet-v2 architecture. Another key for succeeding the task was the use of data augmentation with usual transformations such as random cropping, horizontal flipping, rotation, for increasing artificially the number of training samples and helping the CNNs to generalize better. Mario MNB team added two more interesting transformations with slight modifications of the color saturation and lightness, and they correlated the intensity of these transformations with the diminution of the learning rate during training to let the CNNs see patches closer to the original image at the end of each training process. Last but not least, Mario MNB is the only team who extended the test images with similar transformations: each image from a given test observation was augmented with 4 more transformed images.

The most recent model race: We can make more comments about the choice of the CNN architectures: one can suppose that the most recent models such as inception-resnet-v2 or inception-v4 should lead to better results than older ones such as AlexNet, VGGNet and GoogleNet. For instance, the runs with GoogleNet and VGGNet by Sabanci [2], or with a PReLU version of inception-v1 by the PlantNet team, or with the historical AlexNet architecture by the UPB HES SO team [44] performed the worst results. However, one can notice that the "winning" team used also numerous GoogLeNet models, while the old VGGNet used in UM run 2 gave quite high and intermediate results around a MRR of 0.8.

This highlights how much the training strategies are important and how ensemble classifiers, bagging and data augmentation can greatly improve the performance even without the most recent architectures from the state of the art. Besides the use of ensemble of classifiers, some teams also tried to propose modifications of existing models. KDE TUT, in particular, modified the architecture of the first convolutional layers of ResNet-50 and report consistent improvements in their validation experiments [18]. CMP also reported slight improvements on the inception-resnet-v2 by using a maxout activation function instead of RELU. The UM team proposed an original architecture called Hybrid Generic-Organ learned on the trusted dataset (UM Run 1). Unfortunately, it performed worst than a standard VGGNet model learned on the noisy dataset (UM Run 2). This can be partially explained by the fact that the HBO-CNN model need tagged images (flower, fruit, leaf,...), a missing information for the noisy dataset and partially available for the trusted dataset.

The GPU race: Like discussed above, best performances were obtained with ensembles of very deep networks (up to 17 learned models for the CMP team) learned over millions of images produced with data augmentation techniques. In the case of the best run Mario MNB Run 4, test images were also augmented so that the prediction of a single image finally relies on the combination of 60 probability distributions (5 patches × 12 models). Overall, the best performing system requires a huge GPU consumption so that their use in data intensive contexts is limited by cost issues (*e.g.* the Pl@ntNet mobile application accounts for millions of users). A promising solution towards this issue could be to rely on knowledge distilling [20]. Knowledge distilling consists in transferring the generalization ability of a cumbersome model to a small model by using the class probabilities produced by the cumbersome model as soft targets for training the small model. Alternatively, more efficient architectures and learning procedures should be devised.

3   Task2: BirdCLEF

The general public as well as professionals like park rangers, ecological consultants and of course ornithologists are potential users of an automated bird song identifying system. A typical professional use would be in the context of wider initiatives related to ecological surveillance or biodiversity conservation. Using audio records rather than bird pictures is justified [4,5,45,46]since birds are in fact not that easy to photograph and calls and songs have proven to be easier to collect and have been found to be species specific.

The 2017 edition of the task shares similar objectives and scenarios with the previous edition: (i) the identification of a particular bird species from a recording of one of its sounds, and (ii) the recognition of all species vocalising in so-called "soundscapes" that can contain up to several tens of birds vocalising. The first scenario is aimed at developing new automatic and interactive identification tools, to help users and experts to assess species and populations from field recordings obtained with directional microphones. The soundscapes, on the other side, correspond to a much more passive monitoring scenario in which any multi-directional audio recording device could be used without or with very light user's involvement. These (possibly crowdsourced) passive acoustic monitoring scenarios could scale the amount of annotated acoustic biodiversity records by several orders of magnitude.

3.1 Data and Task Description

As the soundscapes appeared to be very challenging in 2015 and 2016 (with an accuracy below 15%), new soundscape recordings containing time-coded bird species annotations were integrated in the test set (so as to better understand what makes state-of-the-art methods fail on such contents). This new data was specifically created for BirdCLEF thanks to the work of three people: Paula Caycedo Rosales (ornithologist from the Biodiversa Foundation of Colombia and Instituto Alexander von Humboldt, Xeno-Canto member), Hervé Glotin (bio-accoustician, co-author of this paper) and Lucio Pando (field guide and ornithologist). In total, about 6,5 h of audio recordings were collected and annotated in the form of time-coded segments with associated species name. This data is composed of two main subsets:

Peru soundscapes, about 2 h (1:57:08) 32 annotated segments: recorded in the summer of 2016 with the support of Amazon Explorama Lodges within the BRILA-SABIOD project. These recordings have been realized in the jungle canopy at 35 m high (the highest point of the area), and at the level of the Amazon river, in the Peruvian basin. The recordings are sampled at 96 kHz, 24 bits PCM, stereo, dual −12 dB, using multiple systems: TASCAM DR, SONY PMC10, Zoom H1.

Colombia soundscapes, about 4,5 h (4:25:55), 1990 annotated segments: These documents were annotated by Paula Caycedo Rosales, ornithologist from the Biodiversa Foundation of Colombia and an active Xeno-Canto member.

In addition to these newly introduced records, the test set still contained the 925 soundscapes and 8,596 single species recordings of BirdCLEF 2016 (collected by the members of Xeno-Canto[12] network, see [16] for more details).

As for the training data, we consistently enriched the training set of the 2016 edition of the task, in particular to cover the species represented in the newly introduced time-coded soundscapes. Therefore, we extended the covered geographical area to the union of Brazil, Colombia, Venezuela, Guyana, Suriname, French Guiana, Bolivia, Ecuador and Peru, and collected all Xeno-Canto records in these countries. We then kept only the 1500 species having the most recordings so as to get sufficient training samples per species (48,843 recordings in total). The training set has a massive class imbalance with a minimum of four recordings for *Laniocera*

---

[12] http://www.xeno-canto.org/contributors.

rufescens and a maximum of 160 recordings for *Henicorhina leucophrys*. Recordings are associated to various metadata such as the type of sound (call, song, alarm, flight, etc.), the date, the location, textual comments of the authors, multilingual common names and collaborative quality ratings.

Participants were asked to run their system so as to identify all the actively vocalising birds species in each test recording (or in each test segment of 5 s for the soundscapes). The submission *run files* had to contain as many lines as the total number of identifications, with a maximum of 100 identifications per recording or per test segment). Each prediction had to be composed of a species name belonging to the training set and a normalized score in the range $[0, 1]$ reflecting the likelihood that this species is singing in the segment. The used evaluation metric used was the Mean Average Precision. Up to 4 *run files* per participant could be submitted to allow evaluating different systems or system configurations.

3.2 Participants and Results

78 research groups registered for the BirdCLEF 2017 challenge and downloaded the data. Only 5 of them finally submitted run files and technical reports. Details of the systems and the methods used in the runs are synthesized in the overview working note of the task [17] and further developed in the individual working notes of the participants ([9,10,28,37]). Below we give more details about the 3 systems that performed the best runs.

DYNI UTLN system (Soundception) [37]: This system is based on an adaptation of the image classification model Inception V4 [42] extended with a time-frequency attention mechanism. The main steps of the processing pipeline are (i) the construction of multi-scaled time-frequency representations to be passed as RGB images to the Inception model, (ii) data augmentation (random hue, contrast, brightness, saturation, random crop in time and frequency domain) and (iii) the training phase relying on transfer learning from the initial weights of the Inception V4 model (learned in the visual domain using the ImageNet dataset).

TUCMI system [28]: This system is also based on convolutional neural networks (CNN) but using more classical architectures than the Inception model used by DYNI UTLN. The main steps of the processing pipeline are (i) the construction of magnitude spectrograms with a resolution of 512×256 pixels, which represent five-second chunks of audio signal, (ii) data augmentation (vertical roll, Gaussian noise, Batch Augmentation) and (iii) the training phase relying on either a classical categorical loss with a softmax activation (TUCMI Run 1), or on a set of binary cross entropy losses with sigmoid activations as an attempt to better handle the multi-labeling scenario of the soundscapes (TUCMI Run 2). TUCMI Run 3 is an ensemble of 7 CNN models including the ones of Run 1 and Run 2. TUCMI Run 4 was an attempt to use geo-coordinates and time as a way to reduce the list of species to be recognized in the soundscapes recordings.

Therefore, the occurrences of the eBird initiative were used complementary to the data provided within BirdCLEF. More precisely, only the 100 species having the most occurrences in the Loreto/Peru area for the months of June, July and August were kept in the training set.

Cynapse system [9]**:** This system is based on a multi-modal deep neural network taking audio samples and metadata as input. The audio is fed into a convolutional neural network using four convolutional layers. The additionally provided metadata is processed using fully connected layers. The flattened convolutional layers and the fully connected layer of the metadata were joined and put into a large dense layer. For the sound pre-processing and data augmentation, they used a similar pipeline as the best system of BirdCLEF 2016 described in [8]. The two runs Cynapse Run 2 and 3 mainly differ in the FFT window size used for constructing the time-frequency representation passed as input to the CNN (respectively 512 and 256). Cynapse Run 4 is an average of Cynapse Run 2 and 3.

Figure 2 reports the performance measured for the 18 submitted runs. For each run (*i.e.* each evaluated system), we report the Mean Average Precision for the three categories of queries: traditional mono-directional recordings (the same as the one used in 2016), non time-coded soundscape recordings (the same as the one used in 2016) and the newly introduced time-coded soundscape recordings. To measure the progress over last year, we also plot on the graph the performance of last year's best system [8].

It is remarked that all submitted runs were based on Convolutional Neural Networks (CNN) confirming the supremacy of this approach over previous methods (in particular the ones based on hand-crafted features which were performing the best until 2015). The best MAP of 0.71 (for the single species recordings) was achieved by the best system configuration of DYNI UTLN (Run 1). That rather similar to the MAP of 0.68 achieved last year by [8] but with 50% more species in the training set. Regarding the newly introduced time-coded soundscapes, the best system was also the one of DYNI UTLN (Run 1) whereas it did not introduce any specific features towards solving the multi-labeling issue. The main conclusions we can draw from the results are the following:

The network architecture plays a crucial role: Inception V4 that was known to be the state of the art in computer vision [42] also performed the best within the BirdCLEF 2017 challenge that is much different (time-frequency representations instead of images, a very imbalanced training set, mono- and multi-labeling scenarios, etc.). This shows that its architecture is intrinsically well-suited for a variety of machine-learning tasks across different domains.

The use of ensembles of networks improves the performance consistently: This can be seen through Cynapse Run 4 and TUCMI Run 3 that outperform the other respective runs of these participants.

The use of a multi-labeling training: The use of the binary cross-entropy losses in TUCMI Run 2 did allow a slight performance gain compared to the classical softmax loss in TUCMI Run 1. Unfortunately, this was not enough to compensate the gains due to other factors in DYNI UTLN Run 1 (in particular the network architecture).

The use of metadata was not successful: The attempt of Cynapse did not allow a sufficient performance improvement to compensate the gains due to other factors (in particular the network architecture). TUCMI Run 4, working on the restricted list of the 100 most likely species according to eBird data, did not outperform the other TUCMI Runs on the soundscape test data.

4   Task3: SeaCLEF

The SeaCLEF 2017 task originates from the previous editions (2014 and 2015, 2016) of marine organism identification in visual data for ecological surveillance and biodiversity monitoring. SeaCLEF 2017 significantly extends past editions in the tackled marine organisms species as well in the application tasks. The need of automated methods for sea-related multimedia data and to extend the originally tasks is driven by the recent sprout of marine and ocean observation approaches (mainly imaging - including thermal - systems) and their employment for marine ecosystem analysis and biodiversity monitoring. Indeed in recent years we have assisted an exponential growth of sea-related multimedia data in the forms of images/videos/sounds, for disparate reasoning ranging from fish

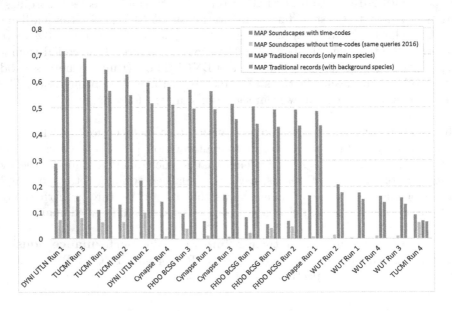

Fig. 2. BirdCLEF 2017 results overview - Mean Average Precision.

biodiversity monitoring, to marine resource managements, to fishery, to educational purposes. However, the analysis of such data is particularly expensive for human operators, thus limiting greatly the impact that the technology may have in understanding and sustainably exploiting the sea/ocean.

4.1   Data and Task Description

The SeaCLEF 2017 challenge was composed of four subtasks and related datasets:

Subtask 1 - Automated Fish Identification and Species Recognition on Coral Reef Videos: The participants have access to a training set consisting of twenty underwater videos in which bounding boxes and fish species labels are provided. Then, providing a test set of 20 videos, the goal of the task is to automatically detect and recognize fish species. The evaluation metrics are the precision, the counting score (CS), and the normalized counting score (NCS). More information about these metrics and the whole set up of the subtask can be found in the LifeCLEF 2016 overview [26].

Subtask 2 - Automated Frame-level Salmon Identification in Videos for Monitoring Water Turbine: The participants have access to a training set consisting of eight underwater videos with frame-level annotations indicating the presence of salmons. Then, providing a test set of 8 videos, the goal of the task is to identify in which frames salmon appear. Such events are pretty rare and salmons are often very small, thus the task mainly pertains detection of rare events involving unclear objects (salmons).

Subtask 3 - Marine Animal Species Recognition using Weakly-Labelled Images and Relevance Ranking: Contrary to the previous subtasks, this one aims at classifying marine animals from 2D images. The main difficulties of the task are: (1) high similarity between species and (2) weak annotations, for training, gathered automatically from the Web and filtered by non-experts. In particular, the training dataset consists of up to 100 images for each considered species (in total 148 fish species). Training images are weakly labelled, i.e., Web images have been retrieved automatically from the Web using marine animal scientific names as query. The retrieved images were then filtered by non-experts who were instructed to only remove images not showing fish/marine animals. Furthermore, the relevance ranking to the query is provided for each crawled image and can be used during training.

Subtask 4 - Whale Individual Recognition: This subtask aims at automatically matching image pairs, over a large set of images, of same individual whales through the analysis of their caudal fins. Indeed, the caudal fin is the most discriminant pattern for distinguishing an individual whale from another. Finding the images that correspond to the same individual whale is a crucial step for further biological analysis (e.g. for monitoring population displacement) and it is currently done manually by human operators (hence a painful, error prone and unscalable process usually known as *photo-identification*). The metric used to evaluate each run is the Average Precision. More information about this metric and the whole set up of the subtask can be found in the LifeCLEF 2016 overview [26].

4.2   Participants and Results

Over 40 research teams registered and downloaded data for the SeaCLEF 2017 challenge. Only seven of the registered participants submitted runs. No one team participated to all the four challenge subtasks and only one team (SIATMMLAB) submitted runs for more than one task (subtask 1, 2 and 3).

Subtask 1 results: Fig. 3 displays the results obtained by the 2 participating groups who submitted a total of 4 runs. Details of the methods of the team SIATMMLAB can be found in [48]. Unfortunately, none of them was able to outperform the baseline method that obtained the best results in LifeCLEF 2015 (by SNUMED [6]).

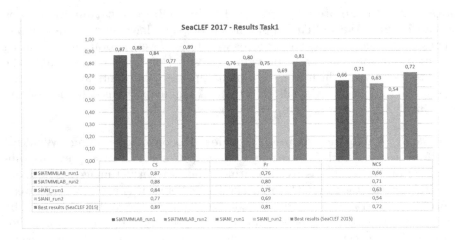

Fig. 3. Results of SeaCLEF subtask 1 - Automated Fish Identification and Species Recognition on Coral Reef Videos.

Subtask 2 results: For this task, only one participant (SIATMMLAB [48]) submitted only one run, achieving the following results: Precision = 0.04, Recall = 0.82 and F-measure = 0.07. The low performance, especially due to false positives, demonstrates the complexity of this task.

Subtask 3 results: Fig. 4 reports the performance in terms of P@1, P@3 and P@5 of the two participant teams (HITSZ and SIATMMLAB) that submitted three runs each. Details of the methods used by the SIATMMLAB can be found in [48].

Subtask 4 results: Results achieved by the three groups who participated to subtask 4 are summarized in Table 1[13]. A first conclusion is that, as last year, using a RANSAC-like spatial consistency checking step is crucial for reaching good performance. The spatial arrangement of the local features is actually a precious information for rejecting the masses of mismatches obtained by simply matching the low level (SIFT) features. The two other elements explaining the big performance gap achieved by BME_DCLab_Run3 was (i) the use of a preliminary segmentation to separate the fin from the background and (ii), the use of a clustering algorithm on top of the image matching graph (to recover many pairs that were missed by the matching process but that can be infer by transitivity).

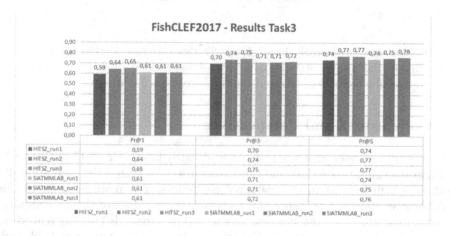

Fig. 4. Results of SeaCLEF subtask 3 - Marine Animal Species Recognition using Weakly-Labelled Images and Relevance Ranking.

For all the first three subtasks, Convolutional Neural Networks based on either Inception or Res-Net were employed. The good performance achieved by CNNs over the subtasks 1 and 3 combined to the low one in subtask 3 confirm, what is already known in the literature, that CNNs work best with large data and that, instead, show limitations in case of detection of rare instances (despite no one of the employed methods discussed any data augmentation technique). Another observation is related to the fact that all methods exploit only visual cues to perform the tasks, which is expected given the nature of the released data. To overcome this limitation, this year we added in a task (subtask 3) another dimension related to image search ranking to support image classification using multimedia data, which, however, was not used by the participants. Finally, the

---

[13] We precise that there was probably a bug in the runfile MLRG_Run2 that performed abnormally low with regard to the used technique.

Table 1. Individual whale identification results (SeaCLEF subtask 4).

Run name	AP	Method description	Paper
BME_DCLab_Run3	0.51	SIFT features matching + RANSAC spatial consistency filtering + random walks on the matches to discover clusters	[7]
ZenithINRIA_Run1	0.39	SIFT features matching (through multi-probe hashing) + RANSAC-like spatial consistency filtering	[27]
BME_DCLab_Run2	0.30	SIFT features matching + re-ranking	[7]
BME_DCLab_Run1	0.30	SIFT features matching	[7]
MLRG_Run2	0.01	Preprocessing using Grabcut Segmentation and Clustering Techniques + SIFT features matching (through FLANN) + re-ranking	[22]

low performance in subtask 3 and 4 clearly demonstrate that some problems, especially in the underwater domain, cannot be tackled merely by brute force learning (or fine-tuning) of low and middle-level visual features.

5   Conclusions and Perspectives

With about 130 research groups who downloaded LifeCLEF 2017 data and 18 of them who submitted runs, the third edition of the LifeCLEF evaluation did confirm a high interest in the evaluated challenges. The main outcome of this collaborative effort is a snapshot of the performance of state-of-the-art computer vision, bio-acoustic and machine learning techniques towards building real-world biodiversity monitoring systems. The results did show that very high identification rates can be reached by the evaluated systems, even on large number of species (up to 10,000 species). The most noticeable progress came from the deployment of new convolutional neural network architectures, confirming the fast growing progress of that techniques. Interestingly, the best performing system on the bird sounds recognition task was based on an the architecture of the image-based CNN Google model (Inception V4). This shows the convergence of the best performing technique whatever the targeted domain. Another important outcome was about the use of noisy Web data to train such deep learning models. The plant task confirmed that doing so allows achieving very good performance, even better than the one obtained with validated data. The combination of both noisy and clean data enabled even better performance gains, up to an amazing accuracy of 92% for the 10 K species plant challenge. Despite these impressive results, there is still a large room of improvements for several of the evaluated challenges including: (i) the soundscapes for birds monitoring, (ii) the underwater imagery for fish monitoring, and (iii) the photo-identification of whale individuals.

Acknowledgements. The organization of the PlantCLEF task is supported by the French project Floris'Tic (Tela Botanica, INRIA, CIRAD, INRA, IRD) funded in the context of the national investment program PIA. The organization of the BirdCLEF task is supported by the Xeno-Canto foundation for nature sounds as well as the French CNRS project SABIOD.ORG and EADM MADICS, and Floris'Tic. The annotations of some soundscape were prepared with regreted wonderful Lucio Pando at Explorama Lodges, with the support of Pam Bucur, Marie Trone and H. Glotin. The organization of the SeaCLEF task is supported by the Ceta-mada NGO and the French project Floris'Tic.

References

1. Affouard, A., Goeau, H., Bonnet, P., Lombardo, J.C., Joly, A.: Pl@ntnet app. in the era of deep learning. In: 5th International Conference on Learning Representations (ICLR 2017), 24–26 April 2017, Toulon, France (2017)
2. Atito, S., Yanikoglu, B., Aptoula, E.: Plant identification with large number of classes: Sabanciu-gebzetu system in plantclef 2017. In: Working Notes of CLEF 2017 (Cross Language Evaluation Forum) (2017)
3. Baillie, J., Hilton-Taylor, C., Stuart, S.N.: 2004 IUCN red list of threatened species: a global species assessment. Iucn (2004)
4. Briggs, F., Lakshminarayanan, B., Neal, L., Fern, X.Z., Raich, R., Hadley, S.J., Hadley, A.S., Betts, M.G.: Acoustic classification of multiple simultaneous bird species: A multi-instance multi-label approach. J. Acoust. Soc. Am. **131**, 4640 (2012)
5. Cai, J., Ee, D., Pham, B., Roe, P., Zhang, J.: Sensor network for the monitoring of ecosystem: Bird species recognition. In: 3rd International Conference on Intelligent Sensors, Sensor Networks and Information, ISSNIP 2007 (2007)
6. Choi, S.: Fish identification in underwater video with deep convolutional neural network: snumedinfo at lifeclef fish task 2015. In: Working Notes of CLEF 2015 (Cross Language Evaluation Forum) (2015)
7. Papp, D., Mogyorósi, F., Szücs, G.: Image matching for individual recognition with SIFT, RANSAC and MCL. In: Working Notes of CLEF 2017 (Cross Language Evaluation Forum) (2017)
8. Sprengel, E., Martin Jaggi, Y.K., Hofmann, T.: Audio based bird species identification using deep learning techniques. In: Working Notes of CLEF 2016 (Cross Language Evaluation Forum) (2016)
9. Fazekas, B., Schindler, A., Lidy, T.: A multi-modal deep neural network approach to bird-song identication. In: Working Notes of CLEF 2017 (Cross Language Evaluation Forum) (2017)
10. Fritzler, A., Koitka, S., Friedrich, C.M.: Recognizing bird species in audio files using transfer learning. In: Working Notes of CLEF 2017 (Cross Language Evaluation Forum) (2017)
11. Gaston, K.J., O'Neill, M.A.: Automated species identification: why not? Philos. Trans. R. Soc. Lond. B Biol. Sci. **359**(1444), 655–667 (2004)
12. Goëau, H., Bonnet, P., Joly, A.: Plant identification based on noisy web data: the amazing performance of deep learning (lifeclef 2017). In: Working Notes of CLEF 2017 (Cross Language Evaluation Forum) (2017)
13. Goëau, H., Bonnet, P., Joly, A., Bakic, V., Barthélémy, D., Boujemaa, N., Molino, J.F.: The imageclef 2013 plant identification task. In: CLEF 2013, Valencia (2013)

14. Goëau, H., Bonnet, P., Joly, A., Boujemaa, N., Barthélémy, D., Molino, J.F., Birnbaum, P., Mouysset, E., Picard, M.: The imageclef 2011 plant images classification task. In: CLEF 2011 (2011)

15. Goëau, H., Bonnet, P., Joly, A., Yahiaoui, I., Barthélémy, D., Boujemaa, N., Molino, J.F.: Imageclef 2012 plant images identification task. In: CLEF 2012, Rome (2012)

16. Goëau, H., Glotin, H., Planqué, R., Vellinga, W.P., Joly, A.: Lifeclef bird identification task 2016. In: CLEF 2016 (2016)

17. Goëau, H., Glotin, H., Planqué, R., Vellinga, W.P., Joly, A.: Lifeclef bird identification task 2017. In: CLEF 2017 (2017)

18. Hang, S.T., Aono, M.: Residual network with delayed max pooling for very large scale plant identification. In: Working Notes of CLEF 2017 (Cross Language Evaluation Forum) (2017)

19. He, K., Zhang, X., Ren, S., Sun, J.: Deep residual learning for image recognition. In: Proceedings of the IEEE Conference on Computer Vision and Pattern Recognition, pp. 770–778 (2016)

20. Hinton, G., Vinyals, O., Dean, J.: Distilling the knowledge in a neural network. arXiv preprint arxiv:1503.02531 (2015)

21. Ioffe, S., Szegedy, C.: Batch normalization: Accelerating deep network training by reducing internal covariate shift. CoRR abs/1502.03167 (2015). http://arxiv.org/abs/1502.03167

22. Jaisakthi, S., Mirunalini, P., Jadhav, R.: Automatic whale matching system using feature descriptor. In: Working Notes of CLEF 2017 (Cross Language Evaluation Forum) (2017)

23. Joly, A., Bonnet, P., Goëau, H., Barbe, J., Selmi, S., Champ, J., Dufour-Kowalski, S., Affouard, A., Carré, J., Molino, J.F., et al.: A look inside the pl@ntnet experience. Multimedia Syst. 22(6), 751–766 (2016)

24. Joly, A., Goëau, H., Bonnet, P., Bakić, V., Barbe, J., Selmi, S., Yahiaoui, I., Carré, J., Mouysset, E., Molino, J.F., et al.: Interactive plant identification based on social image data. Ecol. Inform. 23, 22–34 (2014)

25. Joly, A., Goëau, H., Bonnet, P., Bakic, V., Molino, J.F., Barthélémy, D., Boujemaa, N.: The imageclef plant identification task 2013. In: International Workshop on Multimedia Analysis for Ecological Data (2013)

26. Joly, A., Goëau, H., Glotin, H., Spampinato, C., Bonnet, P., Vellinga, W.P., Champ, J., Planqué, R., Palazzo, S., Müller, H.: Lifeclef 2016: multimedia life species identification challenges. In: International Conference of the Cross-Language Evaluation Forum for European Languages 2016 (2016)

27. Joly, A., Lombardo, J.C., Champ, J., Saloma, A.: Unsupervised individual whales identification: spot the difference in the ocean. In: Working Notes of CLEF 2016 (Cross Language Evaluation Forum) (2016)

28. Kahl, S., Wilhelm-Stein, T., Hussein, H., Klinck, H., Kowerko, D., Ritter, M., Eibl, M.: Large-scale bird sound classification using convolutional neural networks. In: CLEF 2017 (2017)

29. Krause, J., Sapp, B., Howard, A., Zhou, H., Toshev, A., Duerig, T., Philbin, J., Fei-Fei, L.: The unreasonable effectiveness of noisy data for fine-grained recognition. In: Leibe, B., Matas, J., Sebe, N., Welling, M. (eds.) ECCV 2016. LNCS, vol. 9907, pp. 301–320. Springer, Cham (2016). doi:10.1007/978-3-319-46487-9_19

30. Krizhevsky, A., Sutskever, I., Hinton, G.E.: Imagenet classification with deep convolutional neural networks. In: Advances in Neural Information Processing Systems, pp. 1097–1105 (2012)

31. Kumar, N., Belhumeur, P.N., Biswas, A., Jacobs, D.W., Kress, W.J., Lopez, I.C., Soares, J.V.B.: Leafsnap: a computer vision system for automatic plant species identification. In: Fitzgibbon, A., Lazebnik, S., Perona, P., Sato, Y., Schmid, C. (eds.) ECCV 2012. LNCS, pp. 502–516. Springer, Heidelberg (2012). doi:10.1007/978-3-642-33709-3_36

32. Lasseck, M.: Image-based plant species identification with deep convolutional neural networks. In: Working Notes of CLEF 2017 (Cross Language Evaluation Forum) (2017)

33. Lee, D.J., Schoenberger, R.B., Shiozawa, D., Xu, X., Zhan, P.: Contour matching for a fish recognition and migration-monitoring system. In: Optics East, pp. 37–48. International Society for Optics and Photonics (2004)

34. Lee, S.H., Chang, Y.L., Chan, C.S.: Lifeclef 2017 plant identification challenge: classifying plants using generic-organ correlation features. In: Working Notes of CLEF 2017 (Cross Language Evaluation Forum) (2017)

35. Ludwig, A.R., Piorek, H., Kelch, A.H., Rex, D., Koitka, S., Friedrich, C.M.: Improving model performance for plant image classification with filtered noisy images. In: Working Notes of CLEF 2017 (Cross Language Evaluation Forum) (2017)

36. Nilsback, M.E., Zisserman, A.: Automated flower classification over a large number of classes. In: Proceedings of the Indian Conference on Computer Vision, Graphics and Image Processing, December 2008

37. Sevilla, A., Glotin, H.: Audio bird classification with inception v4 joint to an attention mechanism. In: Working Notes of CLEF 2017 (Cross Language Evaluation Forum) (2017)

38. Silvertown, J., Harvey, M., Greenwood, R., Dodd, M., Rosewell, J., Rebelo, T., Ansine, J., McConway, K.: Crowdsourcing the identification of organisms: a case-study of ispot. ZooKeys **480**, 125 (2015)

39. Simonyan, K., Zisserman, A.: Very deep convolutional networks for large-scale image recognition. CoRR abs/1409.1556 (2014)

40. Šulc, M., Matas, J.: Learning with noisy and trusted labels for fine-grained plant recognition. In: Working Notes of CLEF 2017 (Cross Language Evaluation Forum) (2017)

41. Sullivan, B.L., Aycrigg, J.L., Barry, J.H., Bonney, R.E., Bruns, N., Cooper, C.B., Damoulas, T., Dhondt, A.A., Dietterich, T., Farnsworth, A., et al.: The ebird enterprise: an integrated approach to development and application of citizen science. Biol. Conserv. **169**, 31–40 (2014)

42. Szegedy, C., Ioffe, S., Vanhoucke, V., Alemi, A.: Inception-v4, inception-resnet and the impact of residual connections on learning. arXiv preprint arXiv:1602.07261 (2016)

43. Szegedy, C., Liu, W., Jia, Y., Sermanet, P., Reed, S., Anguelov, D., Erhan, D., Vanhoucke, V., Rabinovich, A.: Going deeper with convolutions. In: Proceedings of the IEEE Conference on Computer Vision and Pattern Recognition, pp. 1–9 (2015)

44. Toma, A., Stefan, L.D., Ionescu, B.: Upb hes so @ plantclef 2017: automatic plant image identification using transfer learning via convolutional neural networks. In: Working Notes of CLEF 2017 (Cross Language Evaluation Forum) (2017)

45. Towsey, M., Planitz, B., Nantes, A., Wimmer, J., Roe, P.: A toolbox for animal call recognition. Bioacoustics **21**(2), 107–125 (2012)

46. Trifa, V.M., Kirschel, A.N., Taylor, C.E., Vallejo, E.E.: Automated species recognition of antbirds in a mexican rainforest using hidden markov models. J. Acoust. Soc. Am. **123**, 2424 (2008)

47. Wilson, E.O.: The encyclopedia of life. Trends Ecol. Evol. **18**(2), 77–80 (2003)
48. Zhuang, P., Xing, L., Liu, Y., Guo, S., Qiao, Y.: Marine animal detection and recognition with advanced deep learning models. In: Working Notes of CLEF 2017 (Cross Language Evaluation Forum) (2017)

Overview of PAN'17
Author Identification, Author Profiling, and Author Obfuscation

Martin Potthast[1](✉), Francisco Rangel[2,5], Michael Tschuggnall[3],
Efstathios Stamatatos[4], Paolo Rosso[5], and Benno Stein[1]

[1] Web Technology and Information Systems,
Bauhaus-Universität Weimar, Weimar, Germany
pan@webis.de
[2] Autoritas Consulting, S.A., Valencia, Spain
[3] Department of Computer Science, University of Innsbruck, Innsbruck, Austria
[4] Department of Information and Communication Systems Engineering,
University of the Aegean, Samos, Greece
[5] PRHLT Research Center, Universitat Politècnica de València, Valencia, Spain
http://pan.webis.de

Abstract. The PAN 2017 shared tasks on digital text forensics were
held in conjunction with the annual CLEF conference. This paper gives
a high-level overview of each of the three shared tasks organized this
year, namely author identification, author profiling, and author obfuscation. For each task, we give a brief summary of the evaluation data,
performance measures, and results obtained. Altogether, 29 participants
submitted a total of 33 pieces of software for evaluation, whereas 4 participants submitted to more than one task. All submitted software has
been deployed to the TIRA evaluation platform, where it remains hosted
for reproducibility purposes.

1 Introduction

Digital text forensics is a key area for the application of technologies that analyze
writing style. For decades, scientists with various backgrounds, ranging from
linguistics over natural language processing to computer security have conducted
research to quantify and reliably analyze the writing style of a given text. A
general goal is to find a kind of "style fingerprint", which would render its author
personally identifiable when other pieces of writing known to be written by
the same author are at hand. Collectively termed author identification, several
subordinate tasks have been identified and extensively studied under this goal.
Besides identifying authors, also the question came up whether authors who
share a personal trait also expose this fact via shared writing style characteristics.
If true, the author of a text of unknown authorship may still be identified via
circumstantial evidence, by narrowing down the list of candidates to those whose
profiles match personal traits predicted for the author of the text in question.

© Springer International Publishing AG 2017
G.J.F. Jones et al. (Eds.): CLEF 2017, LNCS 10456, pp. 275–290, 2017.
DOI: 10.1007/978-3-319-65813-1_25

Predicting one or many of the traits of a text's author is hence called author profiling.

However, despite decades of research the problem of reliably extracting writing style fingerprints from text is only partially solved, and, the problem of identifying style markers that are reliably correlated with personal traits is even far from being solved. At the same time, recent advances in automatic text generation based on deep neural networks have opened the door to mixtures of human-generated and machine-generated texts, rendering future writing style analyses more difficult. In this regard, also the vulnerability of writing style analysis to targeted attacks is being investigated, since text synthesis technology may be applied to alter the writing style of a text in such a way that its author cannot be reliably identified anymore, or, similarly, that wrong personal traits are predicted from it. Any systematic attempt to alter a text in such a way is called author obfuscation. Note that the question whether author identification and profiling dominate author obfuscation or vice versa is open.

At PAN, we have been addressing all of these tasks head-on—in particular, by organizing shared tasks for each of them for the past couple of years. While the specific variants of the tasks in question have changed significantly throughout the years, the underlying goal of getting to the "principles" of writing style technologies and their application remains the same. In the 2017 edition of PAN, we focus on (1) author clustering and style break detection, two tasks that belong to author identification, (2) gender and native language prediction, which belong to author profiling, and (3) author masking as a specific case of author obfuscation. Participation and interest from scientists worldwide has been strong throughout the years, and again a total of 33 teams participated in the current edition. In what follows, we briefly review each shared task and the achieved results.

2   Author Identification

In certain authorship analysis tasks, a given document could be written by multiple authors. In such cases, it is necessary to decompose the document into authorial components [14]. For instance, in intrinsic plagiarism detection, the main part of the document is assumed to be by the alleged author, while the rest of the document has been taken by other authors. In author diarization, several authors collaborated to write a document and the contribution of each one of them should be detected [30]. Tackling such a problem requires to master two basic sub-problems: to properly segment a document into stylistically homogeneous parts, and, to group these parts by authorship. The current edition of PAN focuses on exactly these two tasks, here called *style breach detection* and *author clustering*.

2.1   Style Breach Detection

The style breach detection task at PAN 2017 attaches to a series of subtasks of previous PAN events that focused on intrinsic characteristics of text documents.

Including tasks like intrinsic plagiarism detection [17] or author diarization [29], one commonality is that the style of authors has to be extracted and quantified in some way in order to tackle the specific problem types. In a similar way, an intrinsic analysis of the writing style is also key to approach the PAN 2017 style breach detection task, which can be summarized as follows: Given a document, determine whether it is multi-authored, and, if in-fact it is multi-authored, find the borders where authors switch.

From a different perspective, the detection of style breaches, i.e., locating borders between different authors, can be seen as a special case of a general text segmentation problem. However, a crucial difference to existing segmentation approaches is the following: While the latter focus on detecting switches of topics or stories (e.g., [11,15,27]), the aim of style breach detection is to identify borders based on style, disregarding the content. In contrast to the author clustering task described in Sect. 2.2, the goal is to only find borders; it is irrelevant to identify or cluster authors of segments.

Evaluation Datasets. To evaluate the approaches, distinct training and test data sets have been provided, which are based on the Webis-TRC-12 data set [20]. The original corpus contains documents on 150 topics used at the TREC Web Tracks from 2009–2011 [4], whereby professional writers were hired and asked to search for a given topic and to compose a single document from the search results. From these documents, the respective data sets have been generated randomly by varying several configurations:

- number of borders (0–8, 0 for single-author documents, i.e., containing no style breaches)
- number of collaborating authors (1–5)
- average segment length ($\sim$ 30–2500 words)
- document length ($\sim$ 200–6000 words)
- allow borders either only at the end or within paragraphs
- either uniformly or randomly distribute borders with respect to segment lengths

Parts of the original corpus have already been used and published, and we ensured that the test documents have been created from previously unpublished documents only. Overall, the number of documents in the training data set is 187, whereas the test data set contains 99 documents.

Performance Measures. The performance of the submitted algorithms have been measured with two common metrics used in the field of text segmentation. The *WindowDiff* metric [16] proposed for general text segmentation evaluation is computed, because it is widely used for similar problems. It calculates an error rate between 0 and 1 for predicting borders (0 indicates a perfect prediction), by penalizing near-misses less than other/complete misses or extra borders. Depending on the problem types and data sets used, text segmentation approaches report near-perfect windowDiff values of less than 0.01, while on the other side, the error rate exceeds values of 0.6 and higher under certain circumstances [6]. A more recent adaption of the WindowDiff metric is the *WinPR* metric [28]. It enhances WindowDiff by computing the common information retrieval measures precision (WinP) and recall (WinR) and thus allows to give a more detailed, qualitative statement about the prediction. Internally, WinP and WinR are computed based on the calculation of true and false positives/negatives of border positions, respectively. It also assigns higher scores, if predicted borders are closer to the real border position.

Both metrics have been computed on word-level, whereby the participants were asked to provide character positions (i.e., the tokenization was delegated to the evaluator script). For the final ranking of all participating teams, the F-score of WinPR (WinF) is employed.

Results. This year, five teams participated in the style breach detection task, whereas three of them submitted their software to TIRA [7,18]. An overview in the nutshell: Karaś, Śpiewak & Sobecki use bags of 3-grams in combination with other common stylometric features that serve as input for a statistical test that determines whether or not two consecutive paragraphs are similar in style. Khan uses feature vectors based on word lists and other basic lexical metrics; borders are then detected by applying a distance function on sliding windows. Finally, Safin & Kuznetsova compute high-dimensional vectors for sentences, using a skip-thought-model [12], which can be seen as a word embedding technique operating on sentences as atomic units. The vectors then serve as input for an outlier detection analysis, similar to identifying suspicious sentences in intrinsic plagiarism detection.

To be able to compare the results, two simple baselines have been computed: *BASELINE-rnd* randomly places 0–10 borders on arbitrary positions inside a document. As a variant, *BASELINE-eq* also decides on a random basis how many borders should be placed (also 0–10), but then places the borders uniformly, i.e., such that all resulting segments are of equal size with respect to tokens contained. Both baselines have been computed based on the average of 100 runs.

The final results of the three submitting teams are presented in Table 1, which shows the average value of each computed measure (note that by doing so WinF is not the result of computing the F-score on the presented WinP

Table 1. Style breach detection results. Participants are ranked according to their WinF score.

Rank	Participant	WinP	WinR	WinF	WindowDiff
1	Karaś, Śpiewak & Sobecki	0.315	0.586	**0.323**	0.546
–	BASELINE-eq	0.337	**0.645**	0.289	0.647
2	Khan	**0.399**	0.487	0.289	**0.480**
3	Safin & Kuznetsova	0.371	0.543	0.277	0.529
–	BASELINE-rnd	0.302	0.534	0.236	0.598

and WinR values, but rather the average of the individual WinF scores). Karaś et al. could surpass the baseline equalizing the segment sizes in case of WinF, whereas the baseline using completely random positions could be exceeded by all participants. In comparison to WindowDiff, all approaches perform better than both baselines, whereby Khan achieves the best result. Finally, fine-grained sub performances depending on the data set configuration, e.g., the number of borders, are presented in the respective overview paper of this task [31].

2.2   Author Clustering

Given a small set of short (paragraph-length) documents D, the task is to group them by authorship. More specifically, we adopt the following two scenarios:

- *Complete clustering.* Each document $d \in D$ has to be assigned to exactly one of k clusters, where each cluster corresponds to a distinct author; k is not given.
- *Authorship-link ranking.* Pairs of documents by the same author (authorship-links), $(d_i, d_j) \in D \times D$, have to be extracted and ranked in decreasing order of confidence (a score belonging to [0,1]).

All documents within a clustering problem are single-authored, written in the same language, and belong to the same genre. However, topic and text-length may vary. The main difference when compared to the corresponding PAN-2016 task is that the documents are short, including a few sentences only. This makes the task harder since text-length is crucial when attempting to extract stylometric information.

Evaluation Datasets. The datasets used for training and evaluation were extracted from the corresponding PAN-2016 corpora [30]. They include clustering problems in three languages (English, Dutch, and Greek) and two genres (articles and reviews). Each PAN-2016 text was segmented into paragraphs; all paragraphs with less than 100 characters or more than 500 characters were discarded. In each clustering problem, documents by the same authors were selected randomly from all original documents. This means that paragraphs of the same original document or other documents (by the same author) may be grouped. The only exception in this process was the Dutch reviews corpus since its texts were already short (one paragraph each). For this special case, the PAN-2017 datasets were built using the PAN-2016 procedure.

Table 2 shows the details about the training and test datasets. Most of the clustering problems include 20 documents (paragraphs) by an average of 6 authors. In each clustering problem there is an average of about 50 authorship links, whereas the largest cluster contains about 8 documents. Each document has an average of about 50 words. Note that in the case of Dutch reviews these figures deviate from the norm (documents are longer and authorship links are less).

An important factor to each clustering problem is the *clusteriness ratio* $r = k/N$, where N is the size of D. When r is high, most documents belong to single-item clusters, and there are only few authorship links. When r is low, most documents belong to multi-item clusters, and there are plenty of authorship links. In this new corpus r ranges between 0.1 and 0.5 in both training and test datasets, in contrast to the PAN-2016 corpus where $r \geq 0.5$ [30].

Table 2. The author clustering corpus. Average clusteriness ratio (r), number of documents (N), number of authors (k), number of authorship links, maximum cluster size (maxC), and words per document are given.

	Language	Genre	Problems	r	N	k	Links	maxC	Words
Training	English	articles	10	0.3	20	5.6	57.3	9.2	52.6
	English	reviews	10	0.3	19.4	6.1	45.4	8.2	62.2
	Dutch	articles	10	0.3	20	5.3	61.6	9.8	51.8
	Dutch	reviews	10	0.4	18.2	6.5	19.7	4.0	140.6
	Greek	articles	10	0.3	20	6.0	38.0	6.7	48.2
	Greek	reviews	10	0.3	20	6.1	41.6	7.5	39.4
Test	English	articles	20	0.3	20	5.7	59.3	9.5	52.5
	English	reviews	20	0.3	20	6.4	43.5	7.9	65.3
	Dutch	articles	20	0.3	20	5.7	49.4	8.3	49.3
	Dutch	reviews	20	0.4	18.4	7.1	19.3	4.1	152.0
	Greek	articles	20	0.3	19.9	5.2	59.6	9.6	46.6
	Greek	reviews	20	0.3	20	6.0	42.2	7.6	37.1

Evaluation Framework. The same evaluation measures introduced in PAN-2016 are used here [30]. For the complete clustering scenario, Bcubed Recall, Bcubed Precision, and Bcubed F-score are calculated. These are among the best extrinsic clustering evaluation measures [1]. With respect to the authorship-link ranking scenario, Mean Average Precision (MAP), R-precision, and P@10 are used to estimate the ability of systems to rank high correct results.

In order to understand the complexity of tasks and the effectiveness of participant systems, we used a set of baseline approaches and applied them to the evaluation datasets. The baseline methods range from naive to strong and will allow to estimate weaknesses and strengths of participant approaches. More specifically, the following baseline methods were used:

– *BASELINE-Random.* k is randomly chosen from $[1, N]$ and then each $d \in D$ is randomly assigned to one cluster. Authorship links are extracted from the produced clusters and a random score is calculated. The average performance of this method over 30 repetitions is reported. This naive approach can only serve as an indication of the lowest performance.

- *BASELINE-Singleton.* This method sets $k = N$, i.e., all documents are from different authors. It forms singleton clusters and it is used only for the complete clustering scenario. This simple method was found very effective in PAN-2016 datasets, and its performance increases with r [30].
- *BASELINE-Cosine.* Each document is represented by the normalized frequencies of all words occurring at least 3 times in the given collection of documents. Then, for each pair of documents the cosine similarity is used as an authorship-link score. This simple method was found hard-to-beat in PAN-2016 evaluation campaign [30].
- *BASELINE-PAN16.* This is the top-performing method submitted to the corresponding PAN-2016 task. It is based on a character-level recurrent neural network and it is a modification of an effective authorship verification approach [2].

Results. We received 6 software submissions that were evaluated in TIRA experimentation platform [7,18]. Table 3 shows the overall evaluation results for both complete clustering and authorship-link ranking on the entire test dataset. The elapsed runtime of each submission is also reported. As can be seen, the method of Gómez-Adorno et al. [8] achieves the best results in both scenarios. Actually, this is the top-performing method taking into account all but one evaluation measures (BCubed precision). By definition, BASELINE-Singleton achieves perfect Bcubed precision since it provides single-item clusters exclusively. BASELINE-PAN16 also tends to optimize precision since it was tuned for another corpus with much higher clusteriness ratio. Within the submitted methods, the approaches of García et al. [5] and Kocher & Savoy [13] are the best ones in terms of Bcubed precision. However, the winning approach of Gómez-Adorno et al. [8] is the only

Table 3. Overall evaluation results in author clustering (mean values for all clustering problems). Participants are ranked according to Bcubed F-score.

Participant	Complete clustering			Authorship-link ranking			Runtime
	B^3 F	B^3 rec.	B^3 prec.	MAP	RP	P@10	
Gómez-Adorno et al.	**0.573**	**0.639**	0.607	**0.456**	**0.417**	**0.618**	00:02:06
García et al.	0.565	0.518	0.692	0.381	0.376	0.535	00:15:49
Kocher & Savoy	0.552	0.517	0.677	0.396	0.369	0.509	00:00:42
Halvani & Graner	0.549	0.589	0.569	0.139	0.251	0.263	00:12:25
Alberts	0.528	0.599	0.550	0.042	0.089	0.284	00:01:46
BASELINE-PAN16	0.487	0.331	0.987	0.443	0.390	0.583	50:17:49
Karaś et al.	0.466	0.580	0.439	0.125	0.218	0.252	00:00:26
BASELINE-Singleton	0.456	0.304	1.000	–	–	–	–
BASELINE-Random	0.452	0.339	0.731	0.024	0.051	0.209	–
BASELINE-Cosine	–	–	–	0.308	0.294	0.348	–

one that achieves both Bcubed recall and precision higher than 0.6. All submitted methods but one surpass baseline approaches in the complete clustering scenario. On the other hand, in the authorship-link ranking scenario, BASELINE-PAN16 is very competitive while BASELINE-Cosine surpasses half of submissions. More detailed evaluation results are presented in [31].

In general, almost all submitted approaches are quite efficient and can process all evaluation datasets quickly. The approaches of García et al. [5] and Halvani & Graner [10] are relatively slower compared to the other submissions but, however, much more efficient than BASELINE-PAN16. The most successful approaches use low-level (character or lexical) features [5,8,10,13]. Relatively simple clustering algorithms, like hierarchical agglomerative clustering [8] or β-compact graph-based clustering [5] provided the best results in the complete clustering scenario. A more detailed survey of submissions is included in [31].

3   Author Profiling

Author profiling aims at classifying authors in different classes depending on their sociolect aspects, namely how they share language. This allows the identification of author traits such as age, gender, native language, language variety, or personality type. Author profiling is growing in interest, specially due to the rise of social media, where authors may hide personal information, or even lie. Author profiling may help to improve marketing segmentation, security, and it allows the use of the language as evidence in possible cases of abuse or harassing messages.

In previous editions at PAN we have mainly focused on age and gender identification in different genres or in a cross-genre environment. The Author Profiling shared task at PAN'17 focuses on the following aspects:

- *Gender and language variety identification.* As in previous editions, the task contains gender prediction. Instead of age identification, the aim this year is at discriminating among different varieties of the same languages (also known as dialects).
- *Demographic idiosyncrasies.* This is the first time the gender dimension is studied together with the language variety, which may provide insights on the difficulty of the task depending on geographical and cultural idiosyncrasies.
- *Multilinguality.* Participants are provided with data in Arabic, English, Spanish and Portuguese.

3.1   Evaluation Framework

The evaluation data has been collected from Twitter in four different languages, namely Arabic, English, Spanish and Portuguese. The authors have been annotated with their gender and language variety. The gender annotation has been carried out with the help of dictionaries of proper nouns, and the variety has been based on the geographical retrieval of the tweets. For each author, we considered exactly 100 tweets. The dataset is balanced by gender and variety. There are 500 authors per variety and gender. The dataset has been split in a 60/40 proportion with 300 authors for training and 200 for test. The corresponding languages and varieties are shown in Table 4, together with the total number of authors for each subtask.

Table 4. Languages and varieties. There are 500 authors per variety and gender, 300 for training and 200 for test. Each author contains 100 tweets.

(AR) Arabic	(EN) English	(ES) Spanish	(PT) Portuguese
Egypt	Australia	Argentina	Brazil
Gulf	Canada	Chile	Portugal
Levantine	Great Britain	Colombia	
Maghrebi	Ireland	Mexico	
	New Zealand	Peru	
	United States	Spain	
		Venezuela	
4,000	6,000	7,000	2,000

For evaluation, the accuracy for variety, gender and joint identification per language is calculated. Then, we average the results obtained per language (Eq. 1).

$$\overline{gender} = \frac{gender\_ar + gender\_en + gender\_es + gender\_pt}{4}$$

$$\overline{variety} = \frac{variety\_ar + variety\_en + variety\_es + variety\_pt}{4} \quad (1)$$

$$\overline{joint} = \frac{joint\_ar + joint\_en + joint\_es + joint\_pt}{4}$$

The final ranking is calculated as the average of the previous values (Eq. 2):

$$ranking = \frac{\overline{gender} + \overline{variety} + \overline{joint}}{3} \quad (2)$$

In order to understand the complexity of the subtasks in each language and with the aim at comparing the performance of the participants approaches, we propose the following baselines:

- *BASELINE-stat.* It is a statistical baseline that emulates the random choice. This baseline depends on the number of classes: 2 in case of gender identification, and from 2 to 7 in case of variety identification.
- *BASELINE-bow.* This method represents documents as a bag-of-words with the 1,000 most common words in the training set, weighting by absolute frequency of occurrence. The texts are preprocessed in order to lowercase words, remove punctuation signs and numbers, and remove the stop words for the corresponding language.

Table 5. Joint accuracies per language and global ranking as average per language of gender, variety and joint identification.

Ranking	Team	Global	Arabic	English	Spanish	Portuguese
1	nissim17	**0.8361**	**0.6831**	**0.7429**	**0.8036**	0.8288
2	martinc17	0.8285	0.6825	0.7042	0.7850	0.8463
3	miranda17	0.8258	0.6713	0.7267	0.7621	0.8425
4	miura17	0.8162	0.6419	0.6992	0.7518	**0.8575**
5	lopezmonroy17	0.8111	0.6475	0.7029	0.7604	0.8100
6	markov17	0.8097	0.6525	0.7125	0.7704	0.7750
7	poulston17	0.7942	0.6356	0.6254	0.7471	0.8188
8	sierraloaiza17	0.7822	0.5694	0.6567	0.7279	0.8113
	BASELINE-LDR	0.7325	0.5888	0.6357	0.6943	0.7763
9	romanov17	0.7653	0.5731	0.6450	0.6846	0.7775
10	benajiba17	0.7582	0.5688	0.6046	0.7021	0.7525
11	schaetti17	0.7511	0.5681	0.6150	0.6718	0.7300
12	kodiyan17	0.7509	0.5688	0.6263	0.6646	0.7300
13	zampieri17	0.7498	0.5619	0.5904	0.6764	0.7575
14	kheng17	0.7176	0.5475	0.5704	0.6400	0.6475
15	ganesh17	0.6881	0.5075	0.4713	0.5614	0.7300
16	kocher17	0.6813	0.5206	0.4650	0.4971	0.7575
17	akhtyamova17	0.6270	0.2875	0.4333	0.5593	0.6675
	BASELINE-bow	0.6195	0.1794	0.4713	0.5561	0.7588
18	khan17	0.4952	0.3650	0.1900	0.2189	0.5488
	BASELINE-stat	0.2991	0.1250	0.0833	0.0714	0.2500
19	ribeirooliveira17	0.2087	-	-	-	0.7538
20	alrifai17	0.1701	0.5638	-	-	-
21	bouzazi17	0.1027	-	0.2479	-	-
22	castrocastro17	0.0695	-	0.1017	-	-

– *BASELINE-LDR* [23]. This method represents documents on the basis of the probability distribution of occurrence of their words in the different classes.

3.2 Results

This year 22[1] have been the teams who participated in the shared task. In this section a summary of the obtained results is shown. In Table 5 the overall

---

[1] In the five editions of the author profiling shared task we have had respectively 21 (2013: age and gender identification [24]), 10 (2014: age and gender identification in different genre social media [22]), 22 (2015: age and gender identification and personality recognition in Twitter [21]), 22 (2016: cross-genre age and gender identification [26]) and 22 (2017: gender and language variety identification [25]) participating teams.

Table 6. Best results per language and task.

Language	*Joint*	Gender	Variety
Arabic	0.6831	0.8031	0.8313
English	0.7429	0.8233	0.8988
Spanish	0.8036	0.8321	0.9621
Portuguese	0.8575	0.8700	0.9838

performance per language and users' ranking are shown. We can observe that the best results were achieved in Portuguese (85.75%), followed by Spanish (80.36%), English (74.29%) and Arabic (68.31%). The difference on accuracy among languages is very significant. Most of the participants obtained better results than both baselines. However, in case of Portuguese only 9 teams outperformed the bag-of-words baseline, showing the power of simple words to discriminate among varieties and genders in that language. On the contrary, this baseline shows its inefficiency in case of Arabic, where the accuracy drops to values close to the statistical baseline.

In Table 6 the best results per language and task are shown. We can observe that for both the gender and variety subtasks, the best results were achieved in Portuguese, followed by Spanish, English and Arabic. In case of gender identification, the accuracies are between 80.31% in case of Arabic and 87% in case of Portuguese, whereas the difference is higher for language variety identification, where the worst results obtained in Arabic is 83.13% (4 varieties), against a 98.38% obtained in Portuguese (2 varieties). Results for Spanish (7 varieties) (96.21%) are close to Portuguese, while in English (6 varieties) they fall to 89.88%. A more in-depth analysis of the results and the different approaches can be found in [25].

4   Author Obfuscation

Author obfuscation is the youngest branch of PAN's main tasks, and perhaps also one of the most difficult ones. Its goal is to attack the approaches to the other tasks by altering their text input in a way that will cause them return an incorrect answer. The difficulty arises not so much from making changes to the texts that have such an effect on the attacked approaches, but to do so in a way so that the text input can still be understood by a human and so that its original message is not twisted beyond recognition. The latter two severely limit the potential changes that can be made, rendering any form of obfuscation a form of style paraphrasing where the goal is to change the writing style of a piece of writing without changing a text's pragmatics.

The author obfuscation task at PAN 2017 concerns author masking, where the specific goal is to attack authorship verification technology. For the latter, the task is to verify whether a given pair of texts has been written by the same author, whereas for the former, the task is to alter the writing style of a designated text from a given pair written by the same author in order to prevent verification algorithms from arriving at just that decision. As a shared task, author masking has been organized for the first time at PAN 2016 [19]. We continue with author masking in much the same way as before. Since the setup did no change significantly, just to be self-contained, the following gives only a brief recap.

4.1 Evaluation Datasets

The evaluation data consist of the English portion of the joint datasets of the PAN 2013–2015 authorship verification tasks, separated by training datasets and test datasets. The datasets cover a broad range of genres, namely computer science textbook excerpts, essays from language learners, horror fiction novel excerpts, and dialog lines from plays. The joint training dataset was handed out to participants, while the joint test dataset was held back and only accessible via the TIRA experimentation platform. The test dataset contains a total of 464 problem instances, each consisting of a to-be-obfuscated text and one or more other texts from the same author. The approaches submitted by participants were supposed to process each problem instance and to return for each of the to-be-obfuscated texts and paraphrased version, perhaps using the remaining texts from the same author to learn what style changes are at least necessary to make the writing styles of the two texts the most dissimilar.

4.2 Performance Measures

We call an obfuscation software

- **safe**, if its obfuscated texts can not be attributed to their original authors anymore,
- **sound**, if its obfuscated texts are textually entailed by their originals, and
- **sensible**, if its obfuscated texts are well-formed and inconspicuous.

Any evaluation of an author obfuscation approach must at least cover these three dimensions, whereas the assessment and quantification of especially the latter two is still an open problem. To cut a long story short, in this shared task, we evaluate the safety of a submitted approach by feeding the obfuscated evaluation dataset that it produces to as many pre-trained authorship verification approaches as possible. Fortunately, with 44 authorship verifiers, the number of such approaches available to us is rather high, allowing for meaningful conclusions. This is made possible due to the fact that TIRA has been employed at PAN since before 2013, so that all authorship verification approaches submitted to the corresponding shared tasks are available to us in working condition. By counting the number of cases where a true positive prediction of an authorship verifier is flipped to a false negative prediction because of applying a to-be-evaluated obfuscator beforehand, we can calculate the relative impact the obfuscator has on the verifier. When doing so not just for one verifier, but for 44 state-of-the-art verifiers, this tells a lot about the ability of the obfuscator to fulfill its purpose of obfuscating the writing style of texts in a way that cannot be defeated by any verifier known to date.

Regarding soundness and sensibleness of an author obfuscation approach, we rely own judgment as well as on peer-review. Here, we grade a selection of Likert scale of 1–5 with regard to sensibleness, and on 3-point scale with regard to soundness. In the past, the participants who participated in peer-evaluation came up with similar grade scales, obtained results commensurate with ours.

4.3 Results

A detailed evaluation of the results of a total of 5 obfuscation approaches, two of which have been submitted this year, and three of which past year can be found in the task overview paper [9].

5 Summary

Altogether, PAN presented its participants again with a set of challenging shared tasks, including new ones as well as "classical" ones which were given new spin. Multilingual as well as multigenre corpora have been prepared, which will henceforth serve as new benchmark datasets for their tasks. At the same time, the software underlying each of the submitted approaches has been collected and hosted on the TIRA experimentation platform, ensuring replicability of results as well as reproducibility, e.g., by allowing for their reevaluation using new datasets as they arrive in the future. In this regard, for future work, we will continue to develop PAN's shared tasks, providing new and challenging datasets as well as inventing new tasks belonging to author identification, author profiling, and author obfuscation.

Acknowledgements. Our special thanks go to all of PAN's participants, to Symanto Group (https://www.symanto.net/) for sponsoring PAN and to Meaning-Cloud (https://www.meaningcloud.com/) for sponsoring the author profiling shared task award. The work at the Universitat Politècnica de València was funded by the MINECO research project SomEMBED (TIN2015-71147-C2-1-P).

References

1. Amigó, E., Gonzalo, J., Artiles, J., Verdejo, F.: A comparison of extrinsic clustering evaluation metrics based on formal constraints. Inf. Retrieval **12**(4), 461–486 (2009)
2. Bagnall, D.: Authorship clustering using multi-headed recurrent neural networks—notebook for PAN at CLEF 2016. In: Balog et al. [3] (2016). http://ceur-ws.org/Vol-1609/
3. Balog, K., Cappellato, L., Ferro, N., Macdonald, C. (eds.): CLEF 2016 Evaluation Labs and Workshop – Working Notes Papers, 5–8 September, Évora, Portugal. CEUR Workshop Proceedings. CEUR-WS.org (2016). http://www.clef-initiative.eu/publication/working-notes

4. Clarke, C.L., Craswell, N., Soboroff, I., Voorhees, E.M.: Overview of the TREC 2009 web track. Technical report, DTIC Document (2009)
5. García, Y., Castro, D., Lavielle, V., Noz, R.M.: Discovering author groups using a β-compact graph-based clustering. In: Cappellato, L., Ferro, N., Goeuriot, L., Mandl, T. (eds.) CLEF 2017 Working Notes. CEUR Workshop Proceedings, CLEF and CEUR-WS.org, September 2017
6. Glavaš, G., Nanni, F., Ponzetto, S.P.: Unsupervised text segmentation using semantic relatedness graphs. In: Association for Computational Linguistics (2016)
7. Gollub, T., Stein, B., Burrows, S.: Ousting ivory tower research: towards a web framework for providing experiments as a service. In: Hersh, B., Callan, J., Maarek, Y., Sanderson, M. (eds.) 35th International ACM Conference on Research and Development in Information Retrieval (SIGIR 2012), pp. 1125–1126. ACM, August 2012
8. Gómez-Adorno, H., Aleman, Y., no, D.V., Sanchez-Perez, M.A., Pinto, D., Sidorov, G.: Author clustering using hierarchical clustering analysis. In: Cappellato, L., Ferro, N., Goeuriot, L., Mandl, T. (eds.) CLEF 2017 Working Notes. CEUR Workshop Proceedings, CLEF and CEUR-WS.org, September 2017
9. Hagen, M., Potthast, M., Stein, B.: Overview of the author obfuscation task at PAN 2017: safety evaluation revisited. In: Cappellato, L., Ferro, N., Goeuriot, L., Mandl, T. (eds.) Working Notes Papers of the CLEF 2017 Evaluation Labs. CEUR Workshop Proceedings, CLEF and CEUR-WS.org, September 2017
10. Halvani, O., Graner, L.: Author clustering based on compression-based dissimilarity scores. In: Cappellato, L., Ferro, N., Goeuriot, L., Mandl, T. (eds.) CLEF 2017 Working Notes. CEUR Workshop Proceedings, CLEF and CEUR-WS.org, September 2017
11. Hearst, M.A.: TextTiling: segmenting text into multi-paragraph subtopic passages. Comput. Linguist. **23**(1), 33–64 (1997)
12. Kiros, R., Zhu, Y., Salakhutdinov, R.R., Zemel, R., Urtasun, R., Torralba, A., Fidler, S.: Skip-thought vectors. In: Advances in Neural Information Processing Systems (NIPS), pp. 3294–3302 (2015)
13. Kocher, M., Savoy, J.: UniNE at CLEF 2017: author clustering. In: Cappellato, L., Ferro, N., Goeuriot, L., Mandl, T. (eds.) CLEF 2017 Working Notes. CEUR Workshop Proceedings, CLEF and CEUR-WS.org, September 2017
14. Koppel, M., Akiva, N., Dershowitz, I., Dershowitz, N.: Unsupervised decomposition of a document into authorial components. In: Lin, D., Matsumoto, Y., Mihalcea, R. (eds.) Proceedings of the 49th Annual Meeting of the Association for Computational Linguistics (ACL), pp. 1356–1364 (2011)
15. Misra, H., Yvon, F., Jose, J.M., Cappe, O.: Text segmentation via topic modeling: an analytical study. In: Proceedings of CIKM 2009, pp. 1553–1556. ACM (2009)
16. Pevzner, L., Hearst, M.A.: A critique and improvement of an evaluation metric for text segmentation. Comput. Linguis. **28**(1), 19–36 (2002)
17. Potthast, M., Eiselt, A., Barrón-Cedeño, A., Stein, B., Rosso, P.: Overview of the 3rd international competition on plagiarism detection. In: Notebook Papers of the 5th Evaluation Lab on Uncovering Plagiarism, Authorship and Social Software Misuse (PAN), Amsterdam, The Netherlands, September 2011
18. Potthast, M., Gollub, T., Rangel, F., Rosso, P., Stamatatos, E., Stein, B.: Improving the reproducibility of PAN's shared tasks: plagiarism detection, author identification, and author profiling. In: Kanoulas, E., Lupu, M., Clough, P., Sanderson, M., Hall, M., Hanbury, A., Toms, E. (eds.) CLEF 2014. LNCS, vol. 8685, pp. 268–299. Springer, Cham (2014). doi:10.1007/978-3-319-11382-1_22

19. Potthast, M., Hagen, M., Stein, B.: Author obfuscation: attacking the state of the art in authorship verification. In: Working Notes Papers of the CLEF 2016 Evaluation Labs. CEUR Workshop Proceedings, CLEF and CEUR-WS.org, September 2016. http://ceur-ws.org/Vol-1609/

20. Potthast, M., Hagen, M., Völske, M., Stein, B.: Crowdsourcing interaction logs to understand text reuse from the web. In: Fung, P., Poesio, M. (eds.) Proceedings of the 51st Annual Meeting of the Association for Computational Linguistics (ACL 13), pp. 1212–1221. Association for Computational Linguistics (2013). http://www.aclweb.org/anthology/p13-1119

21. Rangel, F., Celli, F., Rosso, P., Potthast, M., Stein, B., Daelemans, W.: Overview of the 3rd author profiling task at PAN 2015. In: Cappellato, L., Ferro, N., Jones, G., San Juan, E. (eds.) CLEF 2015 Evaluation Labs and Workshop – Working Notes Papers, 8–11 September, Toulouse, France. CEUR Workshop Proceedings, CEUR-WS.org, September 2015

22. Rangel, F., Rosso, P., Chugur, I., Potthast, M., Trenkmann, M., Stein, B., Verhoeven, B., Daelemans, W.: Overview of the 2nd author profiling task at PAN 2014. In: Cappellato, L., Ferro, N., Halvey, M., Kraaij, W. (eds.) CLEF 2014 Evaluation Labs and Workshop – Working Notes Papers, 15–18 September, Sheffield, UK. CEUR Workshop Proceedings, CEUR-WS.org, September 2014

23. Rangel, F., Rosso, P., Franco-Salvador, M.: A low dimensionality representation for language variety identification. In: 17th International Conference on Intelligent Text Processing and Computational Linguistics, CICLing. LNCS. Springer (2016). arXiv:1705.10754

24. Rangel, F., Rosso, P., Koppel, M., Stamatatos, E., Inches, G.: Overview of the author profiling task at PAN 2013. In: Forner, P., Navigli, R., Tufis, D. (eds.) CLEF 2013 Evaluation Labs and Workshop – Working Notes Papers, 23–26 September, Valencia, Spain (2013)

25. Rangel, F., Rosso, P., Potthast, M., Stein, B.: Overview of the 5th author profiling task at PAN 2017: gender and language variety identification in Twitter. In: Cappellato, L., Ferro, N., Goeuriot, L., Mandl, T. (eds.) Working Notes Papers of the CLEF 2017 Evaluation Labs. CEUR Workshop Proceedings, CLEF and CEUR-WS.org, September 2017

26. Rangel, F., Rosso, P., Verhoeven, B., Daelemans, W., Potthast, M., Stein, B.: Overview of the 4th author profiling task at PAN 2016: cross-genre evaluations. In: Balog et al. [3]

27. Riedl, M., Biemann, C.: TopicTiling: a text segmentation algorithm based on LDA. In: Proceedings of ACL 2012 Student Research Workshop, pp. 37–42. Association for Computational Linguistics (2012)

28. Scaiano, M., Inkpen, D.: Getting more from segmentation evaluation. In: Proceedings of the 2012 Conference of the North American Chapter of the Association for Computational Linguistics: Human Language Technologies. pp. 362–366. Association for Computational Linguistics (2012)

29. Stamatatos, E., Tschuggnall, M., Verhoeven, B., Daelemans, W., Specht, G., Stein, B., Potthast, M.: Clustering by authorship within and across documents. In: Working Notes Papers of the CLEF 2016 Evaluation Labs. CEUR Workshop Proceedings, CLEF and CEUR-WS.org. http://ceur-ws.org/Vol-1609/

30. Stamatatos, E., Tschuggnall, M., Verhoeven, B., Daelemans, W., Specht, G., Stein, B., Potthast, M.: Clustering by authorship within and across documents. In: Working Notes Papers of the CLEF 2016 Evaluation Labs. CEUR Workshop Proceedings, CLEF and CEUR-WS.org, September 2016

31. Tschuggnall, M., Stamatatos, E., Verhoeven, B., Daelemans, W., Specht, G., Stein, B., Potthast, M.: Overview of the author identification task at PAN-2017: style breach detection and author clustering. In: Cappellato, L., Ferro, N., Goeuriot, L., Mandl, T. (eds.) Working Notes Papers of the CLEF 2017 Evaluation Labs. CEUR Workshop Proceedings, CLEF and CEUR-WS.org, September 2017

CLEF 2017 eHealth Evaluation Lab Overview

Lorraine Goeuriot[1], Liadh Kelly[2], Hanna Suominen[3,4,5]([✉]), Aurélie Névéol[6],
Aude Robert[7], Evangelos Kanoulas[8], Rene Spijker[9], João Palotti[10],
and Guido Zuccon[11]

[1] University Grenoble Alpes, CNRS, Grenoble INP, LIG, 38000 Grenoble, France
Lorraine.Goeuriot@imag.fr
[2] ADAPT Centre, Dublin City University, Dublin, Ireland
liadh.kelly@dcu.ie
[3] Data61/CSIRO, The Australian National University (ANU),
Canberra, ACT, Australia
hanna.suominen@anu.edu.au
[4] University of Canberra, Canberra, ACT, Australia
[5] University of Turku, Turku, Finland
[6] LIMSI CNRS UPR 3251 Université Paris-Saclay, 91405 Orsay, France
Aurelie.Neveol@limsi.fr
[7] INSERM - CépiDc, 80 Rue du Général Leclerc,
94276 Le Kremlin-Bicêtre Cedex, France
aude.robert@inserm.fr
[8] Informatics Institute, University of Amsterdam, Amsterdam, Netherlands
E.Kanoulas@uva.nl
[9] Julius Center for Health Sciences and Primary Care, Cochrane Netherlands
and UMC Utrecht, Utrecht, Netherlands
R.Spijker-2@umcutrecht.nl
[10] Vienna University of Technology, Vienna, Austria
palotti@ifs.tuwien.ac.at
[11] Queensland University of Technology, Brisbane, QLD, Australia
g.zuccon@qut.edu.au

Abstract. In this paper we provide an overview of the fifth edition of the
CLEF eHealth evaluation lab. CLEF eHealth 2017 continues our evaluation resource building efforts around the easing and support of patients,
their next-of-kins, clinical staff, and health scientists in understanding,
accessing, and authoring eHealth information in a multilingual setting.
This year's lab offered three tasks: Task 1 on multilingual information
extraction to extend from last year's task on French corpora, Task 2
on technologically assisted reviews in empirical medicine as a new pilot
task, and Task 3 on patient-centered information retrieval (IR) building
on the 2013-16 IR tasks. In total 32 teams took part in these tasks (11
in Task 1, 14 in Task 2, and 7 in Task 3). We also continued the replication track from 2016. Herein, we describe the resources created for these
tasks, evaluation methodology adopted and provide a brief summary of
participants of this year's challenges and results obtained. As in previous

G. Zuccon—In alphabetical order by surname, LG, LK & HS co-chaired the lab. AN
& AR, EK & RS, and JP & GZ led Tasks 1–3, respectively.

G.J.F. Jones et al. (Eds.): CLEF 2017, LNCS 10456, pp. 291–303, 2017.
DOI: 10.1007/978-3-319-65813-1_26

years, the organizers have made data and tools associated with the lab tasks available for future research and development.

Keywords: Evaluation · Entity linking · Information retrieval · Health records · Information extraction · Medical informatics · Systematic reviews · Test-set generation · Text classification · Text segmentation · Self-diagnosis

1   Introduction

This paper presents an overview of the CLEF eHealth 2017 evaluation lab, organized within the Conference and Labs of the Evaluation Forum (CLEF) to support the development of approaches for helping patients, their next-of-kins, and clinical staff in understanding, accessing, and authoring health information in a multilingual setting. This fifth year of the evaluation lab aimed to build upon the resource development and evaluation approaches offered in the previous four years of the lab [5,10,11,19], which focused on patients and their next-of-kins' ease in understanding and accessing health information.

Task 1 addressed *Multi-lingual Information Extraction* (IE) related to diagnosis coding in written text with a focus on unexplored languages corpora, specifically French. English was also offered. This built upon the 2016 task, which analyzed French biomedical text with the IE of causes of death from a corpus of French death reports [15]. This is an essential task in epidemiology, as the determination and analysis of causes of death at a global level informs public health policies. This task was treated as a named entity recognition and normalization task or as a text classification task. Each language could be considered independently, but we encouraged participants to explore multilingual approaches and approaches which could be easily adapted to a new language. Only fully automated means were allowed, that is, human-in-the-loop approaches were not permitted.

Task 2 on *Technology Assisted Reviews in Empirical Medicine* was introduced for the first time in 2017. It was a high-recall Information Retrieval (IR) task that aimed at evaluating search algorithms that seek to identify all studies relevant for conducting a systematic review in empirical medicine. Evidence-based medicine has become an important strategy in health care and policy making. In order to practice evidence-based medicine, it is important to have a clear overview of the current scientific consensus. These overviews are provided in systematic review articles, that summarize all evidence that is published regarding a certain topic (e.g., a treatment or diagnostic test). In order to write a systematic review, researchers have to conduct a search that will retrieve all the documents that are relevant. This is a difficult task, known in the IR domain as the total recall problem. With the reported medical studies expanding rapidly, the need for automation in this process becomes of utmost importance. CLEF 2017 Task 2 had a focus on Diagnostic Test Accuracy (DTA) reviews. Search in this area is generally considered the difficult, and a breakthrough in this field would likely be applicable to other areas as well [12]. The task coordinators considered all 57 systematic reviews conducted by Cochrane[1] experts on DTA studies and published in the Cochrane library[2]. The coordinators of the task managed to reconstruct the MEDLINE Boolean query used for 50 of these systematic reviews. The corpus considered was Document Abstracts and Titles retrieved by these 50 Boolean queries from the medline database (either through Ovid[3] or PubMed[4]). The goal of the participants was to (a) rank the documents returned by the Boolean query studies, and (b) find an optimal threshold that could inform experts when to stop examining documents in the ranked list. Recall, precision, effort, cost of missing studies, and combinations of these metrics were used to assess the quality of the participating systems. The set of relevant titles and abstracts used in the evaluation were directly extracted from the reference section of the systematic reviews.

Task 3, the IR Task, aimed at evaluating the effectiveness of IR systems when searching for health content on the web, with the objective to foster research and development of search engines tailored to health information seeking. This year's IR task continued the growth path identified in 2013, 2014, 2015, and 2016's CLEF eHealth IR challenges [3,4,16,22]. The corpus (ClueWeb12) and the topics used are similar to 2016's. This year new use cases were explored and the pool of assessed documents deepened. The subtasks within the IR challenge were similar to 2016's: ad hoc search, query variation, and multilingual search. A new subtask was also organized, aimed at exploring methods to personalize health search. Query variations were generated based on the fact that there are multiple ways to express a single information need. Translations of the English queries into several langues were also provided. Participants were required to translate the queries back to English and use the English translation to search the collection.

This paper is structured as follows: in Sect. 2 we detail the tasks, evaluation and datasets created; in Sect. 3 we describe the submission and results for each task; and in Sect. 4 we provide conclusions.

2 Materials and Methods

2.1 Text Documents

Task 1 used a corpus of death certificates comprising free-text descriptions of causes of death as reported by physicians in the standardized causes of death forms in France and in the United States. Each document was manually coded by experts with ICD-10 per international WHO standards. The languages of the challenge this year are French and English. Table 1 below provides some statistics on the datasets.

---

[1] http://www.cochrane.org/.
[2] http://www.cochranelibrary.com.
[3] http://ovid.com/site/catalog/databases/901.jsp.
[4] https://www.ncbi.nlm.nih.gov/pubmed/.

Table 1. Descriptive statistics of the causes of death certificates corpus

	FR			EN	
	Train (2006–2012)	Dev (2013)	Test (2014)	Train (2015)	Test (2015)
Documents	65, 844	27, 850	31, 690	13, 330	6, 665
Tokens	1, 176, 994	496, 649	599, 127	88, 530	40, 130
Total ICD codes	266, 808	110, 869	131, 426	39, 334	18, 928
Unique ICD codes	3, 233	2, 363	2, 527	1, 256	900
Unique unseen ICD codes	3, 233	224	266	1, 256	157

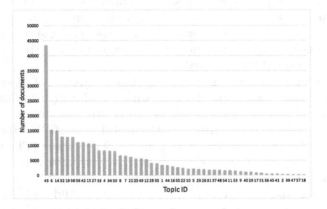

Fig. 1. The distribution of the number of documents across topics in Task 2.

The new technologically assisted reviews in empirical medicine task, Task 2, used a subset of PubMed documents for its challenge to make Abstract and Title Screening more effective. More specifically the PubMed Document Identifiers (PIDs) of potentially relevant PubMed Document abstracts were provided for each training and test topic. The PIDs were collected by the task coordinators by re-running the MEDLINE Boolean query used in the original systematic reviews conducted by Cochrane to search PubMed. A distribution of the number of documents to be ranked by participants per topic can be found in Fig. 1.

The IR challenge, Task 3, once again used the ClueWeb12 B13[5] corpus, first introduced to the CLEF eHealth IR task in 2016. This corpus is a large snapshot of the Web, crawled between February and May 2012. Unlike the Khresmoi dataset [6] used in earlier years of the IR task [3,4,16], ClueWeb12 does not contain only Health On the Net certified pages and pages from a selected list of known health domains, making the dataset more in line with the material current web search engines index and retrieve. ClueWeb12 B13 contains approximately 52.3 million web pages, for a total of 1.95 TB of data, once uncompressed.

For participants who did not have access to the ClueWeb dataset, Carnegie Mellon University granted the organisers permission to make the dataset

---

[5] http://lemurproject.org/clueweb12/index.php.

available through cloud computing instances provided by Microsoft Azure[6]. The Azure instances that were made available to participants for the IR challenge included (1) the Clueweb12 B13 dataset, (2) standard indexes built with the Terrier [13] and the Indri [18] toolkits, (3) additional resources such as a spam list [2], Page Rank scores, anchor texts [7], and urls, made available through the ClueWeb12 website.

2.2 Human Annotations, Queries, and Relevance Assessments

For Task 1, the ICD10 codes were abstracted from the raw lines of death certificate text by professional curators at INSERM over the period of 2006–2014 for the French dataset, and curators at the CDC (Center for Disease Control) in the year 2015 for the American dataset. During this time, curators from both groups also manually built dictionaries of terms associated with ICD10 codes. Several versions of these lexical resources were supplied to participants in addition to the training data. Because of the interface used by curators to perform coding, the data used in the challenge comes in separate files: one file contains the original "raw" text of the death certificates presented line by line, one contains the metadata associated with the certificates at the document level, and one contains the ICD codes assigned to the certificate. As detailed in the task overview, a "raw" version of datasets was distributed for French and English. For French, we also distributed an "aligned" version of the data where the ICD10 codes are reconciled with the specific text line that supported the assignment.

For the technology assisted reviews in empirical medicine task, focusing on title and abstract screening, topics consisted of the Boolean Search from the first step of the systematic review process. Specifically, for each topic the following information was provided:

1. Topic-ID
2. The title of the review, written by Cochrane experts;
3. The Boolean query manually constructed by Cochrane experts;
4. The set of PubMed Document Identifiers (PID's) returned by running the query in MEDLINE.

Twenty of these topics were randomly selected to be used as a training set, while the remaining thirty were used as a test set. The original systematic reviews written by Cochrane experts included a reference section that listed Included, Excluded, and Additional references to medical studies. The union of Included and Excluded references are the studies that were screened at a Title and Abstract level and were considered for further examination at a full content level. These constituted the relevant documents at the abstract level, while the

---

[6] The organisers are thankful to Carnegie Mellon University, and in particular to Jamie Callan and Christina Melucci, for their support in obtaining the permission to redistribute ClueWeb 12. The organisers are also thankful to Microsoft Azure who provided the Azure cloud computing infrastructure that was made available to participants through the Microsoft Azure for Research Award CRM:0518649.

Included references constituted the relevant documents at the full content level. References in the original systematic reviews were collected from a variety of resources, not only MEDLINE. Therefore, studies that were cited but did not appear in the results of the Boolean query were excluded from the label set.

The IR task, Task 3, uses 2016's task topics [22], with the aim to acquire more relevance assessments and improve the collection reusability. The queries consider real health information needs expressed by the general public through posts published in public health web forums. Forum posts were extracted from the 'askDocs' section of Reddit[7], and presented to query creators. Query creators were asked to formulate queries based on what they read in the initial user post. Six query creators with different medical expertise were used for this task. This year, apart from the AdHoc retrieval task (IRTask 1), the query variation task (IRTask 3) introduced in 2016 and the multilingual task (IRTask 4) introduced in 2013, we proposed a personalized search task (IRTask2) in which participants have to personalize the retrieved list of search results so as to match user expertise, measured by how likely the person is to understand the content of a document (with respect to the health information).

Relevance assessments were collected by pooling participants' submitted runs as well as baseline runs. Assessment was performed by paid medical students who had access to the queries, to the documents, and to the relevance criteria drafted by a junior medical doctor that guided assessors in the judgment of document relevance. The relevance criteria were drafted considering the entirety of the forum posts used to create the queries; a link to the forum posts was also provided to the assessors. Along with relevance assessments, readability/understandability and reliability/trustworthiness judgments were also collected for the assessment pool; these were used to evaluate systems across different dimensions of relevance [20,21].

2.3 Evaluation Methods

Task 1. Teams could submit up to two runs for the tasks for each language. System performance was assessed by the precision, recall and F-measure for ICD code extraction at the document level for English and both at the line and document level for French. Evaluation measures were computed overall (for all ICD codes) and for a subset of the codes, called external causes of death, which are of specific interest to public health specialists. Two baselines were also implemented by the organizers and one participating team.

After submitting their result files for the IE challenges, participating teams had one extra week to submit the system used to produce them, or a remote access to the system, along with instructions on how to install and operate the system. Participating teams were also invited to act as analysts to attempt replicating results with the submitted systems. The replication work is still ongoing at the time of writing this paper.

---

[7] https://www.reddit.com/r/AskDocs/.

Task 2. Teams could submit up to eight official runs. System performance was assessed using a Simple Evaluation approach and a Cost-Effective Evaluation approach. The assumption behind the Simple Evaluation approach is the following: The user of your system is the researcher that performs the abstract and title screening of the retrieved articles. Every time an abstract is returned (i.e. ranked) there is an incurred cost/effort, while the abstract is either irrelevant (in which case no further action will be taken) or relevant (and hence passed to the next stage of document screening) to the topic under review. Evaluation measures were: Area under the recall-precision curve, i.e. Average Precision; Minimum number of documents returned to retrieve all R relevant documents; Work Saved over Sampling at different Recall levels; Area under the cumulative recall curve normalized by the optimal area; Recall @ 0% to 100% of documents shown; a number of newly constructed cost-based measures; and reliability [1]. The assumption behind the Cost-Effective Evaluation approach is the following: The user that performs the screening is not the end-user. The user can interchangeably perform abstract and title screening, or document screening, and decide what PID's to pass to the end-user. Every time an abstract is returned the user can either (a) read the abstract (with an incurred cost CA) and decide whether to pass this PID to the end-user, or (b) read the full document (with an incurred cost of CA+CD) and decide whether to pass this PID to the end-user, or (c) directly pass the PID to the end user (with an incurred cost of 0), or (d) directly discard the PID and not pass it to the end user (with an incurred cost of 0). For every PID passed to the end-user there is also a cost attached to it: CA if the abstract passed on is not relevant, and CA+CD if the abstract passed on is relevant (that is, we assume that the end-user completes a two-round abstract and document screening, as usual, but only for the PIDs the algorithm and feedback user decided to be relevant). More details on the evaluation are provided in the Task 2 overview paper [9].

Task 3. For IRTask 1 (Ad-Hoc Search), participants could treat each query individually (without grouping variants together) and submit up to 7 ranked runs with up to 1,000 documents per query for all 300 queries. For IRTask 2 (Personalized Search), participants could submit up to 7 ranked runs with up to 1,000 documents per information need. For IRTask 3 (Query Variations), participants could submit results for each group of queries of a post, i.e. up to 7 ranked runs with up to 1,000 documents per information need. For IRTask 4 (Multilingual Search), participants could again treat each query individually (like in IRTask 1), submitting up to 7 ranked runs with up to 1,000 documents per query for all 300 queries for each language (Czech (CS), French (FR), Hungarian (HU), German (DE), Polish (PL) and Swedish (SV)).

The organizers also generated baseline runs and a set of benchmark systems using popular IR models implemented in Terrier and Indri. System evaluation was conducted using precision at 10 (p@10) and normalised discounted cumulative gain [8] at 10 (nDCG@10) as the primary and secondary measures, respectively. Precision was computed using the binary relevance assessments; nDCG was computed using the graded relevance assessments. A separate evaluation was conducted using the multidimensional relevance assessments (topical relevance, understandability and trustworthiness) following the methods in [20]. For all runs, Rank biased precision (RBP)[8] was computed along with the multidimensional modifications of RBP, namely uRBP (using binary topicality relevance and understandability assessments), uRBPgr (using graded topicality relevance and understandability assessments), u+tRBP (using binary topicality relevance, understandability and trustworthiness assessments) and α-uRBP (using a user expertise parameter α, binary topicality relevance and understandability assessments). More details on this multidimensional evaluation are provided in the Task overview paper [17]. Precision and Mean Average Precision were computed using `trec_eval`; while the multidimensional evaluation (comprising RBP) was performed using `ubire`[9].

3   Results

The number of people who registered their interest in CLEF eHealth tasks was 34, 40, and 43 respectively (and a total of 67 unique teams). In total, 32 teams submitted to the three shared tasks.

Task 1 received considerable interest with 34 registered participants. However, only 11 teams submitted runs, including one team from Australia (UNSW), five teams from France (LIMSI, LIRMM, LITL, Mondeca, and SIBM), two teams from Germany (TUC and WBI), one team from Italy (UNIPD), and one team from Russia (KFU). Five teams also submitted systems to the replication track, and two teams also volunteered to participate in the replication track as analysts. The training datasets were released at the end of January 2017 and the test datasets by 25 April 2017. The ICD-10 coding task submission on French and English death certificates were due by 5 May 2017 and the replication track systems by 12 May 2017.

For the English raw dataset, 9 teams submitted 15 runs (Table 2). For the French raw dataset, 6 teams submitted 7 runs for the raw dataset (Table 3) and 9 runs for the aligned dataset(Table 4). In addition to these official runs, unofficial runs were submitted by the task organizers and by some participants after the test submission deadline[10].

The best performance in official runs was achieved with an F-measure of 0.804 for French and of 0.850 for English. Systems relied both on knowledge based methods, machine learning methods, and sometimes a combination of them. The level of performance observed shows that there is potential for integrating automated assistance in the death certificate coding work flow both in French and in English. We hope that continued efforts towards reproducibility will support the shift from research prototypes to operational production systems. See the Task 1 overview paper for further details [14].

---

[8] The persistence parameter p in RBP was set to 0.8.

[9] https://github.com/ielab/ubire, [20].

[10] See Task 1 paper for details on unofficial runs [14].

Table 2. System performance for ICD10 coding on the English raw test corpus in terms of Precision (P), recall (R), and F-measure (F). The top part of the table displays official runs, while the bottom part displays baseline runs.

		ALL				EXTERNAL		
	Team	P	R	F	Team	P	R	F
	KFU-run1	.893	.811	**.850**	KFU-run1	.584	.357	**.443**
	KFU-run2	.891	**.812**	.850	KFU-run2	.631	.325	.429
	TUC-MI-run1	**.940**	.725	.819	SIBM-run1	.426	.389	.407
	SIBM-run1	.839	.783	.810	LIRMM-run2	.233	**.524**	.323
	TUC-MI-run2	.929	.717	.809	LIRMM-run1	.232	**.524**	.322
	WBI-run1	.616	.606	.611	TUC-MI-run1	.880	.175	.291
Official runs submitted	WBI-run2	.616	.606	.611	TUC-MI-run2	**1.00**	.159	.274
	LIRMM-run1	.691	.514	.589	UNSW-run1	.168	.262	.205
	LIRMM-run2	.646	.527	.580	Unipd-run2	.292	.111	.161
	Unipd-run1	.496	.442	.468	WBI-run1	.246	.119	.160
	UNSW-run1	.401	.352	.375	WBI-run2	.246	.119	.160
	Unipd-run2	.382	.341	.360	Unipd-run1	.279	.095	.142
	UNSW-run2	.371	.328	.348	UNSW-run2	.043	.310	.076
	Mondeca-run1	*invalid format*			Mondeca-run1	*invalid format*		
	average	.670	.582	.622	**average**	.405	.267	.261
	median	.646	.606	.611	**median**	.279	.262	.274
	Frequency baseline	.115	.085	.097	Frequency baseline	0.00	0.00	0.00
	ICD baseline	.029	.007	.011	ICD baseline	0.00	0.00	0.00

Table 3. System performance for ICD10 coding on the French raw test corpus in terms of Precision (P), recall (R), and F-measure (F). A horizontal dash line places the frequency baseline performance. The top part of the table displays official runs, while the bottom part displays baseline runs.

		ALL				EXTERNAL		
	Team	P	R	F	Team	P	R	F
	SIBM-run1	**.857**	**.689**	**.764**	SIBM-run1	**.567**	**.431**	**.490**
	LITL-run2	.666	.414	.510	LIRMM-run1	.443	.367	.401
Official runs	LIRMM-run1	.541	.480	.509	LIRMM-run2	.443	.367	.401
	LIRMM-run2	.540	.480	.508	LITL-run2	.560	.283	.376
	LITL-run1	.651	.404	.499	LITL-run1	.538	.277	.365
	TUC-MI-run2	.044	.026	.033	TUC-MI-run2	.010	.004	.005
	TUC-MI-run1	.025	.015	.019	TUC-MI-run1	.006	.005	.005
	average	.475	.358	.406	**average**	.367	.247	.292
	median	.541	.414	.508	**median**	.443	.283	.376
	Frequency baseline	.339	.237	.279	Frequency baseline	.381	.110	.170

Task 2 also received much interest with 40 registered participants. Of these 14 teams submitted runs, including 1 team from Australia (QUT), 1 team from Canada (Waterloo), 1 team from China (ECNU), 1 team from France (CNRS), 1 team from Greece (AUTH), 1 team from India (IIIT), 1 team from Italy

Table 4. System performance for ICD10 coding on the French aligned test corpus in terms of Precision (P), recall (R), and F-measure (F). A horizontal dash line places the frequency baseline performance. The top part of the table displays official runs, while the bottom part displays baseline runs.

	ALL			EXTERNAL			
Team	P	R	F	Team	P	R	F
SIBM-run1	.835	**.775**	**.804**	SIBM-run1	.534	**.472**	**.501**
WBI-run1	.780	.751	.765	TUC-MI-run2	**.740**	.318	.445
TUC-MI-run2	**.874**	.611	.719	LIRMM-run1	.412	.403	.407
LITL-run1	.612	.550	.579	LIRMM-run2	.412	.403	.407
LIRMM-run1	.506	.530	.518	LITL-run1	.482	.348	.404
LIRMM-run2	.505	.530	.517	LITL-run2	.534	.275	.363
LITL-run2	.646	.402	.495	WBI-run1	.709	.151	.249
TUC-MI-run1	.426	.297	.350	TUC-MI-run1	.218	.119	.154
average	.648	.555	.593	**average**	.505	.311	.366
median	.629	.540	.548	**median**	.508	.333	.406
Frequency baseline	.640	.470	.542	Frequency baseline	.508	.338	.406
ICD baseline	.346	.041	.073	ICD baseline	.000	.000	.000

(The left block rows SIBM-run1 through TUC-MI-run1 are labeled "Official runs".)

(Padua), 1 team from the Netherlands (AMC), 1 team from Singapore (NTU), 1 team from Switzerland (ETH), 3 teams from the United Kingdom (Sheffield, UCL, UOS), and 1 team from the United States (NCSU). The training datasets were released on the 10 March 2017 and the test datasets (with gold standard annotations) by May 2017. Participants submissions were due by 14 May 2017. In total, 14 teams submitted at least one run. See the Task 2 overview paper for further details and the results of the evaluation [9].

Task 3 received much interest with 43 registered participants. Of these 7 teams submitted runs, including 1 team from Australia (QUT), 1 team from Austria (TUW), 1 team from Botswana (UB-Botswana), 1 team from Czech Republic (CUNI), 1 team from Korea (KISTI), 1 team from Portugal (UEvora), and 1 team from Spain (SINAI). Participants submissions were due by 9 June 2017 and the relevance assessments are being collected at the time of writing of this paper. See the Task 3 overview paper for further details and the results of the evaluation [17].

4   Conclusions

In this paper we provided an overview of the CLEF eHealth 2017 evaluation lab. In recent year's the CLEF eHealth lab has offered a recurring contribution to the creation and dissemination of test collections in the fields of biomedical IR and IE. This edition of CLEF eHealth offered three tasks: Task 1 on multilingual IE to extend from last year's task on French corpora, Task 2 on technologically assisted reviews in empirical medicine as a new pilot task, and Task 3 on patient-centred IR extending the 2013–16 IR tasks. We also continued the replication track from 2016 in Task 1. More specifically, Task 1 offered test collections addressing the task of automatic coding using the International Classification of Diseases for death certificates in two languages. Task 2 and Task 3 offered test collections addressing two aspects of biomedical IR: high-recall IR over PubMed Abstracts and Titles for the purpose of conducting systematic reviews of Diagnostics Test Accuracy studies (Task 2) and effectiveness, quality, and personalization for health related searches made on the Web (Task 3).

Each task's test collections offered a specific task definition, implemented in a dataset distributed together with an implementation of relevant evaluation metrics to allow for direct comparability of the results reported by systems evaluated on the collections. The established CLEF eHealth IE and IR tasks (Task 1 and Task 3) used a traditional shared task model evaluation approach again this year whereby a community-wide evaluation is executed in a controlled setting: participants have access to test data at the same time, following which no further updates to systems are allowed and following submission of the outputs from their frozen IE or IR system to the task organiser, their results are evaluated blindly by an independent third party who reports label results for all participants. With our new pilot IR task (Task 2) we aspire to offering means to conduct cross comparable relevance feedback loops, with plans to introduce a newer form of shared evaluation next year through the use of a live evaluation service.

The CLEF eHealth lab has matured and established its presence during its five iterations in 2013–2017. In total, 67 unique teams registered their interests and 32 teams took part in the 2017 tasks (11 in Task 1, 14 in Task 2, and 7 in Task 3). In comparison, in 2016, 2015, 2014, and 2013, the number of team registrations was 116, 100, 220, and 175, respectively and the number of participating teams was 20, 20, 24, and 53 [5,10,11,19]. Given the significance of the tasks, all test collections and resources associated with the lab have been made available to the wider research community through our CLEF eHealth website[11].

Acknowledgements. The CLEF eHealth 2017 evaluation lab has been supported in part by (in alphabetical order) the ANR, the French National Research Agency, under grant CABeRneT ANR-13-JS02-0009-01, the CLEF Initiative and Data61. We are also thankful to the people involved in the annotation, query creation, and relevance assessment exercise. Last but not least, we gratefully acknowledge the participating teams' hard work. We thank them for their submissions and interest in the lab.

References

1. Cormack, G.V., Grossman, M.R.: Engineering quality and reliability in technology-assisted review. In: Proceedings of the 39th International ACM SIGIR Conference on Research and Development in Information Retrieval, SIGIR 2016, pp. 75–84. ACM, New York (2016). http://doi.acm.org/10.1145/2911451.2911510
2. Cormack, G.V., Smucker, M.D., Clarke, C.L.: Efficient and effective spam filtering and re-ranking for large web datasets. Inf. Retrieval **14**(5), 441–465 (2011)

---

[11] https://sites.google.com/site/clefehealth/.

3. Goeuriot, L., Jones, G.J., Kelly, L., Leveling, J., Hanbury, A., Müller, H., Salantera, S., Suominen, H., Zuccon, G.: ShARe, CLEF eHealth Evaluation Lab 2013, Task 3: Information retrieval to address patients' questions when reading clinical reports. CLEF 2013 Online Working. Notes 8138 (2013)

4. Goeuriot, L., Kelly, L., Lee, W., Palotti, J., Pecina, P., Zuccon, G., Hanbury, A., Gareth J.F. Jones, H.M.: ShARe/CLEF eHealth Evaluation Lab 2014, Task 3: user-centred health information retrieval. In: CLEF 2014 Evaluation Labs and Workshop: Online Working Notes, Sheffield, UK (2014)

5. Goeuriot, L., Kelly, L., Suominen, H., Hanlen, L., Névéol, A., Grouin, C., Palotti, J., Zuccon, G.: Overview of the CLEF eHealth evaluation lab 2015. In: Mothe, J., Savoy, J., Kamps, J., Pinel-Sauvagnat, K., Jones, G.J.F., SanJuan, E., Cappellato, L., Ferro, N. (eds.) CLEF 2015. LNCS, vol. 9283, pp. 429–443. Springer, Cham (2015). doi:10.1007/978-3-319-24027-5_44

6. Hanbury, A., Müller, H.: Khresmoi - multimodal multilingual medical information search. In: Medical Informatics Europe 2012 (MIE 2012), Village of the Future (2012)

7. Hiemstra, D., Hauff, C.: Mirex: Mapreduce information retrieval experiments. arXiv preprint arXiv:1004.4489

8. Järvelin, K., Kekäläinen, J.: Cumulated gain-based evaluation of IR techniques. ACM Trans. Inf. Syst. 20(4), 422–446 (2002)

9. Kanoulas, E., Li, D., Azzopardi, L., Spijker, R.: Overview of the CLEF technologically assisted reviews in empirical medicine. In: Working Notes of Conference and Labs of the Evaluation (CLEF) Forum. CEUR Workshop Proceedings (2017)

10. Kelly, L., Goeuriot, L., Suominen, H., Névéol, A., Palotti, J., Zuccon, G.: Overview of the CLEF eHealth evaluation lab 2016. In: Fuhr, N., Quaresma, P., Gonçalves, T., Larsen, B., Balog, K., Macdonald, C., Cappellato, L., Ferro, N. (eds.) CLEF 2016. LNCS, vol. 9822, pp. 255–266. Springer, Cham (2016). doi:10.1007/978-3-319-44564-9_24

11. Kelly, L., Goeuriot, L., Suominen, H., Schreck, T., Leroy, G., Mowery, D.L., Velupillai, S., Chapman, W.W., Martinez, D., Zuccon, G., Palotti, J.: Overview of the ShARe/CLEF eHealth evaluation lab 2014. In: Kanoulas, E., Lupu, M., Clough, P., Sanderson, M., Hall, M., Hanbury, A., Toms, E. (eds.) CLEF 2014. LNCS, vol. 8685, pp. 172–191. Springer, Cham (2014). doi:10.1007/978-3-319-11382-1_17

12. Leeflang, M.M., Deeks, J.J., Takwoingi, Y., Macaskill, P.: Cochrane diagnostic test accuracy reviews. Syst. Rev. 2(1), 82 (2013)

13. Macdonald, C., McCreadie, R., Santos, R.L., Ounis, I.: From puppy to maturity: experiences in developing terrier. In: Proceeding of OSIR at SIGIR, pp. 60–63 (2012)

14. Névéol, A., Anderson, R.N., Cohen, K.B., Grouin, C., Lavergne, T., Rey, G., Robert, A., Zweigenbaum, P.: CLEF eHealth 2017 multilingual information extraction task overview: ICD10 coding of death certificates in English and French. In: CLEF 2017 Online Working Notes. CEUR-WS (2017)

15. Névéol, A., Cohen, K.B., Grouin, C., Hamon, T., Lavergne, T., Kelly, L., Goeuriot, L., Rey, G., Robert, A., Tannier, X., Zweigenbaum, P.: Clinical information extraction at the CLEF eHealth evaluation lab 2016. In: CLEF 2016 Evaluation Labs and Workshop: Online Working Notes. CEUR-WS (2016)

16. Palotti, J., Zuccon, G., Goeuriot, L., Kelly, L., Hanburyn, A., Jones, G.J., Lupu, M., Pecina, P.: CLEF eHealth evaluation lab 2015, Task 2: retrieving information about medical symptoms. In: CLEF 2015 Online Working Notes. CEUR-WS (2015)

17. Palotti, J., Zuccon, G., Jimmy, Pecina, P., Lupu, M., Goeuriot, L., Kelly, L., Hanbury, A.: CLEF 2017 task overview: the IR task at the eHealth evaluation lab. In: Working Notes of Conference and Labs of the Evaluation (CLEF) Forum. CEUR Workshop Proceedings (2017)
18. Strohman, T., Metzler, D., Turtle, H., Croft, W.B.: Indri: a language model-based search engine for complex queries. In: Proceedings of the International Conference on Intelligent Analysis, vol. 2, pp. 2–6. Citeseer (2005)
19. Suominen, H., et al.: Overview of the ShARe/CLEF eHealth evaluation lab 2013. In: Forner, P., Müller, H., Paredes, R., Rosso, P., Stein, B. (eds.) CLEF 2013. LNCS, vol. 8138, pp. 212–231. Springer, Heidelberg (2013). doi:10.1007/978-3-642-40802-1_24
20. Zuccon, G.: Understandability biased evaluation for information retrieval. In: Advances in Information Retrieval, pp. 280–292 (2016)
21. Zuccon, G., Koopman, B.: Integrating understandability in the evaluation of consumer health search engines. In: Medical Information Retrieval Workshop at SIGIR 2014, p. 32 (2014)
22. Zuccon, G., Palotti, J., Goeuriot, L., Kelly, L., Lupu, M., Pecina, P., Mueller, H., Budaher, J., Deacon, A.: The IR task at the CLEF eHealth evaluation lab 2016: user-centred health information retrieval. In: CLEF 2016 Evaluation Labs and Workshop: Online Working Notes, CEUR-WS, September 2016

CLEF 2017 Microblog Cultural Contextualization Lab Overview

Liana Ermakova[1], Lorraine Goeuriot[3], Josiane Mothe[2], Philippe Mulhem[3], Jian-Yun Nie[4], and Eric SanJuan[5(✉)]

[1] LISIS (UPEM, INRA, ESIEE, CNRS), Université de Lorraine, Nancy, France
liana.ermakova@univ-lorraine.fr
[2] IRIT, UMR5505 CNRS, ESPE, Université de Toulouse, Toulouse, France
josiane.mothe@irit.fr
[3] LIG, Université de Grenoble, Grenoble, France
[4] RALI, Université de Montréal, Québec, Canada
[5] LIA, Université d'Avignon, Avignon, France
eric.sanjuan@univ-avignon.fr

Abstract. MC2 CLEF 2017 lab deals with how cultural context of a microblog affects its social impact at large. This involves microblog search, classification, filtering, language recognition, localization, entity extraction, linking open data, and summarization. Regular Lab participants have access to the private massive multilingual microblog stream of *The Festival Galleries* project. Festivals have a large presence on social media. The resulting mircroblog stream and related URLs is appropriate to experiment advanced social media search and mining methods. A collection of 70,000,000 microblogs over 18 months dealing with cultural events in all languages has been released to test multilingual content analysis and microblog search. For content analysis topics were in any language and results were expected in four languages: English, Spanish, French, and Portuguese. For microblog search topics were in four languages: Arabic, English, French and Spanish, and results were expected in any language.

1 Introduction: From Microblog to Cultural Contextualization

Microblog Contextualization was introduced as a Question Answering task of INEX 2011 [1]. The main idea was to help Twitter users to understand a tweet by providing some context associated to it. It has evolved in a Focus IR task over WikiPedia [2].

The CLEF 2016 Cultural Microblg Contextualization Workshop considered specific cultural twitter feeds [3]. In this context restricted context implicit localization and language identification appeared to be important issues. It also required identifying implicit timelines over long periods. The MC2 CLEF 2017 lab has been centered on Cultural Contextualization based on Microblog feeds. It dealt with how cultural context of a microblog affects its social impact

© Springer International Publishing AG 2017
G.J.F. Jones et al. (Eds.): CLEF 2017, LNCS 10456, pp. 304–314, 2017.
DOI: 10.1007/978-3-319-65813-1_27

at large [4]. This involved microblog search, classification, filtering, language recognition, localization, entity extraction, linking open data, and summarization. Regular Lab participants had access to the private massive multilingual microblog stream of *The Festival Galleries* project[1]. Festivals have a large presence on social media. The resulting microblog stream and related URLs were appropriate to experiment advanced social media search and mining methods.

The overall usage scenario for the lab has been centered on festival attendees:

- an insider attendee who receives a microblog about the cultural event which he will participate in will need context to understand it (microblogs often contain implicit information).
- a participant in a specific location wants to know what is going on in surrounding events related to artists, music, or shows that he would like to see. Starting from a list of bookmarks in the Wikipedia app, the participant will seek for a short list of microblogs summarizing the current trends about related cultural events. We hypothesize that she/he is more interested in microblogs from insiders than outsiders or officials.

These scenari lead to three tasks lab participants could answer to:

- Content analysis
- Microblog search
- Timeline illustration

These tasks are detailed in Sects. 3 to 5. Section 2 depicts the data used in the various tasks and Sect. 6 describes the evaluation. Finally Sect. 7 draws some conclusions.

2 Data

The lab gave access to registered participants to a massive collection of microblogs and URLs related to cultural festivals in the world.

It allows researchers in IR (Information Retrieval) and NLP (Natural Language Processing) to experiment a broad variety of multilingual microblog search techniques (Wikipedia entity search, automatic summarization, language identification, text localization, etc.).

A personal login was required to acces the data. Once registered on CLEF each registered team can obtain up to 4 extra individual logins by writing to admin@talne.eu. This collection is still accessible on demand. Any usage requires to make a reference to the following paper: "L. Ermakova, L. Goeuriot, J. Mothe, P. Mulhem, J.-Y. Nie, and E. SanJuan, CLEF 2017 Microblog Cultural Contextualization Lab Overview, Proceedings of Experimental IR Meets Multilinguality, Multimodality, and Interaction 8th International Conference of the CLEF Association, CLEF 2017, LNCS 10439, Dublin, Ireland, September 11–14, 2017". Updates will be frequently posted on the lab website[2].

---

[1] http://www.agence-nationale-recherche.fr/?Project=ANR-14-CE24-0022.

[2] https://mc2.talne.eu/lab/.

An Indri index with a web interface is available to query the whole set of microblogs. Online Indri indexes are available in English, Spanish, French, and Potuguses for Wikipedia search.

2.1 Microblog Collection

The document collection is an updated extension of the microblog stream presented at the CLEF 2016 workshop [5] (see also [6]).

It was provided to registered participants by ANR GAFES project[3]. It consists in a pool of more than 50 M unique microblogs from different sources with their meta-information as well as ground truth for the evaluation.

The microblog collection contains a very large pool of public posts on Twitter using the keyword "festival" since June 2015. These microblogs were collected using private archive services based on streaming API. The average of unique microblog posts (i.e. without re-tweets) between June and September is 2,616,008 per month. The total number of collected microblog posts after one year (from May 2015 to May 2016) is 50,490,815 (24,684,975 without re-posts). These microblog posts are available online on a relational database with associated fields.

Because of privacy issues, they cannot be publicly released but can be analyzed inside the organization that purchased these archives and among collaborators under privacy agreement. The MC2 lab provides this opportunity to share this data among academic participants. These archives can be indexed, analyzed and general results acquired from them can be published without restriction.

2.2 Linked Web Pages

66% of the collected microblog posts contain *Twittert.co* compressed URLs. Sometimes these URLs refer to other online services like *adf.ly, cur.lv, dlvr.it, ow.ly* that hide the real URL. We used the spider mode of the GNU *wget* tool to get the real URL, this process required multiple DNS requests.

The number of unique uncompressed URLs collected in one year is 11,580,788 from 641,042 distinct domains.

2.3 Wikipedia XML Corpus for Summary Generation

Wikipedia is under Creative Commons license, and its content can be used to contextualize tweets or to build complex queries referring to Wikipedia entities.

We have extracted an average of 10 million XML documents from Wikipedia per year since 2012 in the four main Twitter languages: English (en), Spanish (es), French (fr), and Portuguese (pt).

These documents reproduce in an easy-to-use XML structure the content of the main Wikipedia pages: title, abstract, section, and subsections as well as Wikipedia internal links. Other content such as images, footnotes and external

---

[3] http://www.agence-nationale-recherche.fr/?Projet=ANR-14-CE24-0022.

links is stripped out in order to obtain a corpus easier to process using standard NLP tools.

By comparing contents over the years, it was possible to detect long term trends.

3 Content Analysis

The content analysis task has been inspired by [7–9].

Given a stream of microblogs, the task consists in:

- filtering microblogs dealing with festivals;
- language(s) identification;
- event localization;
- author categorization (official account, participant, follower or scam);
- Wikipedia entity recognition and translation in four target languages: English, Spanish, Portuguese, and French.
- automatic summarization of linked Wikipedia pages in the four target languages.

Each item has been evaluated independently, however, language identification could impact Wikipedia linking and the resulting summaries. The filtering and author categorization subtasks were inspired by the filtering and priority tasks at RepLab 2014 [10].

Opinion mining was not initially considered, however two participants did apply binary opinion classifiers and detection of controversies on the provided corpus. It appears that like for the Reputation task in RepLab [10], microblog content needs to be contextualized and expanded to accurately identify opinions, especially for cultural events where nouns and adjectives considered as negative can reflect highly positive opinions for specific communities.

Language(s) identification is challenging over short contents that tend to mix several languages. Festival names over tweets often appear in English but the rest of the content can be in any other languages. Moreover festival attendees tend to add terms from various dialects to highlight the local context.

Event localization requires external resources. For large festivals WikiPedia often contains the information and it can be retrieved based on state-of-the-art QA approaches. However for small events it is necessary to query the public web or social networks. In this lab the 18 months feed of tweets about festivals allows to search for microblogs by festival organizers about venues.

The two subtasks Wikipedia Entity Recognition and Automatic Summarization refer to previous experiments around Tweet Contextualization [2]. Most efficient methods proceed in two steps: (1) retrieve most relevant Wikipedia pages, (2) propose a multidocument summary of them. WikiFying tweets is complex due to the lexical gap between tweets and Wikipedia pages. Extracting summaries looked easier by aggregating sentences from pages but ensuring and evaluating readability is an issue, specially on languages with less resources than English.

4   Microblog Search

Given a cultural query about festivals in Arabic, English, French, or Spanish, the task is to search for the 64 most relevant microblogs in a collection covering 18 months of news about festivals in all languages.

Queries have been extracted from resources suggested by participants.

Arabic and English queries were extracted from the Arab Spring Microblog corpus [11]. We considered the content of all the tweets dealing with festivals during the Arab Spring period and the task consisted in searching for traces of these festivals or artists in the lab corpus two years after. The use case was to follow up artists involved in the Arab spring festivals two or three years later. Indeed, most of the festivals before the Arab spring were relying on tourism and have been stopped after 2014, so they don't appear as festivals in the MC2 corpus of microblogs which spans from 2015 to 2016. However, most famous artists and film makers involved in Arabic festival have been invited in European and Canadian Festivals in 2015. The microblog search task in English and Arabic consisted in retrieving those indirectly related microblogs.

The difficulty of this task relies in generating a query based on a microblog content. Tweets are too short to apply standard name entity extraction algorithms but they are too long to be considered as queries for an IR system without a robust preprocessing that removes empty words and keeps only informative terms. Corpus and queries being encoded in utf8, systems can handle multiple languages, however language stop word lists and lemmatizers are specific to each language. For Arabic another difficulty appeared with dialects. Dealing with them require extra linguistic resources.

French queries were extracted from the VodKaster Micro Film Reviews [12]. VodKaster is a French social network about films. Users can post and share micro reviews in French about movies as they watch them. They can score films but also reviews written by others. We extracted as queries all micro reviews dealing with festivals during the period of the lab corpus (2015–2016). Most of them where posted from phones by festival attendees. Film micro reviews are easier to process than tweets because most of them contain well formed sentences. Searching for related tweets could be improved by considering the date of the micro review to identify the film festival. However microblogs about other festivals mentioning the same films or actors were also considered as relevant.

Spanish queries are a representative sample of sentences dealing with festivals from the Mexican newspaper *La jornada*[4]. We considered all the sentences from the newspaper mentioning a festival and extracted a random sample from this pool. These are well formed sentences easy to analyze but much harder to contextualize. Extracting queries about these sentences often requires to find the source article and neighboring sentences. The use case was that a reader highlights a sentence while reading the newspaper and sees related microblogs. Like for Arabic, it was also necessary to deal with the multiple variants of Spanish language.

---

[4] http://www.jornada.unam.mx.

A language model index powered by Indri and accessible through a web API has been provided. To deal with reposts, there was one document by user grouping all his/her posts including the reposts. Each document has an XML structure (cf. Fig. 1). Figure 2 gives an example of such XML document.

```
<!ELEMENT xml (f, m)+>
<!ELEMENT f ($\#$ user\_id)>
<!ELEMENT m (i, u, l, c d, t)>
<!ELEMENT i ($\#$ microblog\_id)>
<!ELEMENT u ($\#$ user)>
<!ELEMENT l ($\#$ ISO\_language\_code)>
<!ELEMENT c ($\#$ client)>
<!ELEMENT d ($\#$ date)>
<!ELEMENT t ($\#$ PCDATA)>
```

Fig. 1. XML DTD for microblog search

```
<xml><f>20666489</f>
 <m><i>727389569688178688</i>
  <u>soulsurvivornl</u>
  <l>en</l>
  <c>Twitter for iPhone</c>
  <d>2016-05-03</d>
  <t>RT @ndnl: Dit weekend begon het Soul Surivor Festival.</t>
 </m>
 <m><i>727944506507669504</i>
  <u>soulsurvivornl</u>
  <l>en</l>
  <c>Facebook</c>
  <d>2016-05-04</d>
  <t>Last van een festival-hangover?</t>
 </m>
</xml>
```

Fig. 2. An example of document for microblog search

This XML structure permits to work with complex queries like:

```
\# combine[m](
  Instagram.c es.l  \# 1(2016 05).d conduccin
  \# syn(pregoneros pregonero) \# syn(festivales festival))
```

This query will look for microblogs ([m]) posted from Instagram (.c) using Spanish locale (.l) in May 2016 (.d) dealing with pregonero(s) and festival(es).

5   Timeline Illustration

The goal of this task was to retrieve all relevant tweets dedicated to each event of a festival, according to the program provided. We were really looking here at a kind of "total recall" retrieval, based on initial artists' names and names, dates, and times of shows.

For this task, we focused on 4 festivals. Two French Music festivals, one French theater festival and one Great Britain theater festival:

- Vielles Charrues 2015;
- Transmusicales 2015;
- Avignon 2016;
- Edinburgh 2016.

Each topic was related to one cultural event. In our terminology, one event is one occurrence of a show (theater, music, ...). Several occurrences of the same show correspond then to several events (e.g. plays can be presented several times during theater festivals). More precisely, one topic is described by: one id, one festival name, one title, one artist (or band) name, one timeslot (date/time begin and end), and one venue location.

An excerpt from the topic list is:

```
<topic>
  <id>5</id>
  <title></title>
  <artist>Klangstof</artist>
  <festival>transmusicales</festival>
  <startdate>04/12/16-17:45</startdate>
  <enddate>04/12/16-18:30</enddate>
  <venue>UBU</venue>
</topic>
```

The id was an integer ranging from 1 to 664. We see from the excerpt above that, for a live music show without any specific title, the title field was empty. The artist name was a single artist, a list of artist names, an artistic company name or orchestra name, as they appear in the official programs of the festivals.

The festival labels were:

- *charrues* for Vielles Charrues 2015;
- *transmusicales* for Transmusicales 2015;
- *avignon* for Avignon 2016;
- *edinburgh* for Edinburgh 2016.
- For the date/time fields, the format is: *DD/MM/YY-HH:MM*.
- The *venue* is a string corresponding to the name of the location, given by the official programs.

If the start or end time is unknown, they are replaced with: *DD/MM/YY-xx:xx*. If the day is unknown, the date format is the following: *-HH:MM* (day is omitted).

Participants were required to use the full dataset to conduct their experiments.

The runs were expected to respect the classical TREC top files format. Only the top 1000 results for each query run must be given. Each retrieved document is identified using its tweet id. The evaluation is achieved on a subset of the full set of topics, according to the richness of the results obtained. The official evaluation measures were interpolated precision at 1% and recall values at 5, 10, 25, 50 and 100 documents.

6 Results

Overall, 53 teams involving 72 individuals registered to the lab. Among them 12 teams from Brazil (1), France (4), Tunisia (2), Mexico (1), India (1), and Mongolia (1) submitted 42 valid runs. 11 teams submitted a working note to CLEF 2017.

6.1 Evaluation

For *Content Analysis*, q-rels based on pooling from participant submissions appeared to be unstable due to the variety of languages involved in this task and the variety of participants' approaches that could be efficient on different subsets of languages. Results were to be provided in four different languages; however, the submitted runs were extremely multilingual. Reaching stable q-rels by pooling would have required to stratify by language on input and output which leads to very sparse matrices of results. All results had to be extracted from the four WikiPedias in English, Spanish, French and Portuguese to have a common document ground base, but even the WikiPedia appeared to be highly redundant with multiple pages with similar content referring to the same cultural event from different perspectives.

By contrast, extensive textual references by organizers manually built on a reduced random subset of topics using the one powered by Indri provided to participants[5] and the aggregator DuckDuckGo[6] runs and on the four targeted different languages appeared to be more stable to rank runs based on token overlapping following the same methodology as in [2].

For Multilingual Microblog Search, we applied the same methodology based on textual references instead of document q-rels. Seven trilingual annotators fluently speaking 13 languages (Arabic, Hebrew, Euskadi, Catalan, Mandarin Chinese, English, French, German, Italian, Portuguese, Russian, Spanish and Turkish) produced an initial textual reference. This reference was extended to

---

[5] http://tc.talne.eu.

[6] https://ducduckgo.com.

Corean, Japanese and Persian based on Google translate. However this automatic extension appeared to be noisy and had to be dropped out from the reference. Only results in one of the assessors language could then be evaluated.

For *Timeline Illustration* it was anticipated that re-tweets would be excluded from the pools. But the fact that it was recall-oriented task lead participants to return all retweets. Excluding retweets would have disqualified recall oriented runs that missed one original tweet. Moreover it emerged during the evaluation that retweets are often more interesting than original ones. Indeed original ones are often posted by festival organizers meanwhile reposts by individuals are more informative about festival attendees participation.

Therefore, building a set of document q-rels for time-line illustration was a two step process.

First, tweet relevance on original tweets from baselines (each participant was asked to provide a baseline) has been assessed on a 3-level scale:

- Not relevant: the tweet is not related to the topic.
- Partially relevant: the tweet is somehow related to the topic (e.g. the tweet is related to the artist, song, play but not to the event, or is related to a similar event with no possible way to check if they are the same).
- Relevant: the tweet is related to the event.

Secondly, the q-rels were expanded to any microblog containing the text of one previously assessed as relevant this way, the q-rels were expanded to all reposts. Participant runs have then be ranked using treceval program provided by NIST TREC[7]. All measures have been provided since they lead to different rankings.

6.2   Participant Approaches

Among the 12 active participants, 6 teams participated to Content Analysis but only one (LIA) managed to produce multilingual contextual summaries in four languages on all queries (microblogs without urls mixing more than 30 languages) and only one managed to deal with the localization task (Syllabs). 5 teams participated to the multilingual microblog search but none managed to process the four sets of queries. All did process the English set, three could process French queries, one Arabic queries and one Spanish queries. Building realable multilingual stop word lists was a major issue and required linguistic expertise. 4 teams participated to the timeline illustrations task but only one outperformed the BM25 baseline. The main issue was to identify microblogs related to one of the four festivals chosen by organizers. This selection couldn't be only based on festival names since some relevant microblogs didn't include the festival hashtag, neither on the dates since microblogs about videos posted by festivals later on after the event were considered as relevant.

The most effective approaches have been:

- Language Identification: Syllabs enterprise based on linguistic resources on Latin languages.

---

[7] http://trec.nist.gov/trec_eval/.

- Entity Extraction: FELTS system based on string matching over very large lexicons.
- MultiLingual Contextualization: LIA team based on automatic multidocument summarization using Deep Learning.
- MIcroblog Search: LIPAH based on LDA query reformulation for Language Model.
- Timeline Illustration: IITH using BM25 and DRF based on artist name, festival name, top hashtags of each event features.

7   Conclusion

Dealing with a massive multilingual multicultural corpus of microblogs reveals the limits of both statistical and linguistic approaches. Raw utf8 text needs to be indexed without chunking. Synonyms and ambiguous terms over multiple languages have to be managed at query level. This requires positional index but the usage of utf8 encoding makes them slow. It also requires linguistic resources for each language or for specific cultural events. Therefore language and festival recognition appeared to be the key points of MC2 CLEF 2017 official tasks.

The CLEF 2017 MC2 also expanded from a regular IR evaluation task to a task search. Almost all participants used the data and infrastructure to deal with problematics beyond the initial scope of the lab. For example:

- the LSIS-EJCAM team used this data to analyze the role of social media in propagating controversies.
- the ISAMM team experimented opinion polarity detection in Twitter data combining sequence mining and topic modeling.
- the *My Local Influence* and U3ICM team experimented using sociological needs to characterize profiles and contents for Microblog search.

Researchers interested in using MC2 Lab data and infrastructure, but who didn't participate to the 2017 edition, can apply untill march 2019 to get access to the data and baseline system for their academic institution by contacting `eric.sanjuan@talne.eu`. Once the application accepted, they will get a personal private login to access lab resources for research purposes.

References

1. SanJuan, E., Moriceau, V., Tannier, X., Bellot, P., Mothe, J.: Overview of the INEX 2011 question answering track (QA@INEX). In: Geva, S., Kamps, J., Schenkel, R. (eds.) INEX 2011. LNCS, vol. 7424, pp. 188–206. Springer, Heidelberg (2012). doi:10.1007/978-3-642-35734-3_17
2. Bellot, P., Moriceau, V., Mothe, J., SanJuan, E., Tannier, X.: INEX tweet contextualization task: evaluation, results and lesson learned. Inf. Process. Manage. **52**(5), 801–819 (2016). doi:10.1016/j.ipm.2016.03.002

3. Ermakova, L., Goeuriot, L., Mothe, J., Mulhem, P., Nie, J., SanJuan, E.: Cultural micro-blog contextualization 2016 workshop overview: data and pilot tasks. In: Working Notes of CLEF 2016 - Conference and Labs of the Evaluation forum, Évora, Portugal, 5–8 September 2016, pp. 1197–1200 (2016)

4. Murtagh, F.: Semantic mapping: towards contextual and trend analysis of behaviours and practices. In: Working Notes of CLEF 2016 - Conference and Labs of the Evaluation forum, Évora, Portugal, 5–8 September 2016, pp. 1207–1225 (2016)

5. Goeuriot, L., Mothe, J., Mulhem, P., Murtagh, F., SanJuan, E.: Overview of the CLEF 2016 cultural micro-blog contextualization workshop. In: Fuhr, N., Quaresma, P., Gonçalves, T., Larsen, B., Balog, K., Macdonald, C., Cappellato, L., Ferro, N. (eds.) CLEF 2016. LNCS, vol. 9822, pp. 371–378. Springer, Cham (2016). doi:10.1007/978-3-319-44564-9_30

6. Balog, K., Cappellato, L., Ferro, N., Macdonald, C. (eds.): Working Notes of CLEF 2016 - Conference and Labs of the Evaluation forum, Évora, Portugal, 5–8 September 2016, vol. 1609. CEUR Workshop Proceedings (2016). CEUR-WS.org

7. Hoang, T.B.N., Mothe, J.: Building a knowledge base using microblogs: the case of cultural microblog contextualization collection. In: Working Notes of CLEF 2016 - Conference and Labs of the Evaluation forum, Évora, Portugal, 5–8 September 2016, pp. 1226–1237 (2016)

8. Scohy, C., Chaham, Y.R., Déjean, S., Mothe, J.: Tweet data mining: the cultural microblog contextualization data set. In: Working Notes of CLEF 2016 - Conference and Labs of the Evaluation forum, Évora, Portugal, 5–8 September 2016, pp. 1246–1259 (2016)

9. Pontes, E.L., Torres-Moreno, J., Huet, S., Linhares, A.C.: Tweet contextualization using continuous space vectors: Automatic summarization of cultural documents. In: Working Notes of CLEF 2016 - Conference and Labs of the Evaluation forum, Évora, Portugal, 5–8 September 2016, pp. 1238–1245 (2016)

10. Amigó, E., Carrillo-de-Albornoz, J., Chugur, I., Corujo, A., Gonzalo, J., Meij, E., Rijke, M., Spina, D.: Overview of RepLab 2014: author profiling and reputation dimensions for online reputation management. In: Kanoulas, E., Lupu, M., Clough, P., Sanderson, M., Hall, M., Hanbury, A., Toms, E. (eds.) CLEF 2014. LNCS, vol. 8685, pp. 307–322. Springer, Cham (2014). doi:10.1007/978-3-319-11382-1_24

11. Features Extraction To Improve Comparable Tweet corpora Building, JADT (2016)

12. Cossu, J.V., Gaillard, J., Juan-Manuel, T.M., El Bèze, M.: Contextualisation de messages courts: l'importance des métadonnées. In: EGC 2013 13e Conférence Francophone sur l'Extraction et la Gestion des connaissances, Toulouse, France, January 2013

Overview of ImageCLEF 2017: Information Extraction from Images

Bogdan Ionescu[1](✉), Henning Müller[2], Mauricio Villegas[3], Helbert Arenas[4],
Giulia Boato[5], Duc-Tien Dang-Nguyen[6], Yashin Dicente Cid[2],
Carsten Eickhoff[7], Alba G. Seco de Herrera[8], Cathal Gurrin[6], Bayzidul Islam[9],
Vassili Kovalev[10], Vitali Liauchuk[10], Josiane Mothe[4], Luca Piras[11],
Michael Riegler[12], and Immanuel Schwall[7]

[1] University Politehnica of Bucharest, Bucharest, Romania
bionescu@alpha.imag.pub.ro
[2] University of Applied Sciences Western Switzerland (HES-SO),
Sierre, Switzerland
[3] SearchInk, Berlin, Germany
[4] Institut de Recherche en Informatique de Toulouse, Toulouse, France
[5] University of Trento, Trento, Italy
[6] Dublin City University, Dublin, Ireland
[7] ETH Zurich, Zurich, Switzerland
[8] National Library of Medicine, Bethesda, MD, USA
[9] Technische Universität Darmstadt, Darmstadt, Germany
[10] United Institute of Informatics Problems, Minsk, Belarus
[11] University of Cagliari, Cagliari, Italy
[12] Simula Research Laboratory, Lysaker, Norway

Abstract. This paper presents an overview of the ImageCLEF 2017
evaluation campaign, an event that was organized as part of the CLEF
(Conference and Labs of the Evaluation Forum) labs 2017. ImageCLEF
is an ongoing initiative (started in 2003) that promotes the evaluation
of technologies for annotation, indexing and retrieval for providing information access to collections of images in various usage scenarios and
domains. In 2017, the 15th edition of ImageCLEF, three main tasks were
proposed and one pilot task: (1) a LifeLog task about searching in LifeLog
data, so videos, images and other sources; (2) a caption prediction task
that aims at predicting the caption of a figure from the biomedical literature based on the figure alone; (3) a tuberculosis task that aims at
detecting the tuberculosis type from CT (Computed Tomography) volumes of the lung and also the drug resistance of the tuberculosis; and (4)
a remote sensing pilot task that aims at predicting population density
based on satellite images. The strong participation of over 150 research
groups registering for the four tasks and 27 groups submitting results
shows the interest in this benchmarking campaign despite the fact that
all four tasks were new and had to create their own community.

© Springer International Publishing AG 2017
G.J.F. Jones et al. (Eds.): CLEF 2017, LNCS 10456, pp. 315–337, 2017.
DOI: 10.1007/978-3-319-65813-1_28

1   Introduction

20 years ago getting access to large visual data sets for research was a problem and open data collections that could be used to compare algorithms of researchers were rare. Now it is getting easier to access data collections but it is still hard to obtain annotated data with a clear evaluation scenario and strong baselines to compare to. Motivated by this, ImageCLEF has for 15 years been an initiative that aims at evaluating multilingual or language independent annotation and retrieval of images [5,15,18,24]. The main goal of ImageCLEF is to support the advancement of the field of visual media analysis, classification, annotation, indexing and retrieval. It proposes novel challenges and develops the necessary infrastructure for the evaluation of visual systems operating in different contexts and providing reusable resources for benchmarking, which is also linked to initiatives such as Evaluation as a Service (EaaS) [11]. Many research groups have participated over the years in these evaluation campaigns and even more have acquired its datasets for experimentation. The impact of ImageCLEF can also be seen by its significant scholarly impact indicated by the substantial numbers of its publications and their received citations [22].

There are other evaluation initiatives that have had a close relation with ImageCLEF. LifeCLEF [14] was formerly an ImageCLEF task. However, due to the need to assess technologies for automated identification and understanding of living organisms using data not only restricted to images, but also videos and sound, it was decided to be organised independently from ImageCLEF. Other CLEF labs linked to ImageCLEF, in particular the medical task, are: CLEFeHealth [10] that deals with processing methods and resources to enrich difficult-to-understand eHealth text and the BioASQ [3] tasks from the Question Answering lab that targets biomedical semantic indexing and question answering but is now not a lab anymore. Due to their medical topic, the organisation is coordinated in close collaboration with the medical tasks in ImageCLEF.

This paper presents a general overview of the ImageCLEF 2017 evaluation campaign[1], which as usual was an event organised as part of the CLEF labs[2]. Section 2 presents a general description of the 2017 edition of ImageCLEF, commenting about the overall organisation and participation in the lab. Followed by this are sections dedicated to the four tasks that were organised this year. Section 3 explains all details on the life logging task; Sect. 4 details the caption prediction task; Sect. 5 describes the two subtasks for the tuberculosis challenge and the pilot task on remote sensing data is described in Sect. 6.

For the full details and complete results, the readers should refer to the corresponding task overview papers [2,6,7,9]. The final section of this paper concludes by giving an overall discussion, and pointing towards the challenges ahead and possible new directions for future research.

---

[1] http://imageclef.org/2017/.
[2] http://clef2017.clef-initiative.eu/.

2   Overview of Tasks and Participation

ImageCLEF 2017 consisted of three main tasks and a pilot task that covered challenges in diverse fields and usage scenarios. In 2016 [25] the tasks were completely different with a handwritten retrieval task, an image annotation task and a medical task with several subtasks. In 2017 the tasks completely changed and only the caption prediction was a subtask already attempted in 2016 but for which no participant submitted results in 2016. The 2017 tasks are the following:

– **ImageCLEFlifelog:** aims at developing systems for lifelogging data retrieval and summarization, so for persons automatically logging their life.
– **ImageCLEFcaption:** addresses the problem of bio-medical image caption prediction from large amounts of training data. Captions can either be created as free text or concepts of the image captions could be detected.
– **ImageCLEFtuberculosis:** targets the challenge of determining the tuberculosis (TB) subtypes and drug resistances automatically from the volumetric image information (mainly related to texture) and based on clinical information that is available such as age, gender, etc.
– **ImageCLEFremote (pilot task):** targets the estimation of the population of a geographical area based on low definition but free earth observation images as provided by Copernicus program.

In order to participate in the evaluation campaign, the groups first had to register either on the CLEF website or from the ImageCLEF website. To actually get access to the datasets, the participants were required to submit a signed End User Agreement (EUA). Table 1 summarizes the participation in Image-CLEF 2017, including the number of registrations and number of signed EUAs, indicated both per task and for the overall lab. The table also shows the number of groups that submitted results (a.k.a. runs) and the ones that submitted a working notes paper describing the techniques used.

The number of registrations could be interpreted as the initial interest that the community has for the evaluation. However, it is a bit misleading because several people from the same institution might register, even though in the end they count as a single group participation. The EUA explicitly requires all groups that get access to the data to participate, even though this is not enforced. Unfortunately, the percentage of groups that submit results is often relatively small. Nevertheless, as observed in studies of scholarly impact [22,23], in subsequent years the datasets and challenges provided by ImageCLEF do get used quite often, which in part is due to the researchers that for some reason were unable to participate in the original event.

After a decrease in participation in 2016, the participation increased well in 2017 and this despite the fact that all four tasks did not have a participating community as all tasks were new and had to create the community from scratch. Still, of the 167 groups that registered and 60 that submitted a valid copyright agreement, only 27 submitted results in the end. The percentage is in line with past years with 20% of the registered groups submitting results and about 50% of those that signed the agreement. The following four sections are dedicated to

Table 1. Key figures of participation in ImageCLEF 2017.

Task	Online registrations	Signed EUA	Groups that subm. results	Submitted working notes
Lifelog	66	21	3	3
Caption	100	43	11	11
Tuberculosis	96	40	9	8
Remote	59	20	4	4
Overall	167	60	27	26

each of the tasks. Only a short overview is reported, including general objectives, description of the tasks and datasets and a short summary of the results.

3   The Lifelog Task

3.1   Motivation and Task Setup

The availability of a large variety of personal devices, such as smartphones, video cameras as well as wearable devices that allow capturing pictures, videos and audio clips in every moment of our life is creating vast archives of personal data where the totality of an individual's experiences, captured multi-modally through digital sensors are stored permanently as a personal multimedia archive. These unified digital records, commonly referred to as *lifelogs*, gathered increasing attention in recent years within the research community. This happened due to the need for and challenge of building systems that can automatically analyse these huge amounts of data in order to categorize, summarize and also query them to retrieve the information that the user may need.

Despite the increasing number of successful related workshops and panels (e.g., iConf 2016[3], ACM MM 2016[4]) lifelogging has rarely been the subject of a rigorous comparative benchmarking exercise as, for example, the new lifelog evaluation task at NTCIR-12[5]. The ImageCLEF 2017 LifeLog task [6] aims to bring the attention of lifelogging to a wide audience and to promote research into some of the key challenges of the coming years. The ImageCLEF 2017 LifeLog task aims to be a comparative evaluation of information access and retrieval systems operating over personal lifelog data. The task consists of two sub-tasks, both allow participation independently. These sub-tasks are:

– Lifelog Retrieval Task (LRT);
– Lifelog Summarization Task (LST).

---

[3] http://irlld2016.computing.dcu.ie/index.html.
[4] http://lta2016.computing.dcu.ie/styled/index.html.
[5] http://ntcir-lifelog.computing.dcu.ie/NTCIR12/.

Lifelog Retrieval Task. The participants had to analyse the lifelog data and according to several specific queries return the correct answers. For example: *Shopping for Wine: Find the moment(s) when I was shopping for wine in the supermarket* or *The Metro: Find the moment(s) when I was riding a metro.* The ground truth for this sub-task was created by extending the queries from the NTCIR-12 dataset, which already provides a sufficient ground truth.

Lifelog Summarization Task. In this sub-task the participants had to analyse all the images and summarize them according to specific requirements. For instance: *Public Transport: Summarize the use of public transport by a user. Taking any form of public transport is considered relevant, such as bus, taxi, train, airplane and boat. The summary should contain all different day-times, means of transport and locations, etc.*

Particular attention had to be paid to the diversification of the selected images with respect to the target scenario. The ground truth for this sub-task was created utilizing crowdsourcing and manual annotations.

3.2 Data Sets Used

The Lifelog dataset consists of data from three lifeloggers for a period of about one month each. The data contains a large collection of wearable camera images (approximately two images per minute), an XML description of the semantic locations (e.g. Starbucks cafe, McDonalds restaurant, home, work) and the physical activities (e.g. walking, transport, cycling), of the lifeloggers at a granularity of one minute. A summary of the data collection is shown in Table 2.

Given the fact that lifelog data is typically visual in nature and in order to reduce the barriers-to-participation, the output of the Caffe CNN-based visual concept detector was included in the test collection as additional meta data.

Table 2. Statistics of lifelog dataset

Number of lifeloggers	3
Size of the collection (Images)	**88,124** images
Size of the collection (Locations)	**130** locations
Number of LRT topics	**36** (16 for devset, 20 for testset)
Number of LsT topics	**15** (5 for devset, 10 for testset)

Topics. Aside from the data, the test collection included a set of topics (queries) that were representative of the real-world information needs of lifeloggers. There were 36 and 15 ad-hoc search topics representing the challenge of retrieval for the LRT task and the challenge of summarization for the LST task, respectively.

Evaluation Methodology. For the *Lifelog Retrieval Task* evaluation metrics based on NDCG (Normalized Discounted Cumulative Gain) at different depths were used, i.e., $NDCG@N$, where N varies based on the type of the topics, for the recall oriented topics N was larger (>20), and for the precision oriented topics N was smaller N (5, 10 or 20).

In the *Lifelog Summarization Task* classic metrics were deployed:

- Cluster Recall at $X(CR@X)$ – a metric that assesses how many different clusters from the ground truth are represented among the top X results;
- Precision at $X(P@X)$ – measures the number of relevant photos among the top X results;
- F1-measure at $X(F1@X)$ – the harmonic mean of the previous two.

Various cut off points were considered, e.g., $X = 5, 10, 20, 30, 40, 50$. Official ranking metrics this year was the **F1-measure@10** or images, which gives equal importance to diversity (via $CR@10$) and relevance (via $P@10$).

Participants were also encouraged to undertake the sub-tasks in an interactive or automatic manner. For interactive submissions, a maximum of five minutes of search time was allowed per topic. In particular, the organizers would like to emphasize methods that allowed interaction with real users (via Relevance Feedback (RF), for example), i.e., beside of the best performance, the way of interaction (like number of iterations using RF), or innovation level of the method (for example, new way to interact with real users) has been evaluated.

3.3   Participating Groups and Runs Submitted

We received 18 runs submitted from 3 teams from Singapore, Romania, and a multi-nation team from Ireland, Italy, and Norway. The submitted runs are summarized in Table 3.

3.4   Results

We received approaches from fully automatic to fully manual paradigms, from using a single information provided by the task to using all information as well as extra resources. In Table 4, we report the runs with highest score from each team for both subtasks.

3.5   Lessons Learned and Next Steps

What we learned from the lifelogging task is that multi-modal analysis seems still to be a problem that not many address. Often only one type of data is analysed. For the future it would be important to encourage participants to try out all modalities. This could be achieved by providing pre-extracted features with the data. Apart from that there was a large gap between signed-up teams and submitted runs. We think that this is based on the complexity of the task and the large amount of data that need to be analysed. Supporting participants with pre-extracted features could also help in this case because feature extraction can take much time. Finally, and most importantly, we could show how interesting

Table 3. Submitted runs for ImageCLEFlifelog 2017 task.

Lifelog retrieval dubtask		
Team	Run	Description
Organizers [26]	Baseline	Baseline method, fully automatic
	Segmentation	Apply segmentation and automatic retrieval based on concepts
	Fine-tunning	Apply segmentation and fine-tunning. Using all information
Lifelog summarization subtask		
I2R [17]	Run 1	Parameters learned for maximum F1 score. Using only visual information
	Run 2	Parameters learned for maximum F1 score. Using visual and metadata information
	Run 3	Parameters learned for maximum F1 score. Using metadata
	Run 4	Re-clustering in each iteration; 20% extra clusters. Using visual, metadata and interactive
	Run 5	No re-clustering. 100% extra clusters. Using visual, metadata and interactive
	Run 6	Parameters learned for maximum F1 score. Using visual, metadata, and object detection
	Run 7	Parameters learned for maximum F1 score, w/ and w/o object detection. Using visual, metadata, and object detection
	Run 8	Parameters learned for maximum F1 score. Using visual information and object detection
	Run 9	Parameters learned for maximum precision. Using visual and metadata information
	Run 10	No re-clustering. 20% extra clusters. Using visual, metadata and interactive
UPB [8]	Run 1	Textual filtering and word similarity using WordNet and Retina
Organizers [26]	Baseline	Baseline method, fully automatic
	Segmentation	Apply segmentation and automatic retrieval and diversification based on concepts
	Filtering	Apply segmentation, filtering, and automatic diversification. Using all information
	Fine-tunning	Apply segmentation, fine-tunning, filtering, and automatic diversification. Using all information
	RF	Relevance feedback. Using all information

and challenging lifelog data is and that it holds much research potential, not only in multimedia analysis but also from a system point of view for the performance. For next steps we will enrich the dataset with more data and also look into which pre-extracted features would make sense and what is the best format to share it with our colleagues.

Table 4. ImageCLEFlifelog 2017 results.

Retrieval subtask			Summarization subtask		
Team	Best Run	NDCG	Team	Best run	F1@10
Organizers[a][26]	Fine-Tuning	0.386	I2R [17]	Run 2	0.497
			UPB [8]	Run 1	0.132
			Organizers[a][26]	RF	0.769

[a]Note: Results from the organizers team are just for reference.

4   The Caption Task

Interpreting and summarizing the insights gained from medical images such as radiography or biopsy samples is a time-consuming task that involves highly trained experts and often represents a bottleneck in clinical diagnosis. Consequently, there is a considerable need for automatic methods that can approximate the mapping from visual information to condensed textual descriptions.

4.1   Task Setup

The ImageCLEF 2017 caption task [9] casts the problem of image understanding as a cross-modality matching scenario in which visual content and textual descriptors need to be aligned and concise textual interpretations of medical images are generated. The task works on the basis of a large-scale collection of figures from open access biomedical journal articles from PubMed Central (PMC)[6]. Each image is accompanied by its original caption and a set of extracted UMLS® (Unified Medical Language System®)[7] Concept Unique Identifiers (CUIs), constituting a natural testbed for this image captioning task.

In 2016, ImageCLEFmed [12] proposed a caption prediction subtask. This edition of the biomedical image captioning task at ImageCLEF comprises two subtasks: (1) Concept Detection and (2) Image Caption Prediction. Figure 1 shows an example biopsy image along with its relevant concepts as well as the reference caption.

Concept Detection. As a first step to automatic image captioning and understanding, participating systems are tasked with identifying the presence of relevant biomedical concepts in medical images. Based on the visual image content, this subtask provides the building blocks for the image understanding step by identifying the individual components from which full captions can be composed.

---

[6] PubMed Central (PMC) is a free full-text archive of biomedical and life sciences journal literature at the U.S. National Institute of Health's National Library of Medicine (NIH/NLM) (see http://www.ncbi.nlm.nih.gov/pmc/).

[7] https://www.nlm.nih.gov/research/umls.

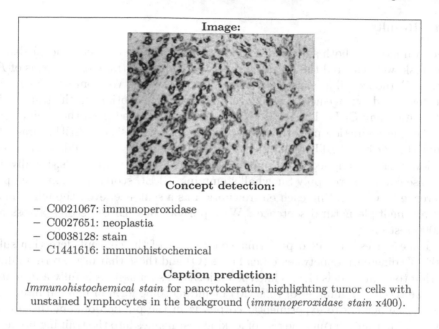

Image:

Concept detection:

- C0021067: immunoperoxidase
- C0027651: neoplastia
- C0038128: stain
- C1441616: immunohistochemical

Caption prediction:

Immunohistochemical stain for pancytokeratin, highlighting tumor cells with unstained lymphocytes in the background (*immunoperoxidase stain* x400).

Fig. 1. Example of an image and the information provided in the training set.

Caption Prediction. On the basis of the concept vocabulary detected in the first subtask as well as the visual information of their interaction in the image, participating systems are tasked with composing coherent natural language captions for the entirety of an image. In this step, rather than the mere coverage of visual concepts, detecting the interplay of visible elements is crucial for recreating the original image caption.

4.2 Dataset

The experimental corpus is derived from scholarly biomedical articles of PMC from which we extract figures and their corresponding captions. The collection is comprised of 184,614 image-caption pairs. This overall set is further split into disjunct training (164,614 pairs), validation (10,000 pairs) and test (10,000 pairs) sets. For the concept detection subtask, we used the QuickUMLS library [21] to identify the CUIs mentioned in the caption text.

4.3 Participating Groups and Submitted Runs

We received a total of 71 runs by 11 individual teams. There was a limit of at most 10 runs per team and subtask and the submissions are roughly evenly split between tasks. The vast majority of participating groups relied on some form of neural network architecture, typically combining convolutional and recurrent layers in order to jointly reason about visual and textual information.

4.4 Results

The evaluation of both subtasks is conducted separately. For the concept detection task, we measured the balanced precision and recall trade-off in terms of F_1 scores. Python's scikit-learn (v0.17.1-2) library is used. We compute micro F_1 per image and average across all images. 393 reference captions in the test set do not contain any CUIs. The respective images are excluded from the evaluation.

Caption prediction performance is assessed on the basis of BLEU scores [19] using the Python NLTK (v3.2.2) default implementation. Candidate captions are lower cased, stripped of all punctuation and English stop words. Finally, to increase coverage, we apply Snowball stemming. BLEU scores are computed per reference image, treating each entire caption as a sentence, even though it may contain multiple natural sentences. We report average BLEU scores across all 10,000 test images.

Table 5 gives a detailed performance overview of the concept detection subtask. We differentiate between official runs (O) and those that use external information to train models (E). While the majority of submissions is fully automatic (A), we received a number of runs (M) including some form of manual intervention. Since our entire experimental corpus is in the public domain, the use of external information runs the risk of leaking test images into the training process. For this reason, the task's official ranking concentrates on those teams that relied only on official material and that are therefore directly comparable. The best official results for this task were obtained by Athens University's Information Processing Laboratory.

The results of the caption prediction subtask can be found in Table 6. While there were no manual submissions to this subtask, here, as well, the use of external information gave teams a considerable, yet difficult-to-compare performance advantage and is therefore excluded from the official team ranking. The best official results were obtained by the Chinese Academy of Sciences' Key Laboratory on Intelligent Information Processing (isia). For additional details regarding the participating teams and their approaches, we refer the reader to the task overview paper [9].

4.5 Lessons Learned and Next Steps

There are several observations that need to be taken into account when analyzing the results presented in the previous section. Most notably, as a consequence of the data source (scholarly biomedical journal articles), the collection contains a considerable amount of noise in the form of compound figures with potentially highly heterogeneous content. In future editions of this task, we plan using a more well-defined source of images such as radiology or biopsy samples in order to reduce the amount of variation in the data.

Second, the CUIs extraction employed to generate ground truth labels is a probabilistic process that introduces its own errors. As a consequence, there are a considerable number of training captions that do not contain any CUIs, making such examples difficult to use for concept detection. In the future, plan to rely

Table 5. Concept detection using official (O) and external (E) resources.

Team	Run	Type	Resources	F_1
NLM	1494012568180	A	E	0.1718
NLM	1494012586539	A	E	0.1648
Aegean AI Lab	1491857120689	A	E	0.1583
Information Processing Laboratory	**1494006128917**	**A**	**O**	**0.1436**
Information Processing Laboratory	1494006074473	A	O	0.1418
Information Processing Laboratory	1494009510297	A	O	0.1417
Information Processing Laboratory	1494006054264	A	O	0.1415
Information Processing Laboratory	1494009412127	A	O	0.1414
Information Processing Laboratory	1494009455073	A	O	0.1394
NLM	1494014122269	A	E	0.1390
Information Processing Laboratory	1494006225031	A	O	0.1365
Information Processing Laboratory	1494006181689	A	O	0.1364
NLM	1494012605475	A	E	0.1228
Information Processing Laboratory	1494006414840	A	O	0.1212
Information Processing Laboratory	1494006360623	A	O	0.1208
AILAB	1493823116836	A	E	0.1208
BMET	1493791786709	A	O	0.0958
BMET	1493791318971	A	O	0.0880
NLM	1494013963830	A	O	0.0880
NLM	1494014008563	A	O	0.0868
BMET	1493698613574	A	O	0.0838
NLM	1494013621939	A	O	0.0811
NLM	1494013664037	A	O	0.0695
Morgan CS	1494060724020	M	O	0.0498
BioinformaticsUA	1493841144834	M	O	0.0488
BioinformaticsUA	1493995613907	M	O	0.0463
mami	1496127572481	M	E	0.0462
Morgan CS	1494049613114	M	O	0.0461
Morgan CS	1494048615677	M	O	0.0434
BioinformaticsUA	1493976564810	M	O	0.0414
Morgan CS	1494048330426	A	O	0.0273
AILAB	1493823633136	A	E	0.0234
AILAB	1493823760708	A	E	0.0215
NLM	1495446212270	A	E	0.0162
MEDGIFT UPB	1493803509469	A	E	0.0028
NLM	1494012725738	A	O	0.0012
mami	1493631868847	M	E	0.0000

Table 6. Caption prediction using official (O) and external (E) resources.

Team	Run	Resources	BLEU
NLM	1494014231230	E	0.5634
NLM	1494081858362	E	0.3317
AILAB	1493825734124	E	0.3211
NLM	1495446212270	E	0.2646
AILAB	1493824027725	E	0.2638
isia	**1493921574200**	**O**	**0.2600**
isia	1493666388885	O	0.2507
isia	1493922473076	O	0.2454
isia	1494002110282	O	0.2386
isia	1493922527122	O	0.2315
NLM	1494038340934	O	0.2247
isia	1493831729114	O	0.2240
isia	1493745561070	O	0.2193
isia	1493715950351	O	0.1953
isia	1493528631975	O	0.1912
AILAB	1493825504037	E	0.1801
isia	1493831517474	O	0.1684
NLM	1494038056289	O	0.1384
NLM	1494037493960	O	0.1131
AILAB	1493824818237	E	0.1107
BMET	1493702564824	O	0.0982
BMET	1493698682901	O	0.0851
BMET	1494020619666	O	0.0826
Biomedical Computer Science Group	1493885614229	E	0.0749
Biomedical Computer Science Group	1493885575289	E	0.0675
BMET	1493701062845	O	0.0656
Biomedical Computer Science Group	1493885210021	E	0.0624
Biomedical Computer Science Group	1493885397459	E	0.0537
Biomedical Computer Science Group	1493885352146	E	0.0527
Biomedical Computer Science Group	1493885286358	E	0.0411
Biomedical Computer Science Group	1493885541193	E	0.0375
Biomedical Computer Science Group	1493885499624	E	0.0365
Biomedical Computer Science Group	1493885708424	E	0.0326
Biomedical Computer Science Group	1493885450000	E	0.0200

on more rigorous (manual and thus expensive) filtering to ensure good concept coverage across training, validation and test data.

Finally, the call for contributions did not make any assumptions about the kinds of strategies participants would rely on. As a consequence, we see a broad range of methods being applied. Evaluation of the results shows that some teams employed methods that were at least partially trained on external resources including Pubmed articles. Since such approaches cannot be guaranteed to have respected our division into training, validation and test folds and might subsequently leak test examples into the training process, future editions of the task will carefully describe the categories of submissions based on the resources used.

5   The Tuberculosis Task

About 130 years after the discovery of Mycobacterium tuberculosis, the disease remains a persistent threat and a leading cause of death worldwide. The greatest disaster that can happen to a patient with tuberculosis (TB) is that the organisms become resistant to two or more of the standard drugs. In contrast to drug sensitive (DS) tuberculosis, its multi-drug resistant (MDR) form is more difficult and expensive to treat. Thus, early detection of the drug resistance (DR) status is of great importance for effective treatment. The most commonly used methods of DR detection are either expensive or take too much time (up to several months). Therefore, there is a need for quick and at the same time cheap methods of DR detection. One of the possible approaches for this task is based on Computed Tomography (CT) image analysis. Another challenging task is automatic detection of TB types using CT volumes.

5.1   Task Setup

Two subtasks were then proposed in the ImageCLEF tuberculosis task 2017 [7]:

– Multi-drug resistance detection (MDR subtask);
– Tuberculosis type classification (TBT subtask).

The goal of the MDR subtask is to assess the probability of a TB patient having resistant form of tuberculosis based on the analysis of a chest CT. For the TBT subtask, the goal is to automatically categorize each TB case into one of the following five types: Infiltrative, Focal, Tuberculoma, Miliary, Fibro-cavernous.

5.2   Dataset

For both subtasks 3D CT images were provided with a size of 512×512 pixels and number of slices varying from 50 to 400. All CT images were stored in NIFTI file format with .nii.gz file extension (g-zipped .nii files). This file format stores raw voxel intensities in Hounsfield units (HU) as well the corresponding image metadata such as image dimensions, voxel size in physical units, slice thickness, etc. For all patients automatically extracted masks of the lungs were provided.

Table 7. Dataset for the MDR subtask.

# Patients	Train	Test
DS	134	101
MDR	96	113
Total patients	230	214

Table 8. Dataset for the TBT subtask.

# Patients	Train	Test
Type 1 (Infiltrative)	140	80
Type 2 (Focal)	120	70
Type 3 (Tuberculoma)	100	60
Type 4 (Miliary)	80	50
Type 5 (Fibro-cavernous)	60	40
Total patients	500	300

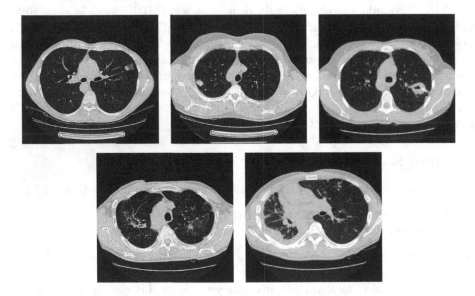

Fig. 2. Examples of the TB types. First row, from left to right: Infiltrative, Focal, and Tuberculoma types. Second row: Miliary, and Fibro-cavernous types.

The dataset for the MDR subtask was composed of 209 MDR and 234 DS patients. The division of the data into training and test sets is shown in Table 7. The TBT task contained 800 patients divided as presented in Table 8. One 2D slice per TB type is shown in Fig. 2.

5.3   Participating Groups and Submitted Runs

In the first year of the task, 9 groups from 6 countries have submitted at least one run to one of the subtask. There were 8 groups participating in the MDR subtask, and 7 in the TBT task. Each group could submit up to 10 runs. Finally 28 and 23 runs were submitted in the MDR and TBT tasks respectively. 5 groups used a deep-learning approach, two were based on graph models encoding local texture features and one build a co-occurrence of adjacent supervoxels. One group did not explain the algorithm.

5.4   Results

The MDR subtask is a 2-class problem. The participants submitted for each patient in the test set the probability of belonging to the MDR group. The Area Under the ROC Curve (AUC) was chosen as the measure to rank results. Accuracy was provided as well. For the TBT subtask, the participants had to submit the tuberculosis category. Since the 5-class problem was not balanced, Cohen's Kappa was used to compare the methods. Again, the accuracy was provided. Tables 9 and 10 show the final results for each run and their rank.

Table 9. Results for the MDR subtask.

Group Name	Run	AUC	ACC	Rank
MedGIFT	MDR_Top1_correct.csv	0.5825	0.5164	1
MedGIFT	MDR_submitted_topBest3_correct.csv	0.5727	0.4648	2
MedGIFT	MDR_submitted_topBest5_correct.csv	0.5624	0.4836	3
SGEast	MDR_LSTM_6_probs.txt	0.5620	0.5493	4
SGEast	MDR_resnet_full.txt	0.5591	0.5493	5
SGEast	MDR_BiLSTM_25_wcrop_probs.txt	0.5501	0.5399	6
UIIP	MDR_supervoxels_run_1.txt	0.5415	0.4930	7
SGEast	MDR_LSTM_18_wcrop_probs.txt	0.5404	0.5540	8
SGEast	MDR_LSTM_21wcrop_probs.txt	0.5360	0.5070	9
MedGIFT	MDR_Top2_correct.csv	0.5337	0.4883	10
HHU DBS	MDR_basecnndo_212.csv	0.5297	0.5681	11
SGEast	MDR_LSTM_25_wcrop_probs.txt	0.5297	0.5211	12
BatmanLab	MDR_submitted_top5.csv	0.5241	0.5164	13
HHU DBS	MDR_basecnndo_113.csv	0.5237	0.5540	14
MEDGIFT UPB	MDR_TST_RUN_1.txt	0.5184	0.5352	15
BatmanLab	MDR_submitted_top4_0.656522.csv	0.5130	0.5024	16
MedGIFT	MDR_Top3_correct.csv	0.5112	0.4413	17
HHU DBS	MDR_basecnndo_132.csv	0.5054	0.5305	18
HHU DBS	MDR_basecnndo_182.csv	0.5042	0.5211	19
HHU DBS	MDR_basecnndo_116.csv	0.5001	0.4930	20
HHU DBS	MDR_basecnndo_142.csv	0.4995	0.5211	21
HHU DBS	MDR_basecnndo_120.csv	0.4935	0.4977	22
SGEast	MDR_resnet_partial.txt	0.4915	0.4930	23
BatmanLab	MDR-submitted_top1.csv	0.4899	0.4789	24
BatmanLab	MDR_SuperVx_Hist_FHOG_rf_0.648419.csv	0.4899	0.4789	25
Aegean Tubercoliosis	MDR_DETECTION_EXPORT2.csv	0.4833	0.4648	26
BatmanLab	MDR_SuperVx_FHOG_rf_0.637994.csv	0.4601	0.4554	27
BioinformaticsUA	MDR_run1.txt	0.4596	0.4648	28

Table 10. Results for the TBT subtask.

Group name	Run	Kappa	ACC	Rank
SGEast	TBT_resnet_full.txt	0.2438	0.4033	1
SGEast	TBT_LSTM_17_wcrop.txt	0.2374	0.3900	2
MEDGIFT UPB	TBT_T_GNet.txt	0.2329	0.3867	3
SGEast	TBT_LSTM_13_wcrop.txt	0.2291	0.3833	4
Image Processing	TBT-testSet-label-Apr26-XGao-1.txt	0.2187	0.4067	5
SGEast	TBT_LSTM_46_wcrop.txt	0.2174	0.3900	6
UIIP	TBT_iiggad_PCA_RF_run_1.txt	0.1956	0.3900	7
MEDGIFT UPB	TBT_...._GoogleNet_10crops_at_different_scales_.txt	0.1900	0.3733	8
SGEast	TBT_resnet_partial.txt	0.1729	0.3567	9
MedGIFT	TBT_Top1_correct.csv	0.1623	0.3600	10
SGEast	TBT_LSTM_25_wcrop.txt	0.1548	0.3400	11
MedGIFT	TBT_submitted_topBest3_correct.csv	0.1548	0.3500	12
BatmanLab	TBT_SuperVx_Hist_FHOG_lr_0.414000.csv	0.1533	0.3433	13
SGEast	TBT_LSTM_37_wcrop.txt	0.1431	0.3333	14
MedGIFT	TBT_submitted_topBest5_correct.csv	0.1410	0.3367	15
MedGIFT	TBT_Top4_correct.csv	0.1352	0.3300	16
MedGIFT	TBT_Top2_correct.csv	0.1235	0.3200	17
BatmanLab	TBT_submitted_bootstrap.csv	0.1057	0.3033	18
BatmanLab	TBT_submitted_top3_0.490000.csv	0.1057	0.3033	19
BatmanLab	TBT_SuperVx_Hist_FHOG_Reisz_lr_0.426000.csv	0.0478	0.2567	20
BatmanLab	TBT_submitted_top2_0.430000.csv	0.0437	0.2533	21
BioinformaticsUA	TBT_run0.txt	0.0222	0.2400	22
BioinformaticsUA	TBT_run1.txt	0.0093	0.1233	23

5.5 Lessons Learned and Next Steps

The results underline the difficulty of both tasks. In the case of the MDR task all participants were close to an AUC of 0.50 that is the performance of a random classifier. When considering the accuracy the results are sometimes worse. The random accuracy for this subtask is 0.5280 and the best participant reached an accuracy of 0.5681. In the TBT subtask the results are more promising. 6 runs achieved a Cohen's Kappa of better than 0.21, threshold to consider a fair agreement between classifications. The random accuracy in this case would be 0.2667 and most of the participants were above this value.

The analysis of the results and the different nature of the methods suggest that the training data did not fully represent the test cases, being fairly small for the diversity of the cases. In the MDR subtask the set of DS patients was composed of patients that may have presented resistance to some drugs, but no all. With more training cases, the groups can be better defined. In a future edition of this task we expect to add the current test set as training and provide new patients for the test set.

6 The Remote (Population Estimation) Task

6.1 Motivation and Task Setup

Before engaging any rescue operation or humanitarian action, NGOs (Non-Governmental Organizations) need to estimate the local population as accurately as possible. Population estimation is fundamental to provide any service for a particular region. While good estimates exists in many parts of the world through accurate census data, this is usually not the case in developing countries.

This pilot task, introduced in 2017, aims at investigating the use of satellite data as a cheaper and quicker process. The task uses Copernicus Sentinel-2 images with resolution between 10 to 60 m.

6.2 Data Sets Used

In this pilot task, participants had to estimate the population for different areas in two regions. To achieve this goal, organizers provided a set of satellite images (Copernicus Sentinel 2)[8]. The boundaries of the areas of interest were provided as shape files. The clipped satellite images were provided as well as the meta data of the original images (before clipping). The data set consists of topographic and geographic information as follows:

- ESRI (Environmental Systems Research Institute) shape files: there is a single shape file by region and the projected shape file of the region has the attributes to represent the various areas the region is composed of.
- Sentinel-2 satellite images: The remote sensing imagery are from the Sentinel-2 platform. The imagery is multi spectral, cloud-free satellite imagery downloaded from Sentinel Data Hub[9]. The images have been clipped to match the bounding box of the areas of interest. The bands for images have different spatial resolutions: 10 m for bands B2 (490 nm), B3 (560 nm) B4 (665 nm) and B8 (84 nm); 20 m for bands B5 (705 nm), B6 (749 nm) B7 (783 nm), B8a (865 nm) B11 (1610 nm) and B12 (2190 nm). For the analysis, participants were encouraged to use Red, Green and Blue bands or in some cases near infrared bands that are 10 m in resolution.
- Meta-data associated to the images: Information regarding the original images is provided in XML files. These files contain information like capture time/date, sensor mode, orbit number, the id of quality files, etc. Further information regarding the Sentinel-2 products, as well as file structure can be found in the Sentinel 2 User handbook[10].

However, participants were allowed to use any other resource they think might help to reach the highest accuracy.

---

[8] The dataset is available on Zenodo with the DOI 10.5281/zenodo.804602 or on demand.
[9] https://scihub.copernicus.eu/dhus/#/home.
[10] https://sentinel.esa.int/documents/247904/685211/Sentinel-2_User_Handbook.

There were 83 areas of interest in the city of Lusaka and 17 in west Uganda for which the population has to be estimated. For 90 of these 100 areas, ground truth provided by NGOs is available, so evaluation considered these areas.

Runs from participants are evaluated against ground truth. For the city of Lusaka, the ground truth comes with a categorical evaluation measure of the population estimation, Good (23 over the 83 areas), Acceptable (37), Doubts (9), High doubts (6) and Unknown (8). For West Uganda, the ground truth corresponds to estimations that are based on a combination of Volunteered Geographic information (VGI) working on BING imagery (2012) with additional ground work. Both have been provided by NGOs.

In our evaluation we use three metrics: (1) Sum of differences, which corresponds to the sum of the absolute value of the difference between ground truth and the estimated population over the areas, (2) Root Mean Square Error (RMSE), which is computed as the square root of the average of squared errors [4], (3) Pearson correlation, and (4) AvgRelDelta, which is the average of the relative deltas as calculated in (1) relative to the population.

The four measures that are detailed in [2], aim at comparing the estimated value against the ground truth. The challenge comprises two areas geographically separated, Uganda, and Zambia. Because of this fact, it was decided to evaluate both areas separately. We evaluate the results on two variables: (1) Population counts, and (2) Dwelling counts. Then each run submitted by the participants has 12 possible metrics. However, not all the participants submitted results for both variables. All the submissions provided estimation for the population, while only two of them provided estimates for both population and dwelling counts.

6.3   Participating Groups and Runs Submitted

Although the pilot task was open to anyone, participants came from local hackathon -like events that were organized within FabSpace 2.0 project (https://fabspace.eu/); see [1] for details. There are four groups participating with their contribution being summarized in Table 11.

Table 11. Participants of the ImageCLEF remote task.

Run	Approach
Darmstadt [13]	Supervised (Maximum Likelihood) by false colour composite and NIR band and unsupervised (K-Means Cluster Analysis by NIR band
Grapes [16]	Supervised classification on Sentinel 2 images coupled with statistical forecasting on historical census data
FABSPACE PL	Pre-processing Sentinel-1 data, creating mask with buildings, mean- shift segmentation process
AndreaDavid [20]	Convolutional Neural Network with Sentinel 2 and open data

6.4   Results

As can be seen in Table 12, the sum of deltas in the prediction over the 90 areas is in the same range for the 3 participants. The correlation is not very high leaving room for improvement. More details are provided in the task overview.

Table 12. Results on the estimation of the population over the 90 areas from the two regions. Detailed results can be found in [2]. Bold font highlights the best result while the italic font highlights the second best.

Participants	Country	Sum Delta	RMSE	Pearson	AvgRelDelta
Darmstadt [13]	Germany	1,493,152	27,495	0.22	*97.89*
Grapes [16]	Greece	1,486,913	34,290	*0.33*	177.55
FABSPACE PL	Poland	1,558,639	31,799	**0.37**	172.84
AndreaDavid [20]	Italy	1,484,088	27,462	0.21	**87.57**

Figure 3 shows an overview of the results submitted by the participating teams. The maps show the prediction errors divided by the ground truth population for each operational zone ($delta = (d_t - s_t)/d_t.100$).

Operational zones where the models severely overestimated the population (the models suggest a higher population) are shown in red or orange. We consider results severely overestimated when the population estimation is over 50% of the actual population. Areas in which the estimation is +/− 50% are depicted in green, while areas in which the models severely underestimated the population are depicted in blue. In this paper, a population estimation would be considered severely underestimated if it is lower than 50% of the actual population.

We can see that there are areas overestimated by all the models: the Industrial area (West), Ngwere (North) and Libala (South). In the case of the industrial areas it seems that the proposed algorithms confused industrial buildings with residential areas. In the case of Ngwere, and Libala, the residential areas have low density, which was incorrectly evaluated by the algorithms.

On the other hand, we can see that there are other areas that are underestimated by all the models: George, Lilanda, Desai, (at the West of the city), Chelston at the East, Chawama and Kuoboka at the South. Most of the models underestimated the population in Makeni, except for the Polish team. This team also differentiated from the rest in an area comprised by Ngombe, Chamba Valley, Kamanga and Kaunda square (North East of the city), providing good estimates with the exception of Chudleigh, which was overestimated by all the teams, except for the Italian team.

In general, all the teams obtained best results in an area near the center of the city, an area roughly defined by the Operational zones, Civic Centre, Rhodes Park and in most cases Northmead (except for the Polish team that did not provide a good result for this zone).

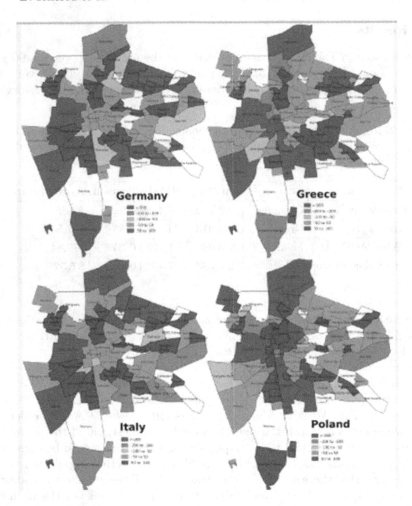

Fig. 3. Overview of the results submitted by the participating teams for the operational zones in Lusaka- Zambia. (Color figure online)

6.5   Lessons Learned and Next Steps

One objective of this pilot task was to evaluate the accuracy of population estimation based on low definition images. The motivation is mainly the availability of such images free of charge, for any place and with a high refresh rate of 5 days) thanks to the European Copernicus program. Participants encountered difficulties mainly linked to the nature of the images. The results show that the accuracy needs to be improved to be fully usable in real applications. The time allowed to solve this task was certainly not sufficient and requires good knowledge of multispectral image analysis, which not all participants had. Thanks to the pilot task we now have several ways to improve the estimation.

7   Conclusions

This paper presents a general overview of the activities and outcomes of the ImageCLEF 2017 evaluation campaign. Four tasks were organised covering challenges in: lifelog retrieval, caption prediction, tuberculosis type and drug resistance detection and of remote sensing.

The participation increased compared to previous years with over 160 registered participants and in the end 27 groups submitting results and a large number of runs. This is remarkable as all four tasks were new and had to create a new community. Whereas several of the participants had participated in the past there was also a large number of groups totally new to ImageCLEF and also collaborations of research groups in several tasks.

Deep Neural Networks were applied for basically all tasks and often led to very good results but this was not true for all tasks as graph-based approaches led to best results in the MDR tuberculosis task. The caption prediction task created a large variety of approaches including using content-based image retrieval to find the visually most similar figures for predicting a caption of an image in addition to the visual content itself. The task also showed that it is important to group the submission based on the resources used as external resources can lead to much better results and a comparison of the techniques needs to be based on the same types of resources used.

ImageCLEF 2017 again brought together an interesting mix of tasks and approaches and we are looking forward to the discussions at the workshop.

Acknowledgements. This research was supported in part by the Intramural Research Program of the National Institutes of Health (NIH), National Library of Medicine (NLM), and Lister Hill National Center for Biomedical Communications (LHNCBC). It is also partly supported European Union's Horizon 2020 Research and Innovation programme under the Grant Agreement n°693210 (FabSpace 2.0).

References

1. Arenas, H., Baker, A., Bialczak, A., Bargiel, D., Becker, M., Gaildrat, V., Carbone, F., Heising, S., Islam, M.B., Lattes, P., Marantos, C., Menou, C., Mothe, J., Nzeh Ngong, A., Paraskevas, I.S., Penalver, M., Sciana, P., Soudris, D.: FabSpaces at population estimation (remote) task - ImageCLEF at the CLEF 2017 Labs. CLEF working notes, CEUR (2017)
2. Arenas, H., Islam, B., Mothe, J.: Overview of the ImageCLEF 2017 population estimation task. In: CLEF 2017 Labs Working Notes. CEUR Workshop Proceedings, CEUR-WS.org, Dublin, Ireland (2017). http://ceur-ws.org
3. Balikas, G., Krithara, A., Partalas, I., Paliouras, G.: BioASQ: a challenge on large-scale biomedical semantic indexing and question answering. In: Müller, H., Jimenez del Toro, O.A., Hanbury, A., Langs, G., Foncubierta Rodríguez, A. (eds.) Multimodal Retrieval in the Medical Domain. LNCS, vol. 9059, pp. 26–39. Springer, Cham (2015). doi:10.1007/978-3-319-24471-6_3
4. Barnston, A.G.: Correspondence among the correlation, RMSE, and heidke forecast verification measures; refinement of the heidke score. Weather Forecast. **7**(4), 699–709 (1992)

5. Clough, P., Müller, H., Sanderson, M.: The CLEF 2004 cross-language image retrieval track. In: Peters, C., Clough, P., Gonzalo, J., Jones, G.J.F., Kluck, M., Magnini, B. (eds.) CLEF 2004. LNCS, vol. 3491, pp. 597–613. Springer, Heidelberg (2005). doi:10.1007/11519645_59

6. Dang-Nguyen, D.T., Piras, L., Riegler, M., Boato, G., Zhou, L., Gurrin, C.: Overview of ImageCLEFlifelog 2017: lifelog retrieval and summarization. In: CLEF 2017 Labs Working Notes. CEUR Workshop Proceedings, CEUR-WS.org, Dublin, Ireland (2017). http://ceur-ws.org

7. Dicente Cid, Y., Kalinovsky, A., Liauchuk, V., Kovalev, V., Müller, H.: Overview of ImageCLEFtuberculosis 2017 - predicting tuberculosis type and drug resistances. In: CLEF 2017 Labs Working Notes. CEUR Workshop Proceedings, CEUR-WS.org, Dublin, Ireland (2017). http://ceur-ws.org

8. Dogariu, M., Ionescu, B.: A textual filtering of hog-based hierarchical clustering of lifelog data, pp. 11–14. CLEF working notes, CEUR (2017)

9. Eickhoff, C., Schwall, I., Seco, G., de Herrera, A., Müller, H.: Overview of Image-CLEF caption 2017 - image caption prediction and concept detection for biomedical images. In: CLEF 2017 Labs Working Notes. CEUR Workshop Proceedings, CEUR-WS.org, Dublin, Ireland (2017). http://ceur-ws.org

10. Goeuriot, L., Kelly, L., Li, W., Palotti, J., Pecina, P., Zuccon, G., Hanbury, A., Jones, G.J.F., Müller, H.: Share/CLEF ehealth evaluation lab 2014, task 3: User-centred health information retrieval. In: Working Notes of CLEF 2014, CEUR (2014)

11. Hanbury, A., Müller, H., Balog, K., Brodt, T., Cormack, G.V., Eggel, I., Gollub, T., Hopfgartner, F., Kalpathy-Cramer, J., Kando, N., Krithara, A., Lin, J., Mercer, S., Potthast, M.: Evaluation-as-a-service: overview and outlook. ArXiv 1512.07454 (2015)

12. Seco, G., de Herrera, A., Schaer, R., Bromuri, S., Müller, H.: Overview of the ImageCLEF 2016 medical task. In: Working Notes of CLEF 2016 (Cross Language Evaluation Forum), September 2016

13. Islam, M.B., Becker, M., Bargiel, D., Ahmed, K.R., Duzak, P., Emana, N.G.: Sentinel-2 satellite imagery based population estimation strategies at FabSpace 2.0 Lab Darmstadt (2017)

14. Joly, A., Goëau, H., Glotin, H., Spampinato, C., Bonnet, P., Vellinga, W.P., Lombardo, J.C., Planqué, R., Palazzo, S., Müller, H.: Lifeclef 2017 lab overview: multimedia species identification challenges. In: Proceedings of CLEF 2017 (2017)

15. Kalpathy-Cramer, J., García Seco de Herrera, A., Demner-Fushman, D., Antani, S., Bedrick, S., Müller, H.: Evaluating performance of biomedical image retrieval systems: overview of the medical image retrieval task at ImageCLEF 2004–2014. Computerized Medical Imaging and Graphics **39**, 55–61 (2015)

16. Koutsouri, K., Skepetari, I., Anastasakis, K., Lappas, S.: Population estimation using satellite imagery. In: CLEF working notes, CEUR. CEUR-WS.org, Dublin, Ireland (2017). http://ceur-ws.org

17. Molino, A.G.D., Mandal, B., Lin, J., Lim, J.H., Subbaraju, V., Chandrasekhar, V.: VC-I2R@ImageCLEF2017: ensemble of deep learned features for lifelog video summarization, pp. 11–14. CLEF working notes, CEUR (2017)

18. Müller, H., Clough, P., Deselaers, T., Caputo, B. (eds.): ImageCLEF - Experimental Evaluation in Visual Information Retrieval, The Springer International Series on Information Retrieval, vol. 32. Springer, Heidelberg (2010)

19. Papineni, K., Roukos, S., Ward, T., Zhu, W.J.: BLEU: a method for automatic evaluation of machine translation. In: Proceedings of the 40th Annual Meeting on Association for Computational Linguistics, pp. 311–318. Association for Computational Linguistics (2002)
20. Pomente, A., Aleandri, D.: Convolutional expectation maximization for population estimation, pp. 11–14. CLEF working notes, CEUR (2017)
21. Soldaini, L., Goharian, N.: QuickUMLS: a fast, unsupervised approach for medical concept extraction. In: MedIR Workshop, SIGIR (2016)
22. Tsikrika, T., Herrera, A.G.S., Müller, H.: Assessing the scholarly impact of Image-CLEF. In: Forner, P., Gonzalo, J., Kekäläinen, J., Lalmas, M., Rijke, M. (eds.) CLEF 2011. LNCS, vol. 6941, pp. 95–106. Springer, Heidelberg (2011). doi:10.1007/978-3-642-23708-9_12
23. Tsikrika, T., Larsen, B., Müller, H., Endrullis, S., Rahm, E.: The scholarly impact of CLEF (2000–2009). In: Forner, P., Müller, H., Paredes, R., Rosso, P., Stein, B. (eds.) CLEF 2013. LNCS, vol. 8138, pp. 1–12. Springer, Heidelberg (2013). doi:10.1007/978-3-642-40802-1_1
24. Villegas, M., et al.: General overview of ImageCLEF at the CLEF 2015 labs. In: Mothe, J., et al. (eds.) CLEF 2015. LNCS, vol. 9283, pp. 444–461. Springer, Cham (2015). doi:10.1007/978-3-319-24027-5_45
25. Villegas, M., et al.: General overview of ImageCLEF at the CLEF 2016 labs. CLEF 2016. LNCS, vol. 9822, pp. 267–285. Springer, Cham (2016). doi:10.1007/978-3-319-44564-9_25
26. Zhou, L., Piras, L., Rieger, M., Boato, G., Dang-Nguyen, D.T., Gurrin, C.: Organizer team at ImageCLEFlifelog 2017: baseline approaches for lifelog retrieval and summarization, pp. 11–14. CLEF working notes, CEUR (2017)

Overview of the CLEF 2017 Personalised Information Retrieval Pilot Lab (PIR-CLEF 2017)

Gabriella Pasi[1], Gareth J.F. Jones[3], Stefania Marrara[1(✉)], Camilla Sanvitto[1], Debasis Ganguly[2], and Procheta Sen[3]

[1] University of Milano Bicocca, Milan, Italy
stefania.marrara@disco.unimib.it
[2] IBM Research Labs, Dublin, Ireland
[3] Dublin City University, Dublin, Ireland

Abstract. The Personalised Information Retrieval Pilot Lab (PIR-CLEF 2017) provides a forum for the exploration of evaluation of personalised approaches to information retrieval (PIR). The Pilot Lab provides a preliminary edition of a Lab task dedicated to personalised search. The PIR-CLEF 2017 Pilot Task is the first evaluation benchmark based on the Cranfield paradigm, with the potential benefits of producing evaluation results that are easily reproducible. The task is based on search sessions over a subset of the ClueWeb12 collection, undertaken by 10 users by using a clearly defined and novel methodology. The collection provides data gathered by the activities undertaken during the search sessions by each participant, including details of relevant documents as marked by the searchers. The intention of the collection is to allow research groups working on PIR to both experience with and provide feedback about our proposed PIR evaluation methodology with the aim of launching a more formal PIR Lab at CLEF 2018.

1 Introduction

The primary aim of the PIR-CLEF laboratory is to provide a framework for the evaluation of Personalised Information Retrieval (PIR). PIR systems are aimed at enhancing traditional IR systems to better satisfy the user's information needs by providing search results that are not only relevant to the query but also to the user who submitted the query. In order to provide a personalised service, a PIR system maintains information about the users and their preferences and interests, which are usually inferred through a variety of interactions of the user with the system. This information is then represented in a user model, which is used to either improve the user's query or to re-rank a set of retrieved results list so that documents that are more relevant to the user are presented in the top positions of the list.

Existing work on the evaluation of PIR has generally relied on a user-centered approach, mostly based on user studies; this approach involves real users undertaking search tasks in a supervised environment. While this methodology has

G.J.F. Jones et al. (Eds.): CLEF 2017, LNCS 10456, pp. 338–345, 2017.
DOI: 10.1007/978-3-319-65813-1_29

the advantage of enabling the detailed study of the activities of real users, it has the significant drawback of not being easily reproducible, thus greatly limiting the scope for algorithmic exploration. Among existing IR benchmark tasks based on the Cranfield paradigm, the closest experiment to the PIR task is the TREC Session track[1] conducted annually between 2010 and 2014. However, this track focused only on stand-alone search sessions, where a "session" is a continuous sequence of query reformulations on the same topic, along with any user interaction with the retrieved results in service of satisfying a specific topical information need, for which no details of the searcher undertaking the task are available. Thus, the TREC Session track did not exploit any user model to personalise the search experience, nor did it allow user actions over multiple search session to be taken into consideration in the ranking of the search output.

The PIR-CLEF 2017 Pilot Task provides search data from a single search session with personal details of the user undertaking the search session. A full edition of this task will gather data across multiple sessions to enable the construction and exploitation of persistent user behaviour data collected from the user across the multiple search sessions.

Since this was a pilot activity we encouraged participants to attempt the task using existing algorithms and to explore new ideas. We also welcomed contributions to the workshop examining the specification and contents of the task and the provided dataset.

The PIR-CLEF 2017 workshop at the CLEF 2017 Conference has grouped together researchers working in PIR and related topics to explore the development of new methods for evaluation in PIR. In particular, the PIR-CLEF workshop has provided a pilot Lab task based on a test collection that has been generated by using the methodology described in [1]. A pilot evaluation using this collection was run to allow research groups working on PIR to both experience with and provide feedback about our proposed PIR evaluation methodology.

The remainder of this paper is organised as follows: Sect. 2 outlines existing related work, Sect. 3 provides an overview of the PIR-CLEF 2017 Pilot task, Sect. 4 discusses the metrics available for the evaluation of the Pilot task, and Sect. 5 concludes the paper.

2   Related Work

Recent years have seen increasing interest in the study of contextualisation in search: in particular, several research contributions have addressed the task of personalising search by incorporating knowledge of user preferences into the search process [2]. This user-centred approach to search has raised the related issue of how to properly evaluate search results in a scenario where relevance is strongly dependent on the interpretation of the individual user. To this purpose several user-based evaluation frameworks have been developed, as discussed in [3].

A key issue when seeking to introduce personalisation into the search process is the evaluation of the effectiveness of the proposed method. A first category of

---

[1] http://trec.nist.gov/data/session.html.

approaches aimed at evaluating personalised search systems attempts to perform a user-centred evaluation provided a kind of extension to the laboratory based evaluation paradigm. The TREC Interactive track [4] and the TREC HARD track [5] are examples of this kind of evaluation framework, which aimed at involving users in interactive tasks to get additional information about them and the query context being formulated. The evaluation was done by comparing a baseline run ignoring the user/topic metadata with another run considering it.

The more recent TREC Contextual Suggestion track [6] was proposed with the purpose of investigating search techniques for complex information needs that are highly dependent on context and user's interests. Participants in the track are given, as input, a set of geographical contexts and a set of user profiles that contain a list of attractions the user has previously rated. The task is to produce a list of ranked suggestions for each profile-context pair by exploiting the given contextual information. However, despite these extensions, the overall evaluation is still system controlled and only a few contextual features are available in the process.

TREC also introduced a Session track [7] whose focus was to exploit user interactions during a query session to incrementally improve the results within that session. The novelty of this task was the evaluation of system performance over entire sessions instead of a single query.

For all these reasons, the problem of defining a standard approach to the evaluation of personalised search is a hot research topic, which needs effective solutions. A first attempt to create a collection in support of PIR research was done in the FIRE Conference held in 2011. The personalised and Collaborative Information Retrieval track [8] was organised with the aim of extending a standard IR ad-hoc test collection by gathering additional meta-information during the topic development process to facilitate research on personalised and collaborative IR. However, since no runs were submitted to this track, only preliminary studies have been carried out and reported using it.

3   Overview of the Task

The goal of the PIR-CLEF 2017 Pilot Task was to investigate the use of a laboratory-based method to enable a comparative evaluation of PIR. The pilot collection used during PIR-CLEF 2017 was created with the cooperation of volunteer users, and was organized into two sequential phases:

– *Data gathering.* This phase involved the volunteer users carrying out a task-based search session during which a set of activities performed by the user were recorded (e.g., formulated queries, bookmarked documents, etc.). Each search session was composed of a phase of query development, refinement and modification, and associated search with each query on a specific topical domain selected by the user, followed by a relevance assessment phase where the user indicated the relevance of documents returned in response to each query and a short report writing activity based on the search activity undertaken.

– *Data cleaning and preparation.* This phase took place once the data gathering had been completed, and did not involve any user participation. It consisted of filtering and elaborating the information collected in the previous phase in order to prepare a dataset with various kinds of information related to the specific user's preferences. In addition, a bag-of-words representation of the participant's user profile was created to allow comparative evaluation of PIR algorithms using the same simple user model.

For the PIR-CLEF 2017 Pilot Task we made available the user profile data and raw search data produced by guided search sessions undertaken by 10 volunteer users as detailed in Sect. 3.1.

The aim of the task was to use the provided information to improve the ranking of a search results list over a baseline ranking of documents judged relevant to the query by the user who entered the query.

The Pilot Task data were provided in csv format to registered participants in the task. Access to the search service for the indexed subset of the ClueWeb12 collection was provided by Dublin City University via an API.

3.1   Dataset

For the PIR-CLEF 2017 Pilot Task we made available both user profile data and raw search data produced by guided search sessions undertaken by 10 volunteer users. The data provided included the submitted queries, the baseline ranked lists of documents retrieved in response to each query by using a standard search system, the items clicked by the user in the result list, and the documents relevance assessments provided by the user on a 4-grade scale. Each session was performed by the user on a topic of her choice selected from a provided list of broad topics, and search was carried out over a subset of the ClueWeb12 web collection.

The data has been extracted and stored in csv format as detailed in the following. In particular 7 csv files were provided in a zip folder. The file *user's session* (csv1) contains the information about each phase of the query sessions performed by each user. Each row of the csv contains:

– username: the user who performed the session
– query_session: id of the performed query session
– category: the top level search domain of the session
– task: the description of the search task fulfilled by the user
– start_time: starting time of the query session
– close_time: closing time of the search phase
– evaluated_time, closing time of the assessment phase
– end_time: closing time of the topic evaluation and the whole session.

The file *user's log* (csv2) contains the search logs of each user, i.e. every search event that has been triggered by a user's action. The file row contains:

– username: the user who performed the session
– query_session: id of the query session within the search was performed

- category: the top level search domain
- query_text: the submitted query
- document_id: the document on which a particular action was performed
- rank: the retrieval rank of the document on which a particular action is performed
- action_type: the type of the action executed by the user (query submission, open_document, close_document, bookmark)
- time_stamp: the timestamp of the action.

The file *user's assessment* (csv3) contains the relevance assessments of a pool of documents with respect to every single query developed by each user to fulfill the given task:

- username: the user who performed the session
- query_session: id of the query session within the evaluation was performed
- query_text: the query on which the evaluation is based
- document_id: the document id for which the evaluation was provided
- rank: the retrieval rank of the document on which a particular action is performed
- relevance_score: the relevance of the document to the topic (1 off-topic, 2 not relevant, 3 somewhat relevant, 4 relevant).

The file *user's info* (csv4) contains some personal information about the users:

- username
- age_range
- gender
- occupation
- native_language.

The file *user's topic* (csv5) contains the TREC-style final topic descriptions about the user's information needs that were developed in the final step of each search session:

- username, the user who formulated the topic
- query_session, id of the query session which the topic refers to
- title, a small phrase defining the topic provided by the user
- description, a detailed sentence describing the topic provided by the user
- narrative, a description of which documents are relevant to the topic and which are not, provided by the user

The file *simple user profile* (csv6a) for each user contains the following information (simple version - the applied indexing included tokenization, shingling, and index terms weighting):

- username: the user whose interests are represented
- category: the search domain of interest
- a list of triples constituted by:
  - a term: a word or n-grams related to the user's searches

- a normalised_score: term weight computed as the mean of the term frequencies in the user's documents of interests, where term frequency is the ratio of the number of occurrences of the term in a document and the number of occurrences of the most frequent term in the same document.

The file *complex user profile* (csv6b) contains, for each user, the same information provided in csv6a, with the difference that the applied indexing was enriched by also including stop word removal:

- username, the user whose interests are represented
- category, the search domain of interest
- a list of triples constituted by:
  - term, a word or a set of words related to the user's searches
  - normalised_score,

Participants had the possibility to contribute to the task in two different ways:

- the two user profile files (csv6a and csv6b) provide the bag-of words profiles of the 10 users involved in the experiment, extracted by applying different indexing procedures to the documents. The user's log file (cvs2) contains for each user all the queries she formulated during the query session. The participant could compare the results obtained by applying their personalisation algorithm on these queries with the results obtained and evaluated by the users on the same queries (and included in the user assessment file csv3). The search had to be carried out on the ClueWeb12 collection, by using the API provided by DCU. Then, by using the 4-graded scale evaluations of the documents (relevant, somewhat relevant, non relevant, off topic) provided by the users and contained in the user assessment file csv3, it was possible to compute Average Precision (AP) and Normalized Discounted Cumulative Gain (NDCG). Note that documents that do not appear in csv3 were considered non-relevant. As further explained in Sect. 4, these metrics were computed both globally (in the literature they are just AP and NDCG) and for each user query individually, by then taking the mean.
- The challenge here was to use the raw data provided in csv1, csv2, csv3, csv4, and csv5 to create user profiles. A user profile is a formal representation of the user interests and preferences; the more accurate the representation of the user model, the higher is the probability to improve the search process. In the approaches proposed in the literature, user profiles are formally represented as bags of words, as vectors, or as conceptual taxonomies, generally defined based on external knowledge resources (such as the WordNet and the ODP - Open Directory Project). The task request here was more research oriented: are the provided information sufficient to create a useful profile? Which information is missing? The outcome here was a report up to 6 pages discussing the theme of user information to profiling aims, by proposing possible integrations of the provided data and by suggesting a way to collect them in a controlled Cranfield style experiment.

Since this was a pilot activity we encouraged participants to be involved in this task by using existing or new algorithms and/or to explore new ideas. We also welcomed contributions that make an analysis of the task and/or of the dataset.

4   Performance Measures

The discussion about performance measures of personalised search was one of the core workshop topics at CLEF 2017. Well known information retrieval metrics, such as Average Precision (AP) and Normalized Discounted Cumulative Gain (NDCG) can be considered to benchmark the participants' systems. Other measures must be investigated, and possibly new metrics have to be defined.

5   Conclusions and Future Work

In this work the PIR-CLEF 2017 personalised Information Retrieval Pilot Task was presented. This task is the preliminary edition of a lab dedicated to the theme of personalised search that is planned to officially start at CLEF 2018. This is the first evaluation benchmark in this field based on the Cranfield paradigm, with the significant benefit of producing results easily reproducible. The PIR-CLEF 2017 workshop has provided a pilot Lab task based on a test collection that has been generated by using a well defined methodology. A pilot evaluation using this collection has been run to allow research groups working on personalised IR to both experience with and provide feedback about our proposed PIR evaluation methodology.

References

1. Sanvitto, C., Ganguly, D., Jones, G.J.F., Pasi, G.: A laboratory-based method for the evaluation of personalised search. In: Proceedings of the Seventh International Workshop on Evaluating Information Access (EVIA 2016), a Satellite Workshop of the NTCIR-12 Conference, Tokyo, 7 June 2016
2. Pasi, G.: Issues in personalising information retrieval. IEEE Intell. Inform. Bull. 11(1), 3–7 (2010)
3. Tamine-Lechani, L., Boughanem, M., Daoud, M.: Evaluation of contextual information retrieval effectiveness: overview of issues and research. Knowl. Inf. Syst. 24(1), 1–34 (2009)
4. Harman, D.: Overview of the fourth text retrieval conference (TREC-4). In Harman, D.K. (ed.) TREC, vol. Special Publication 500-236. National Institute of Standards and Technology (NIST) (1995)
5. Allan, J.: HARD track overview in TREC 2003: high accuracy retrieval from documents. In: Proceedings of The Twelfth Text REtrieval Conference (TREC 2003), Gaithersburg, pp. 24–37 (2003)
6. Dean-Hall, A., Clarke, C.L.A., Kamps, J., Thomas, P., Voorhees, E.M.: Overview of the TREC 2012 contextual suggestion track. In: Voorhees and Bucklan
7. Carterette, B., Kanoulas, E., Hall, M.M., Clough, P.D.: Overview of the TREC 2014 session track. In: Proceedings of the Twenty-Third Text REtrieval Conference (TREC 2014), Gaithersburg (2014)
8. Ganguly, D., Leveling, J., Jones, G.J.F.: Overview of the Personalized and Collaborative Information Retrieval (PIR) Track at FIRE-2011. In: Majumder, P., Mitra, M., Bhattacharyya, P., Subramaniam, L.V., Contractor, D., Rosso, P. (eds.) FIRE 2010-2011. LNCS, vol. 7536, pp. 227–240. Springer, Heidelberg (2013). doi:10.1007/978-3-642-40087-2_22

9. Villegas, M., Puigcerver, J.A., Toselli, H., Sanchez, J.A., Vidal, E.: Overview of the image CLEF 2016 handwritten scanned document retrieval task. In: Proceedings of CLEF (2016)

10. Robertson, S.: A new interpretation of average precision. In: Proceedings of the International ACM SIGIR Conference on Research and Development in Information Retrieval (SIGIR 2008), pp. 689–690. ACM, New York (2008)

eRISK 2017: CLEF Lab on Early Risk Prediction on the Internet: Experimental Foundations

David E. Losada[1]([✉]), Fabio Crestani[2], and Javier Parapar[3]

[1] Centro Singular de Investigación en Tecnoloxías da Información (CiTIUS),
Universidade de Santiago de Compostela, Santiago de Compostela, Spain
david.losada@usc.es
[2] Faculty of Informatics, Universitá della Svizzera italiana (USI),
Lugano, Switzerland
fabio.crestani@usi.ch
[3] Information Retrieval Lab, University of A Coruña, A Coruña, Spain
javierparapar@udc.es

Abstract. This paper provides an overview of eRisk 2017. This was
the first year that this lab was organized at CLEF. The main purpose
of eRisk was to explore issues of evaluation methodology, effectiveness
metrics and other processes related to early risk detection. Early detection technologies can be employed in different areas, particularly those
related to health and safety. The first edition of eRisk included a pilot
task on early risk detection of depression.

1 Introduction

The main goal of eRisk was to instigate discussion on the creation of reusable
benchmarks for evaluating early risk detection algorithms, by exploring issues
of evaluation methodology, effectiveness metrics and other processes related to
the creation of test collections for early risk detection. Early detection technologies can be employed in different areas, particularly those related to health and
safety. For instance, early alerts could be sent when a predator starts interacting
with a child for sexual purposes, or when a potential offender starts publishing
antisocial threats on a blog, forum or social network. eRisk wants to pioneer
a new interdisciplinary research area that would be potentially applicable to a
wide variety of profiles, such as potential paedophiles, stalkers, individuals with
a latent tendency to fall into the hands of criminal organisations, people with
suicidal inclinations, or people susceptible to depression.

Early risk prediction is a challenging and increasingly important research
area. However, this area lacks systematic experimental foundations. It is therefore difficult to compare and reproduce experiments done with predictive algorithms running under different conditions.

Citizens worldwide are exposed to a wide range of risks and threats and many
of these hazards are reflected on the Internet. Some of these threats stem from
criminals such as stalkers, mass killers or other offenders with sexual, racial,
religious or culturally related motivations. Other worrying threats might even

G.J.F. Jones et al. (Eds.): CLEF 2017, LNCS 10456, pp. 346–360, 2017.
DOI: 10.1007/978-3-319-65813-1_30

come from the individuals themselves. For instance, depression may lead to an eating disorder such as anorexia or even to suicide.

In some of these cases early detection and appropriate action or intervention could reduce or minimise these problems. However, the current technology employed to deal with these issues is essentially reactive. For instance, some specific types of risks can be detected by tracking Internet users, but alerts are triggered when the victim makes his disorders explicit, or when the criminal or offending activities are actually happening. We argue that we need to go beyond this late detection technology and foster research on innovative early detection solutions able to identify the states of those at risk of becoming perpetrators of socially destructive behaviour, and the states of those at risk of becoming victims. Thus, we also want to stimulate the development of algorithms that computationally encode the process of becoming an offender or a victim.

It has been shown that the words people use can reveal important aspects of their social and psychological worlds [16]. There is substantial evidence linking natural language to personality, social and situational fluctuations. This is of particular interest to understand the onset of a risky situation and how it reflects the linguistic style of the individuals involved. However, a major hurdle that has to be overcome is the lack of evaluation methodologies and test collections for early risk prediction. In this lab we intended to take the first steps towards filling this gap. We understand that there are two main classes of early risk prediction:

- **Multiple actors.** We include in the first category cases where there is an external actor or intervening factor that explicitly causes or stimulates the problem. For instance, sexual offenders use deliberate tactics to contact vulnerable children and engage them in sexual exploitation. In such cases, early warning systems need to analyse the interactions between the offender and the victim and, in particular, the language of both. The process of predation is known to happen in five phases [11], namely: gaining access, deceptive trust development, grooming, isolation, and approach. Therefore, systems can potentially track conversations and alert about the onset of a risky situation. Initiatives such as the organisation of a sexual predation identification challenge in CLEF [6] (under the PAN lab on Uncovering Plagiarism, Authorship and Social Software Misuse) have fostered research on mining conversations and identifying predatory behaviour. However, the focus was on identifying sexual predators and predatory text. There was no notion of early warning. We believe that predictive algorithms such as those developed under this challenge [13,14] could be further evaluated from an early risk prediction perspective. Another example of risk provoked by external actions is terrorist recruitment. There is currently massive online activity aiming at recruiting young people –particularly, teenagers– for joining criminal networks. Excellent work in this area has been done by the AI Lab of the University of Arizona. Among many other things, this team has created a research infrastructure called "the Dark Web" [19], that is available to social science researchers, computer and information scientists, and policy and security analysts. It permits to study a wide range of social and organizational phenomena of criminal networks. The

Dark Web Forum Portal enables access to critical international jihadist and other extremist web forums. Scanlon and Gerber [17] have analyzed messages from the Dark Web portal forums to perform a two-class categorisation task, aiming at distinguish recruiting posts from non-recruiting posts. Again, the focus was not on early risk prediction because there was not notion of time or sequence of events.

- **Single actor.** We include in this second category cases where there is not an explicit external actor or intervening factor that causes or stimulates the problem. The risk comes "exclusively" from the individual. For instance, depression might not be caused or stimulated by any intervention or action made by external individuals. Of course, there might be multiple personal or contextual factors that affect –or even cause– a depression process (and, as a matter of fact, this is usually the case). However, it is not feasible to have access to sources of data associated to all these external conditions. In such cases, the only element that can be analysed is the language of the individual. Following this type of analysis, there is literature on the language of people suffering from depression [2,3,12,15], post-traumatic stress disorder [1,4], bipolar disorder [7], or teenage distress [5]. In a similar vein, other studies have analysed the language of school shooters [18], terrorists [9], and other self-destructive killers [8].

The two classes of risks described above might be related. For instance, individuals suffering from major depression might be more inclined to fall prey to criminal networks. From a technological perspective, different types of tools are likely needed to develop early warning systems that alert about these two types of risks.

Early risk detection technologies can be adopted in a wide range of domains. For instance, it might be used for monitoring different types of activism, studying psychological disorder evolution, early-warning about sociopath outbreaks, or tracking health-related problems in Social Media.

Essentially, we can understand early risk prediction as a process of sequential evidence accumulation where alerts are made when there is enough evidence about a certain type of risk. For the single actor type of risk, the pieces of evidence could come in the form of a chronological sequence of entries written by a tormented subject in Social Media. For the multiple actor type of risk, the pieces of evidence could come in the form of a series of messages interchanged by an offender and a victim in a chatroom or online forum.

To foster discussion on these issues, we shared with the participants of the lab the test collection presented at CLEF in 2016 [10]. This CLEF 2016 paper discusses the creation of a benchmark on depression and language use that formally defines an early risk detection framework and proposes new effectiveness metrics to compare algorithms that address this detection challenge. The framework and evaluation methodology has the potential to be employed by many other research teams across a wide range of areas to evaluate solutions that infer behavioural patterns –and their evolution– in online activity. We therefore invited eRisk participants to engage in a pilot task on early detection of depression, which is described in the next section.

2   Pilot Task: Early Detection of Depression

This was an exploratory task on early risk detection of depression. The challenge consists of sequentially processing pieces of evidence and detect early traces of depression as soon as possible. The task is mainly concerned about evaluating Text Mining solutions and, thus, it concentrates on texts written in Social Media. Texts should be processed in the order they were created. In this way, systems that effectively perform this task could be applied to sequentially monitor user interactions in blogs, social networks, or other types of online media.

The test collection for this pilot task is the collection described in [10]. It is a collection of writings (posts or comments) from a set of Social Media users. There are two categories of users, depressed and non-depressed, and, for each user, the collection contains a sequence of writings (in chronological order). For each user, his collection of writings has been divided into 10 chunks. The first chunk contains the oldest 10% of the messages, the second chunk contains the second oldest 10%, and so forth.

The task was organized into two different stages:

- **Training stage.** Initially, the teams that participated in this task had access to a training stage where we released the whole history of writings for a set of training users. We provided all chunks of all training users, and we indicated what users had explicitly mentioned that they have been diagnosed with depression. The participants could therefore tune their systems with the training data. This training dataset was released on Nov 30th, 2016.
- **Test stage.** The test stage consisted of 10 sequential releases of data (done at different dates). The first release consisted of the 1st chunk of data (oldest writings of all test users), the second release consisted of the 2nd chunk of data (second oldest writings of all test users), and so forth. After each release, the participants had one week to process the data and, before the next release, each participating system had to choose between two options: (a) emitting a decision on the user (i.e. depressed or non-depressed), or (b) making no decision (i.e. waiting to see more chunks). This choice had to be made for each user in the collection. If the system emitted a decision then its decision was considered as final. The systems were evaluated based on the correctness of the decisions and the number of chunks required to make the decisions (see below). The first release was done on Feb 2nd, 2017 and the last (10th) release was done on April 10th, 2017.

Table 1 reports the main statistics of the train and test collections. Both collections are unbalanced (more non-depression cases than depression cases). The number of subjects is not very high, but each subject has a long history of writings (on average, we have hundreds of messages from each subject). Furthermore, the mean range of dates from the first to the last submission is quite wide (more than 500 days). Such wide chronology permits to study the evolution of the language from the oldest piece of evidence to the most recent one.

Table 1. Main statistics of the train and test collections

	Train		Test	
	Depressed	*Control*	*Depressed*	*Control*
Num. subjects	83	403	52	349
Num. submissions (posts & comments)	30,851	264,172	18,706	217,665
Avg num. of submissions per subject	371.7	655.5	359.7	623.7
Avg num. of days from first to last submission	572.7	626.6	608.31	623.2
Avg num. words per submission	27.6	21.3	26.9	22.5

2.1   Error Measure

We employed ERDE, an error measure for early risk detection defined in [10]. This was an exploratory task and the evaluation was tentative. As a matter of fact, one of the goals of eRisk 2017 was to identify the shortcomings of the collection and error metric.

ERDE is a metric for which the fewer writings required to make the alert, the better. For each user we proceed as follows. Given a chunk of data, if a system does not emit a decision then it has access to the next chunk of data (i.e. more writings from the same user). But the system gets a penalty for *late emission*.

Standard classification measures, such as the F-measure, could be employed to assess the system's output with respect to golden truth judgments that inform us about what subjects are really positive cases. However, standard classification measures are time-unaware and, therefore, we needed to complement them with new measures that reward early alerts.

ERDE stands for *early risk detection error* and it takes into account the correctness of the (binary) decision and the delay taken by the system to make the decision. The delay was measured by counting the number (k) of distinct textual items seen before giving the answer. For example, imagine a user u that has 25 writings in each chunk. If a system emitted a decision for user u after the second chunk of data then the delay k was set to 50 (because the system needed to see 50 pieces of evidence in order to make its decision).

Another important factor is that, in many application domains, data are unbalanced (many more negative cases than positive cases). This was also the case in our data (many more non-depressed individuals). Hence, we also needed to weight different errors in a different way.

Consider a binary decision d taken by a system with delay k. Given golden truth judgments, the prediction d can be a true positive (TP), true negative (TN), false positive (FP) or false negative (FN). Given these four cases, the ERDE measure is defined as:

$$ERDE_o(d,k) = \begin{cases} c_{fp} & \text{if } d = \text{positive AND ground truth} = \text{negative (FP)} \\ c_{fn} & \text{if } d = \text{negative AND ground truth} = \text{positive (FN)} \\ lc_o(k) \cdot c_{tp} & \text{if } d = \text{positive AND ground truth} = \text{positive (TP)} \\ 0 & \text{if } d = \text{negative AND ground truth} = \text{negative (TN)} \end{cases}$$

How to set c_{fp} and c_{fn} depends on the application domain and the implications of FP and FN decisions. We will often face detection tasks where the number of negative cases is several orders of magnitude greater than the number of positive cases. Hence, if we want to avoid building trivial classifiers that always say no, we need to have $c_{fn} >> c_{fp}$. We fixed c_{fn} to 1 and set c_{fp} according to the proportion of positive cases in the test data (e.g. we set c_{fp} to 0.1296). The factor $lc_o(k)(\in [0,1])$ encodes a cost associated to the delay in detecting true positives. We set c_{tp} to c_{fn} (i.e. c_{tp} was set to 1) because late detection can have severe consequences (i.e. late detection is equivalent to not detecting the case at all).

The function $lc_o(k)$ is a monotonically increasing function of k:

$$lc_o(k) = 1 - \frac{1}{1 + e^{k-o}} \tag{1}$$

The function is parameterised by o, which controls the place in the X axis where the cost grows more quickly (Fig. 1 plots $lc_5(k)$ and $lc_{50}(k)$).

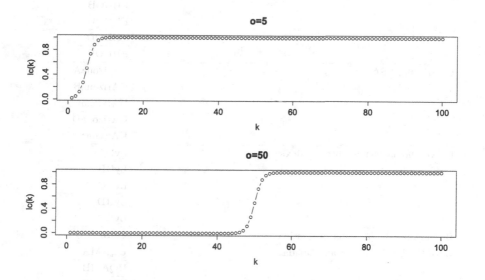

Fig. 1. Latency cost functions: $lc_5(k)$ and $lc_{50}(k)$

Observe that the latency cost factor was introduced only for the true positives. We understand that late detection is not an issue for true negatives. True negatives are non-risk cases that, in practice, would not demand early intervention. They just need to be effectively filtered out from the positive cases. Algorithms should therefore focus on early detecting risk cases and detecting non-risk cases (regardless of when these non-risk cases are detected).

All cost weights are in $[0, 1]$ and, thus, ERDE is in the range $[0, 1]$. Systems had to take one decision for each subject and the overall error is the mean of the p ERDE values.

2.2   Results

We received 30 contributions from 8 different institutions. Table 2 shows the institutions that contributed to eRisk and the labels associated to their runs. Each team could contribute up to five different variants.

Table 2. Participating institutions and submitted results

Institution	Submitted files
ENSEEIHT, France	GPLA
	GPLB
	GPLC
	GPLD
FH Dortmund, Germany	FHDOA
	FHDOB
	FHDOC
	FHDOD
	FHDOE
U. Arizona, USA	UArizonaA
	UArizonaB
	UArizonaC
	UArizonaD
	UArizonaE
U. Autónoma Metropolitana, Mexico	LyRA
	LyRB
	LyRC
	LyRD
	LyRE
U. Nacional de San Luis, Argentina	UNSLA
U. of Quebec in Montreal, Canada	UQAMA
	UQAMB
	UQAMC
	UQAMD
	UQAME
Instituto Nacional de Astrofísica, Optica y Electrónica, Mexico	CHEPEA
	CHEPEB
	CHEPEC
	CHEPED
ISA FRCCSC RAS, Russia	NLPISA

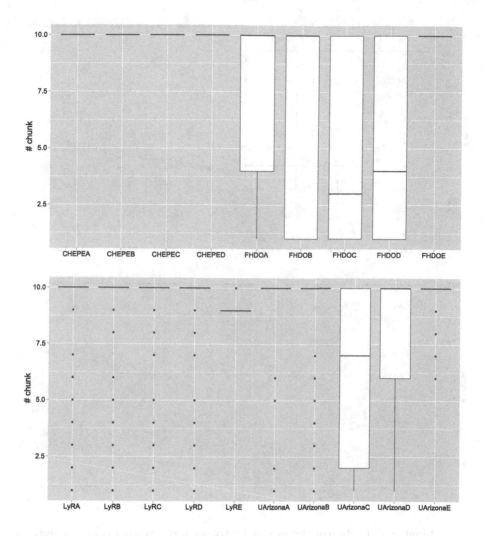

Fig. 2. Number of chunks required by each contributing run in order to emit a decision.

First, let us analyze the behaviour of the algorithms in terms of how quick they were to emit their decisions. Figures 2 and 3 show a boxplot graph of the number of chunks required to make the decisions. The test collection has 401 subjects and, thus, the boxplot associated to each run represents the statistics of 401 cases. Seven variants waited until the last chunk in order to make the decision for all subjects (i.e. no single decision was done before the last chunk). This happened with CHEPEA, CHEPEB, CHEPEC, CHEPED, FHDOE, GPLD, and NLPISA. These seven runs were extremely conservative: they waited to see the whole history of writings for all the individuals and, next, they emitted their decisions (all teams were forced to emit a decision for each user after

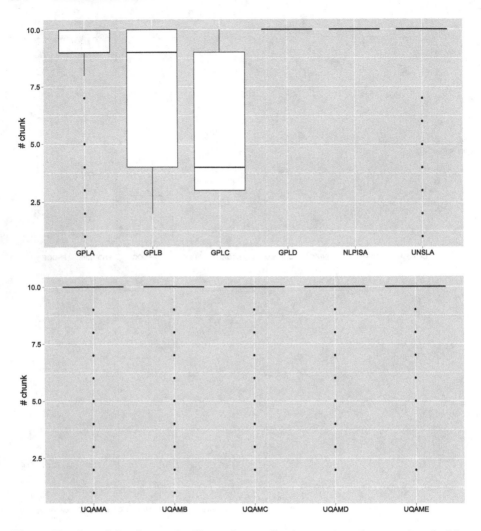

Fig. 3. Number of chunks required by each contributing run in order to emit a decision.

the last chunk). Many other runs –e.g., UNLSA, LyRA, LyRB, LyRC, LyRD, UArizonaA, UArizonaB, UArizonaE, UQAMA, UQAMB, UQAMC, UQAMD, and UQAME– also took most of the decisions after the last chunk. For example, with UNSLA, 316 out of 401 test subjects had a decision assigned after the 10th chunk. Only a few runs were really quick at emitting decisions. Notably, FHDOC had a median of 3 chunks needed to emit a decision.

Figures 4 and 5 represent a boxplot of the number of writings required by each algorithm in order to emit the decisions. Most of the variants waited to see hundreds of writings for each user. Only a few runs (UArizonaC, FHDOC and FHDOD) had a median number of writings analyzed below 100. This was the first year of the task and it appears that most of the teams have concentrated on the

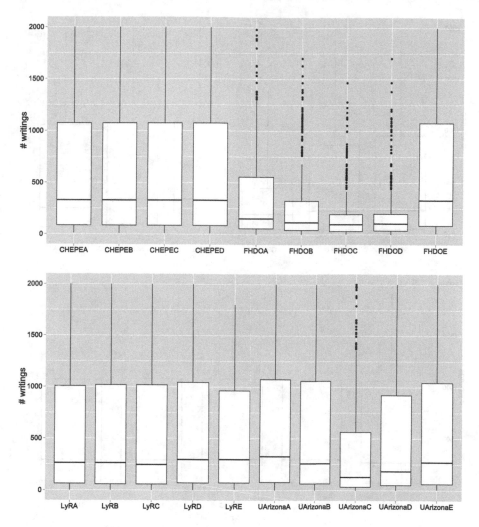

Fig. 4. Number of writings required by each contributing run in order to emit a decision.

effectiveness of their decisions (rather than on the tradeoff between accuracy and delay). The number of writings per subject has a high variance. Some subjects have only 10 or 20 writings, while other subjects have thousands of writings. In the future, it will be interesting to study the impact of the number of writings on effectiveness. Such study could help to answer questions like: was the availability of more textual data beneficial?. Note that the writings were obtained from a wide range of sources (multiple subcommunities from the same Social Network). So, we wonder how well the algorithms perform when a specific user had many offtopic writings.

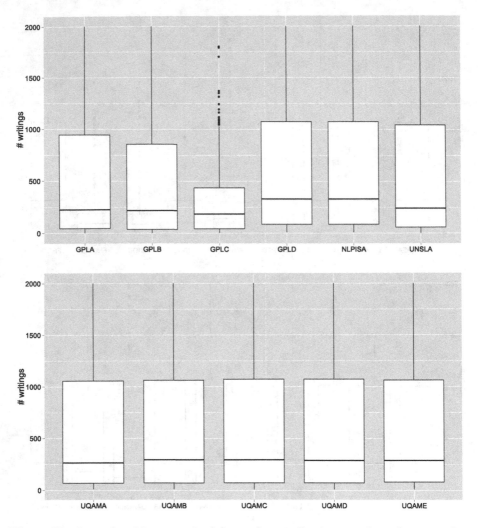

Fig. 5. Number of writings required by each contributing run in order to emit a decision.

A subject whose main (or only) topic of conversation is depression is arguably easier to classify. But the collection contains non-depressed individuals that are active on depression subcommunities. For example, a person that has a close relative suffering from depression. We think that these cases could be false positives in most of the predictions done by the systems. But this hypothesis needs to be validated through further investigations. We will process the system's outputs and analyze the false positives to shed light on this issue.

Figure 6 helps to analyze another aspect of the decisions of the systems. For each group of subjects, it plots the percentage of correct decisions against the number of subjects. For example, the rightmost bar of the upper plot means

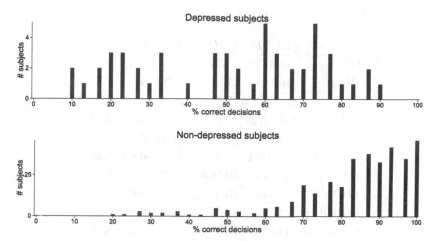

Fig. 6. Number of depressed and non-depressed subjects that had a given percentage of correct decisions.

that 90% of the systems correctly identified one subject as depressed. Similarly, the rightmost bar of the lower plot means that there were 46 non-depressed subjects that were correctly classified by all systems (100% correct decisions). The graphs show that systems tend to be more effective with non-depressed subjects. The distribution of correct decisions for non-depressed subjects has many cases where more than 80% of the systems are correct. The distribution of correct decisions for depressed subjects is flatter, and many depressed subjects are only identified by a low percentage of the systems. Furthermore, there are not depressed subjects that are correctly identified by all systems. However, an interesting point is that no depressed subject has 0% of correct decisions. This means that every depressed subject was classified as such by at least one system.

Let us now analyze the effectiveness results (see Table 3). The first conclusion we can draw is that the task is difficult. In terms of F1, performance is low. The highest F1 is 0.64. This might be related to the way in which the collection was created. The non-depressed group of subjects includes random users of the social networking site, but also a number of users who were active on the depression community and depression fora. There is a variety of such cases but most of them are individuals interested in depression because they have a close relative suffering from depression. These cases could potentially be false positives. As a matter of fact, the highest precision, 0.69, is also relatively low. The lowest $ERDE_5$ was achieved by the FHDO team, which also submitted the runs that performed the best in terms of $F1$ and precision. The run with the lowest $ERDE_{50}$ was submitted by the UNSLA team.

Some systems, e.g. FHDOB, opted for optimizing precision, while other systems, e.g. UArizonaC, opted for optimizing recall. The lowest error tends to be associated with runs with moderate F1 but high precision. For example, FHDO, the run with the lowest $ERDE_5$, is one of the runs that was quicker at makin

Table 3. Results

	$ERDE_5$	$ERDE_{50}$	F1	P	R
GPLA	17.33%	15.83%	0.35	0.22	0.75
GPLB	19.14%	17.15%	0.30	0.18	0.83
GPLC	14.06%	12.14%	0.46	0.42	0.50
GPLD	14.52%	12.78%	0.47	0.39	0.60
FHDOA	12.82%	9.69%	**0.64**	0.61	0.67
FHDOB	**12.70%**	10.39%	0.55	**0.69**	0.46
FHDOC	13.24%	10.56%	0.56	0.57	0.56
FHDOD	13.04%	10.53%	0.57	0.63	0.52
FHDOE	14.16%	12.42%	0.60	0.51	0.73
UArizonaA	14.62%	12.68%	0.40	0.31	0.58
UArizonaB	13.07%	11.63%	0.30	0.33	0.27
UArizonaC	17.93%	12.74%	0.34	0.21	**0.92**
UArizonaD	14.73%	10.23%	0.45	0.32	0.79
UArizonaE	14.93%	12.01%	0.45	0.34	0.63
LyRA	15.65%	15.15%	0.14	0.11	0.19
LyRB	16.75%	15.76%	0.16	0.11	0.29
LyRC	16.14%	15.51%	0.16	0.12	0.25
LyRD	14.97%	14.47%	0.15	0.13	0.17
LyRE	13.74%	13.74%	0.08	0.11	0.06
UNSLA	13.66%	**9.68%**	0.59	0.48	0.79
UQAMA	14.03%	12.29%	0.53	0.48	0.60
UQAMB	13.78%	12.78%	0.48	0.49	0.46
UQAMC	13.58%	12.83%	0.42	0.50	0.37
UQAMD	13.23%	11.98%	0.38	0.64	0.27
UQAME	13.68%	12.68%	0.39	0.45	0.35
CHEPEA	14.75%	12.26%	0.48	0.38	0.65
CHEPEB	14.78%	12.29%	0.47	0.37	0.63
CHEPEC	14.81%	12.57%	0.46	0.37	0.63
CHEPED	14.81%	12.57%	0.45	0.36	0.62
NLPISA	15.59%	15.59%	0.15	0.12	0.21

decisions (see Figs. 2 and 4) and its precision is the highest (0.69). $ERDE_5$ is extremely stringent with delays (after 5 writings, penalties grow quickly, see Fig. 1). This promotes runs that emit few but quick depression decisions. $ERDE_{50}$, instead, gives smoother penalties to delays. This makes that the run with the lowest $ERDE_{50}$, UNSLA, has low precision but relatively high recall (0.79). Such difference between $ERDE_5$ and $ERDE_{50}$ is highly relevant in practice. For example, a mental health agency seeking a tool for automatic screening for depression could set the penalty costs depending on the consequences of late detection of depression.

3   Future Work and Conclusions

This paper provides an overview of eRisk 2017. This was the first year that this lab was organized at CLEF and the lab's activities were concentrated on a pilot task on early risk detection of depression. The task received 30 contributions from 8 different institutions. Being the first year of the task, most teams focused on tuning different classification solutions (depressed vs non-depressed). The tradeoff between early detection and accuracy was not a major concern for most of the participants.

We plan to run eRisk again in 2018. We are currently collecting more data on depression and language, and we plan to expand the lab to other psychological problems. Early detecting other disorders, such as anorexia or post-traumatic stress disorder, would also be highly valuable and could be the focus of some eRisk 2018 subtasks.

Acknowledgements. We thank the support obtained from the Swiss National Science Foundation (SNSF) under the project "Early risk prediction on the Internet: an evaluation corpus", 2015.

We also thank the financial support obtained from the (i) "Ministerio de Economía y Competitividad" of the Government of Spain and FEDER Funds under the research project TIN2015-64282-R, (ii) Xunta de Galicia (project GPC 2016/035), and (iii) Xunta de Galicia – "Consellería de Cultura, Educación e Ordenación Universitaria" and the European Regional Development Fund (ERDF) through the following 2016-2019 accreditations: ED431G/01 ("Centro singular de investigacion de Galicia") and ED431G/08.

References

1. Alvarez-Conrad, J., Zoellner, L.A., Foa, E.B.: Linguistic predictors of trauma pathology and physical health. Appl. Cogn. Psychol. **15**(7), S159–S170 (2001)
2. De Choudhury, M., Counts, S., Horvitz, E.: Social media as a measurement tool of depression in populations. In: Davis, H.C., Halpin, H., Pentland, A., Bernstein, M., Adamic, L.A. (eds.) WebSci, pp. 47–56. ACM (2013)
3. De Choudhury, M., Gamon, M., Counts, S., Horvitz, E.: Predicting depression via social media. In: Kiciman, E., Ellison, N.B., Hogan, B., Resnick, P., Soboroff, I. (eds.) ICWSM. The AAAI Press (2013)
4. Coppersmith, G., Harman, C., Dredze, M.: Measuring post traumatic stress disorder in Twitter. In: Proceedings of the Eighth International Conference on Weblogs and Social Media, ICWSM 2014, Ann Arbor, Michigan, USA, 1–4 June 2014 (2014)
5. Dinakar, K., Weinstein, E., Lieberman, H., Selman, R.L.: Stacked generalization learning to analyze teenage distress. In: Adar, E., Resnick, P., De Choudhury, M., Hogan, B., Oh, A. (eds.) ICWSM. The AAAI Press (2014)
6. Inches, G., Crestani, F.: Overview of the international sexual predator identification competition at PAN-2012. In: Proceedings of the PAN 2012 Lab Uncovering Plagiarism, Authorship, and Social Software Misuse (within CLEF 2012) (2012)

7. Kramer, A.D.I., Fussell, S.R., Setlock, L.D.: Text analysis as a tool for analyzing conversation in online support groups. In: Dykstra-Erickson, E., Tscheligi, M. (eds.) CHI Extended Abstracts, pp. 1485–1488. ACM (2004)
8. Lankford, A.: Précis of the myth of martyrdom: what really drives suicide bombers, rampage shooters, and other self-destructive killers. Behav. Brain Sci. **37**, 351–362 (2014)
9. Leonard, C.H., Annas, G.D., Knoll, J.L., Tørrissen, T.: The case of Anders Behring Breivik - language of a lone terrorist. Behav. Sci. Law **32**, 408–422 (2014)
10. Losada, D.E., Crestani, F.: A test collection for research on depression and language use. In: Fuhr, N., Quaresma, P., Gonçalves, T., Larsen, B., Balog, K., Macdonald, C., Cappellato, L., Ferro, N. (eds.) CLEF 2016. LNCS, vol. 9822, pp. 28–39. Springer, Cham (2016). doi:10.1007/978-3-319-44564-9_3
11. Mcghee, I., Bayzick, J., Kontostathis, A., Edwards, L., Mcbride, A., Jakubowski, E.: Learning to identify internet sexual predation. Int. J. Electron. Commerce **15**(3), 103–122 (2011)
12. Moreno, M.A., Jelenchick, L.A., Egan, K.G., Cox, E., Young, H., Gannon, K.E., Becker, T.: Feeling bad on facebook: depression disclosures by college students on a social networking site, June 2011
13. Parapar, J., Losada, D.E., Barreiro, Á.: Combining psycho-linguistic, content-based and chat-based features to detect predation in chatrooms. J. Univ. Comput. Sci. **20**(2), 213–239 (2014)
14. Parapar, J., Losada, D.E., Barreiro, A.: Approach, learning-based, for the identification of sexual predators in chat logs. In PAN 2012: Lab Uncovering Plagiarism, Authorship, and Social Software Misuse, at Conference and Labs of the Evaluation Forum CLEF. Italy, Rome (2012)
15. Park, M., McDonald, D.W., Cha, M.: Perception differences between the depressed and non-depressed users in Twitter. In: Kiciman, E., Ellison, N.B., Hogan, B., Resnick, P., Soboroff, I. (eds.) ICWSM. The AAAI Press (2013)
16. Pennebaker, J.W., Mehl, M.R., Niederhoffer, K.G.: Psychological aspects of natural language use: our words, our selves. Annu. Rev. Psychol. **54**(1), 547–577 (2003)
17. Scanlon, J.R., Gerber, M.S.: Automatic detection of cyber-recruitment by violent extremists. Secur. Inform. **3**(1), 5 (2014)
18. Veijalainen, J., Semenov, A., Kyppö, J.: Tracing potential school shooters in the digital sphere. In: Bandyopadhyay, S.K., Adi, W., Kim, T., Xiao, Y. (eds.) ISA 2010. CCIS, vol. 76, pp. 163–178. Springer, Heidelberg (2010). doi:10.1007/978-3-642-13365-7_16
19. Zhang, Y., Zeng, S., Huang, C.N., Fan, L., Yu, X., Dang, Y., Larson, C.A., Denning, D., Roberts, N., Chen, H.: Developing a dark web collection and infrastructure for computational and social sciences. In: Yang, C.C., Zeng, D., Wang, K., Sanfilippo, A., Tsang, H.H., Day, M.-Y., Glässer, U., Brantingham, P.L., Chen, H. (eds.) ISI, pp. 59–64. IEEE (2010)

CLEF 2017 Dynamic Search Evaluation Lab Overview

Evangelos Kanoulas[1]([⊠]) and Leif Azzopardi[2]

[1] Informatics Institute, University of Amsterdam, Amsterdam, Netherlands
E.Kanoulas@uva.nl
[2] Computer and Information Sciences, University of Strathclyde, Glasgow, UK
leif.azzopardi@strath.ac.uk

Abstract. In this paper we provide an overview of the first edition of
the CLEF Dynamic Search Lab. The CLEF Dynamic Search lab ran in
the form of a workshop with the goal of approaching one key question:
how can we evaluate dynamic search algorithms? Unlike static search
algorithms, which essentially consider user request's independently, and
which do not adapt the ranking w.r.t the user's sequence of interactions,
dynamic search algorithms try to infer from the user's intentions from
their interactions and then adapt the ranking accordingly. Personalized
session search, contextual search, and dialog systems often adopt such
algorithms. This lab provides an opportunity for researchers to discuss
the challenges faced when trying to measure and evaluate the performance of dynamic search algorithms, given the context of available corpora, simulations methods, and current evaluation metrics. To seed the
discussion, a pilot task was run with the goal of producing search agents
that could simulate the process of a user, interacting with a search system over the course of a search session. Herein, we describe the overall
objectives of the CLEF 2017 Dynamic Search Lab, the resources created
for the pilot task and the evaluation methodology adopted.

1 Introduction

Information Retrieval (IR) research has traditionally focused on serving the best
results for a single query – so-called ad-hoc retrieval. However, users typically
search iteratively, refining and reformulating their queries during a session. IR
systems can still respond to each query in a session independently of the history
of user interactions, or alternatively adopt their model of relevance in the context
of these interactions. A key challenge in the study of algorithms and models
that dynamically adapt their response to a user's query on the basis of prior
interactions is the creation of suitable evaluation resources and the definition
of suitable evaluation metrics to assess the effectiveness of such IR algorithms.
Over the years various initiatives have been proposed which have tried to make
progress on this long standing challenge.

The TREC Interactive track [7], which ran between 1994 and 2002, investigated the evaluation of interactive IR systems and resulted in an early standard-
ization of the experimental design. However, it did not lead to a reusable test

© Springer International Publishing AG 2017
G.J.F. Jones et al. (Eds.): CLEF 2017, LNCS 10456, pp. 361–366, 2017.
DOI: 10.1007/978-3-319-65813-1_31

collection methodology. The High Accuracy Retrieval of Documents (HARD) track [1] followed the Interactive track, with the primary focus on single-cycle user-system interactions. These interactions were embodied in clarification forms which could be used by retrieval algorithms to elicit feedback from assessors. The track attempted to further standardize the retrieval of interactive algorithms, however it also did not lead to a reusable collection that supports adaptive and dynamic search algorithms. The TREC Session Track [3], which ran from 2010 through 2014, made some headway in this direction. The track produced test collections, where included with the topic description was the history of user interactions with a system, that could be used to improve the performance of a given query. While, this mean adaptive and dynamic algorithms could be evaluated for one iteration of the search process, the collection's are not suitable for assessing the quality of retrieval over an entire session. In 2015, the TREC Tasks track [11,12], a different direction was taken, where the test collection provides queries for which all possible sub-tasks did to be inferred, and the documents relevant to those sub-tasks identified. Even though the produced test collections could be used in testing whether a system can help the user to perform a task end-to-end, the focus was not on adapting and learning from the user's interactions as in the case of dynamic search algorithms.

In the related domain of dialogue systems, the advancement of deep learning methods has led to a new generation of data-driven dialog systems. Broadly-speaking, dialog systems can be categorized along two dimensions, (a) goal-driven vs. non-goal-driven, and (b) open-domain vs. closed domain dialog systems. Goal-driven open-domain dialog systems are in par with dynamic search engines: as they seek to provide assistance, advice and answers to a user over unrestricted and diverse topics, helping them complete their task, by taking into account the conversation history. While, a variety of corpora is available for training such dialog systems [10], when it comes to the evaluation, the existing corpora are inappropriate. This is because they only contain a static set of dialogues and any dialog that does not develop in a way similar to the static set cannot be evaluated. Often, the evaluation of goal-driven dialogue systems focuses on goal-related performance criteria, such as goal completion rate, dialogue length, and user satisfaction. Automatically determining whether a task has been solved however is an open problem, while task-completion is not the only quality criterion of interest in the development of dialog systems. Thus, simulated data is often generated by a simulated user [4,5,9]. Given a sufficiently accurate model of how user's converse, the interaction between the dialog system and the user can be simulated over a large space of possible topics. Using such data, it is then possible to deduce the desired metrics. This suggests that a similar approach could be taken in the context of interactive IR. However, while significant effort has been made to render the simulated data as realistic as possible [6,8], generating realistic user simulation models remains an open problem.

2   Lab Overview

Essentially, the CLEF Dynamic Tasks Lab attempts focus attention towards building a bridge between batch TREC-style evaluation methodology and the Interactive Information Retrieval evaluation methodology - so that dynamic search algorithms can be evaluated using re-usable test collections.

The objectives of the lab is threefold:

1. to devise a methodology for evaluating dynamic search algorithms by exploring the role of simulation as a means to create re-usable test collections
2. to develop evaluation metrics that measure the quality during the session (and at different stages) and at the end of the session (overall measures of quality).
3. to develop algorithms that can provide an optimal response in an interactive retrieval setup.

The focus of the CLEF 2017 Dynamic Tasks Lab is to provide a forum that can help foster research around the evaluation of dynamic retrieval algorithms. The Lab, in the form of a Workshop, solicits the submission of two types of papers: (a) position papers, and (b) data papers. Position papers focus on evaluation methodologies for assessing the quality of search algorithms with the user in the loop, under two constraints: any evaluation framework proposed should allow the (statistical) reproducibility of results, and lead to a reusable benchmark collection. Data Papers describe test collections or data sets suitable for guiding the construction of dynamic test collections, tasks and evaluation metrics.

3   Pilot Task

Towards the aforementioned goals of generating simulation data the CLEF 2017 Dynamic Tasks Lab ran a pilot task in the context of developing Task Completion Engines [2] and Intelligent Search Agents [6]. Task Completion Engines and Autonomous Search Agents are being developed to help users in acquire information in order to make a decision and complete a search task. At the same time such Intelligent Search Agents, encode a model of a user, and so present the potential to simulate users submitting queries, which can enable the evaluation of dynamic search algorithms. Such engines/agents need to work with a user to ascertain their information needs, then perform their own searches to dynamically identify relevant material, which will be useful in completing a particular task. For example, consider the task of organizing a wedding. There are many different things that need to be arranged and ordered, e.g. a venue, flowers, catering, gift list, dresses, car hire, hotels, etc. Finding relevant sites and resources requires numerous searches and filtering through many documents/sites. A search agent could help to expedite the process by finding the relevant sites to visit, while a task completion engine would provide a structured interface to help complete the process.

In this year's Dynamic Search Task Track, the task can be interpreted in one of two ways:

1. to generate a series of queries that a search agent would issue to a search engine, in order to compile a set of links useful for the user. This set might be presented to the user or used for further processing by a task completion engine; or
2. to generate a series of query suggestions that a search engine would recommend to the user, and thus the suggested course of interaction.

As a starting point, for building a test collection, we first consider how people look for information required to complete various casual leisure and work tasks. The history of queries and interactions are then used as a reference point during the evaluation to determine if agents/dynamic search algorithms/query suggestion algorithms that can generate queries that are like those posed by people. And thus, see how well human like queries can be generated for suggestions)? Thus the focus of the track is on query generation and models of querying, based on task and interaction history.

3.1   Task Description and Data Sets

Starting with an initial query for a given topic, the task is to generate a series of subsequent or related queries. The data used is TREC ClueWeb Part B test collection and the topics used are sourced from the Session Track 2014 [3]. A fixed search engine was setup, where ClueWeb was indexed using ElasticSearch. The title, url, page contents were indexed, along with the spam rank and page rank of each document. The ElasticSearch API was then provided as the "search engine" that the agent or person is using to undertake each task/topic. From the Session Track 2014 topics, a subset of 26 topics were selected out of the original 50, based on the following criteria: there were four or more queries associated with the topic, where the subsequent interaction on each query lead to identifying at least one TREC relevant document. These were considered, good or useful, queries i.e. they helped identify relevant material. The set of "good" queries were with-held as the relevant set. The TREC topic title, was provided to participants as the initial seed query i.e. the first query in the session.

Interaction data was then provided to provide simulated interaction with queries issued to the search engine. It was anticipated that the simulated clicks could be used by the algorithms to help infer relevance of the documents. This data could be used could be used as (a) a classifier providing relevance decisions regarding observed items in the result list, or (b) as clicks that the user performed when viewing the results of a query (i.e. given a query, assume that this is what the user clicks on, to help infer the next query). A set of judgments/click was generated based on the probability of Session Track users clicks data, conditioned by relevance (i.e. the probability of a click, if then document was TREC Relevant or TREC Non-Relevant).

The task, then, was to provide a list of query suggestions/recommendations along with a list of up to 50 documents.

3.2   Participants

Two teams participated in this Pilot Task: (1) Webis (Web Technology and Information Systems, Bauhaus-Universitat Weimar), and (2) TUW (Vienna University of Technology).

Webis. The general research idea of the Webis team contribution is to evaluate query suggestions in form of key-queries for clicked documents. A key-query for a document set D is a query that returns the documents from D among the top-k ranks. Our query suggestion approach derives key-queries for pairs of documents previously clicked by the user. The assumption then is that the not-already-clicked documents in the top results of the key-queries could also be interesting to the user. The key-query suggestions thus focus on retrieving more documents similar to the ones already clicked.

TUW. The general research idea of the TUW team contribution is to leverage the structure of Wikipedia articles to understand search tasks. Their assumption is that human editors carefully choose meaningful section titles to cover the various aspects of an article. Their proposed search agent explores this fact, being responsible for two tasks: (1) identifying the key Wikipedia articles related to a complex search task, and (2) selecting section titles from those articles. TUW contributed 5 runs, by (a) only using the queries provided; (b) manually choosing which Wikipedia sections to use; (c) automatically choosing the top-5 Wikipedia sections to use; (d) automatically ranking Wikipedia section using word2vec; and (e) automatically ranking Wikipedia sections using word2vec and using a naive Bayes classifier trained on the past qrels to decide if a document was relevant or not.

3.3   Evaluation

Given that evaluation is an open problem, the lab was also open to different ways in which to evaluate this task. Some basic measures that were employed are:

- Query term overlap: how well do the query terms in the suggestions match with the terms used
- Query likelihood: how likely are the queries suggested given the model of relevance.
- Precision and Recall based measures on the set of documents retrieved.
- Suggest your own measure: how to measure the usefulness and value of queries is a rather open question. So how can we evaluate how good a set of queries are in relation to completing a particular task?

During the lab, we will discuss the various challenges of constructing reusable test collections for evaluating dynamic search algorithms and how we can develop appropriate evaluation measures.

Acknowledgements. This work was partially supported by the Google Faculty Research Award program and the Microsoft Azure for Research Award program (CRM:0518163). All content represents the opinion of the authors, which is not necessarily shared or endorsed by their respective employers and/or sponsors. We would also like to thank Dr. Guido Zuccon for setting up the ElasticSearch API.

References

1. Allan, J.: Hard track overview in TREC 2003 high accuracy retrieval from documents. Technical report, DTIC Document (2005)
2. Balog, K.: Task-completion engines: a vision with a plan. In: SCST@ECIR (2015)
3. Carterette, B., Clough, P.D., Hall, M.M., Kanoulas, E., Sanderson, M.: Evaluating retrieval over sessions: the TREC session track 2011–2014. In: Perego, R., Sebastiani, F., Aslam, J.A., Ruthven, I., Zobel, J. (eds.) Proceedings of the 39th International ACM SIGIR Conference on Research and Development in Information Retrieval, SIGIR 2016, Pisa, Italy, 17–21 July 2016, pp. 685–688. ACM (2016). http://doi.acm.org/10.1145/2911451.2914675
4. Georgila, K., Henderson, J., Lemon, O.: User simulation for spoken dialogue systems: learning and evaluation. In: Interspeech pp. 1065–1068 (2006)
5. Jung, S., Lee, C., Kim, K., Jeong, M., Lee, G.G.: Data-driven user simulation for automated evaluation of spoken dialog systems. Comput. Speech Lang. **23**(4), 479–509 (2009). http://dx.doi.org/10.1016/j.csl.2009.03.002
6. Maxwell, D., Azzopardi, L.: Agents, simulated users and humans: An analysis of performance and behaviour. In: Proceedings of the 25th ACM International on Conference on Information and Knowledge Management, CIKM 2016, pp. 731–740 (2016)
7. Over, P.: The TREC interactive track: an annotated bibliography. Inf. Process. Manage. **37**(3), 369–381 (2001)
8. Pääkkönen, T., Kekäläinen, J., Keskustalo, H., Azzopardi, L., Maxwell, D., Järvelin, K.: Validating simulated interaction for retrieval evaluation. Inf. Retrieval J., 1–25 (2017)
9. Pietquin, O., Hastie, H.: A survey on metrics for the evaluation of user simulations. Knowl. Eng. Rev. **28**(01), 59–73 (2013)
10. Serban, I.V., Lowe, R., Henderson, P., Charlin, L., Pineau, J.: A survey of available corpora for building data-driven dialogue systems. CoRR abs/1512.05742 (2015). http://arxiv.org/abs/1512.05742
11. Verma, M., Yilmaz, E., Mehrotra, R., Kanoulas, E., Carterette, B., Craswell, N., Bailey, P.: Overview of the TREC tasks track 2016. In: Voorhees, E.M., Ellis, A. (eds.) Proceedings of The Twenty-Fifth Text REtrieval Conference, TREC 2016, Gaithersburg, Maryland, USA, 15–18 November 2016, vol. Special Publication 500–321. National Institute of Standards and Technology (NIST) (2016). http://trec.nist.gov/pubs/trec25/papers/Overview-T.pdf
12. Yilmaz, E., Verma, M., Mehrotra, R., Kanoulas, E., Carterette, B., Craswell, N.: Overview of the TREC 2015 tasks track. In: Voorhees, E.M., Ellis, A. (eds.) Proceedings of The Twenty-Fourth Text REtrieval Conference, TREC 2015, Gaithersburg, Maryland, USA, 17–20 November 2015, vol. Special Publication 500–319. National Institute of Standards and Technology (NIST) (2015). http://trec.nist.gov/pubs/trec24/papers/Overview-T.pdf

CLEF 2017: Multimodal Spatial Role Labeling (mSpRL) Task Overview

Parisa Kordjamshidi[1]([✉]), Taher Rahgooy[1], Marie-Francine Moens[2],
James Pustejovsky[3], Umar Manzoor[1], and Kirk Roberts[4]

[1] Tulane University, New Orleans, USA
pkordjam@tulane.edu
[2] Katholieke Universiteit Leuven, Leuven, Belgium
[3] Brandies University, Waltham, USA
[4] The University of Texas Health Science Center at Houston, Houston, USA

Abstract. The extraction of spatial semantics is important in many real-world applications such as geographical information systems, robotics and navigation, semantic search, etc. Moreover, spatial semantics are the most relevant semantics related to the visualization of language. The goal of multimodal spatial role labeling task is to extract spatial information from free text while exploiting accompanying images. This task is a multimodal extension of spatial role labeling task which has been previously introduced as a semantic evaluation task in the SemEval series. The multimodal aspect of the task makes it appropriate for the CLEF lab series. In this paper, we provide an overview of the task of multimodal spatial role labeling. We describe the task, sub-tasks, corpora, annotations, evaluation metrics, and the results of the baseline and the task participant.

1 Introduction

The multimodal spatial role labeling task (mSpRL) is a multimodal extension of the spatial role labeling shared task in SemEval-2012 [5]. Although there were proposed extensions of the data and the task in more extensive schemes in Kolomiyets et al. [4] and Pustejovsky et al. [13], the SemEval-2012 data was more appropriate for the goal of incorporating the multimodality aspect. SemEval-2012 annotates CLEF IAPRTC-12 Image Benchmark [1], which includes touristic pictures along with a textual description of the pictures. The descriptions are originally provided in multiple languages though we use the English annotations for the purpose of our research.

The goal of mSpRL is to develop natural language processing (NLP) methods for extraction of spatial information from both images and text. Extraction of spatial semantics is helpful for various domains such as semantic search, question answering, geographical information systems, and even in robotic settings when giving robots navigational instructions or instructions for grabbing and manipulating objects. It is also essential for some specific tasks such as text to

© Springer International Publishing AG 2017
G.J.F. Jones et al. (Eds.): CLEF 2017, LNCS 10456, pp. 367–376, 2017.
DOI: 10.1007/978-3-319-65813-1_32

scene conversion (or vice-versa), scene understanding as well as general information retrieval tasks when using a huge amount of available multimodal data from various resources. Moreover, we have noticed an increasing interest in the extraction of spatial information from medical images that are accompanied by natural language descriptions. The textual descriptions of a subset of images are annotated with spatial roles according to spatial role labeling annotation scheme [7]. We should note that considering the vision and language modalities and combining the two media has become a very popular research challenge nowadays. We distinguish our work and our data from the existing research related to vision and language (inter alia, [3,11]) in considering explicit formal spatial semantics representations and providing direct supervision for machine learning techniques by our annotated data. The formal meaning representation would help to exploit explicit spatial reasoning mechanisms in the future. In the rest of this overview paper, we introduce the task in Sect. 2; we describe the annotated corpus in Sect. 3; the baseline and the participant systems are described in Sect. 4; Sect. 5 reports the results and the evaluation metrics. Finally, we conclude in Sect. 6.

2   Task Description

The task of text-based spatial role labeling (SpRL) [8] aims at mapping natural language text to a formal spatial meaning representation. This formal representation includes specifying spatial entities based on cognitive linguistic concepts and the relationships between those entities, in addition to the type of relationships in terms of qualitative spatial calculi models. The ontology of the target concepts is drawn in Fig. 1 and the concepts are described later in this section. The applied ontology includes a subset of concepts proposed in the scheme described in [7]. We divide this task to three sub-tasks. To clarify these sub-tasks, we use the example of Fig. 2. This figure shows a photograph and a few English sentences that describe it. Given the first sentence *"About 20 kids in traditional clothing and hats waiting on stairs."*, we need to do the following tasks:

- **Sub-task 1:** The first task is to identify the phrases that refer to spatial entities and classify their roles. The spatial roles include (a) spatial indicators, (b) trajectors, (c) landmarks. Spatial indicators indicate the existence of spatial information in a sentence. Trajector is an entity whose location is described and landmark is a reference object for describing the location of a trajector. In the above-mentioned sentence, the location of *about 20 kids* that is the *trajector* has been described with respect to the *the stairs* that is the *landmark* using the preposition *on* that is the *spatial indicator*. These are examples the spatial roles that we aim to extract form the sentence.
- **Sub-task 2:** The second sub-task is to identify the relations/links between the spatial roles. Each spatial relation is represented as a triplet of (spatial-indicator, trajector, landmark). Each sentence can contain multiple relations and individual phrases can even take part in multiple relations. Furthermore, occasionally roles can be implicit in the sentence (i.e., a null item in the

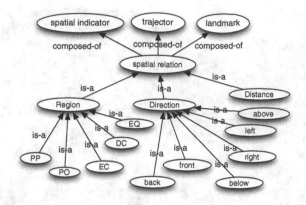

Fig. 1. Given spatial ontology [9]

triplet). In the above example, we have the triplet (kids, on, stairs) that form a spatial relation/link between the three above mentioned roles. Recognizing the spatial relations is very challenging because there could be several spatial roles in the sentence and the model should be able to recognize the right connections. For example (waiting, on, stairs) is a wrong relation here because "kids" is the trajector in this sentence not "waiting".

– **Sub-task 3:** The third sub-task is to recognize the type of the spatial triplets. The types are expressed in terms of multiple formal qualitative spatial calculi models similar to Fig. 1. At the most course-grained level the relations are classified into three categories of topological (regional), directional, or distal. Topological relations are classified according to the well-known RCC (regional connection calculus) qualitative representation. An RCC5 version that is shown in Fig. 1 includes Externally connected (EC), Disconnected (DC), Partially overlapping (PO), Proper part (PP), and Equality (EQ). The data is originally annotated by RCC8 which distinguishes between Proper part (PP), Tangential proper part (TPP) and Inverse tangential proper part inverse (TPPI). For this lab the original RCC8 annotations are used. Directional relations include 6 relative directions: left, right, above, below, back, and front. In the above example, we can state the type of relation between the roles in the triplet (kids, on, stairs) is "above". In general, we can assign multiple types to each relation. This is due to the polysemy of spatial prepositions as well as the difference between the level of specificity of spatial relations expressed in the language compared to formal spatial representation models. However, multiple assignments are not frequently made in our dataset.

The task that we describe here is similar to the specifications that are provided in Kordjamshidi et al. [9], however, the main point of this CLEF lab was to provide an additional resource of information (the accompanying images) and investigate the ways that the images can be exploited to improve the accuracy of the text-based spatial extraction models. The way that the images can be

Fig. 2. "About 20 kids in traditional clothing and hats waiting on stairs. A house and a green wall with gate in the background. A sign saying that plants can't be picked up on the right." (Color figure online)

used is left open to the participants. Previous research has shown that this task is very challenging [8], particularly given the small set of available training data and we aim to investigate if using the images that accompany textual data can improve the recognition of the spatial objects and their relations. Specifically, our hypothesis is that the images could improve the recognition of the type of relations given that the geometrical features of the boundaries of the objects in the images are closer to the formal qualitative representations of the relationships compared to the counterpart linguistic descriptions.

3   Annotated Corpora

The annotated data is a subset of the IAPR TC-12 image Benchmark [1]. It contains 613 text files with a total of 1,213 sentences. The original corpus was available without copyright restrictions. The corpus contains 20,000 images taken by tourists with textual descriptions in up to three languages (English, German, and Spanish). The texts describe objects and their absolute or relative positions in the image. This makes the corpus a rich resource for spatial information. However the descriptions are not always limited to spatial information which makes the task more challenging. The data has been annotated with the roles and relations that were described in Sect. 2, and the annotated data can be used to train machine leaning models to do this kind of extractions automatically. The text has been annotated in previous work (see [6,7]). The role annotations are provided on phrases rather than single words. The statistics about the data is given in Table 1. For this lab, we augmented the textual spatial annotations

Table 1. The statistics of the annotated CLEF-Image Benchmark, some of the spatial relations are annotated with multiple types, e.g., having both region and direction labels.

	Train	Test	All
Sentences	600	613	1213
Trajectors	716	874	1590
Landmarks	612	573	1185
Spatial indicators	666	795	1461
Spatial relations	761	939	1700
Region	560	483	1043
Direction	191	449	640
Distance	40	43	83

with a reference to the aligned images in the xml annotations and fixed some of the annotation mistakes to provide a cleaner version of the data.

4 System Descriptions

We, as organizers of the lab, provided a baseline inspired by previous research for the sake of comparison. The shared task had one official participant who submitted two systems. In this section, we describe the submitted systems and the baseline.

– **Baseline:** For sub-task 1 and classifying each role (Spatial Indicator, Trajector, and Landmark), we created a sparse perceptron binary classifier that uses a set of lexical, syntactical, and contextual features, such as lexical surface patterns, headwords phrases, part-of-speech tags, dependency relations, subcategorization, etc. For classifying the spatial relations, we first trained two binary classifiers on pairs of phrases. One classifier detects Trajector-SpatialIndicator pairs and another detects Landmark-SpatialIndicator pairs. We used the spatial indicator classifier from sub-task 1 to find the indicator candidates and considered all noun phrases as role candidates. Each combination of SpatialRole-SpatialIndicator candidates considered as a pair candidate and the pair classifiers are trained on. We used a number of relational features between the pairs of phrases such as distance, before, etc. to classify them. In the final phase, we combined the predicted phrase pairs that have a common spatial indicator in order to create the final relation/triplet for sub-task 2. for example if (kids, on) pair is classified as Trajector-SpatialIndicator and (stairs,on) is predicted as Landmark-SpatialIndicator then we generate the triplet, (on,kids,stairs) as a spatial triplet since both trajector and landmark relate to the same preposition 'on'. The features of this baseline model are inspired by the work in [9]. For sub-task 3 and training general type and specific value classifiers, we used a very naive pipeline model as the baseline. In this pipeline, the predicted triplets from the last stage are used for training the relations types. For these type classifiers, simply, the phrase features of each argument of the triplets are concatenated and used as features. Obviously, we miss a large number of relations at the stage of spatial relation extraction in sub-task 2 since we depend on its recall.

– **LIP6:** The LIP6 group built a system for sub-task 3 that classifies relation types. For the sub-task 1 and 2, the proposed model in Roberts and Harabagiu [14] was used. Particularly, an implementation of that model in the Saul [10] language/library was applied. For every relation, an embedding is built with available data: the textual relation triplet and visual features from the associated image. Pre-trained word embeddings are used [12] to represent the trajector and landmark and a one-hot vector indicates which spatial indicator is used; the visual features and embeddings from the segmented regions of the trajectors and landmarks are extracted and projected into a low dimensional space. Given those generated embeddings, a linear SVM model is trained to classify the spatial relations and the embeddings remain fixed. Several experiments were made to try various classification modes and discuss the effect of the model parameters, and more particularly the impact of the visual modality. As the best performing model ignores the visual modality, these results highlight that considering multimodal data for enhancing natural language processing is a difficult task and requires more efforts in terms of model design.

5   Evaluation Metrics and Results

About 50% of the data was used as the test set for the evaluation of the systems. The evaluation metrics were precision, recall, and F1-measure, defined as:

$$recall = \frac{TP}{TP + FN}, \quad precision = \frac{TP}{TP + FP}, \quad F1 = \frac{2 * recall * precision}{(recall + precision)}$$

where, TP (true positives) is the number of predicted components that match the ground truth, FP (false positives) is the number of predicted components that do not match the ground truth, and FN (false negatives) is the number of ground truth components that do not match the predicted components. These metrics are used to evaluate the performance on recognizing each type of role, the relations and each type of relation separately. Since the annotations are provided based on phrases, the overlapping phrases are counted as correct predictions instead of exact matchs. The evaluation with exact matching between phrases would provide lower performance than the reported ones. The relation type evaluation for sub-task 3 includes course- and fine-grained metrics. The coarse-grained metric (overall-CG) averages over the labels of region, direction, and distance. The fine-grained metric (overall-FG) shows the performance over

Table 2. Baseline: classic classifiers and linguistically motivated features based on [9]

Label	P	R	F1
SP	94.76	97.74	96.22
TR	56.72	69.56	62.49
LM	72.97	86.21	79.04
Overall	74.36	83.81	78.68
Triplets	75.18	45.47	56.67
Overall-CG	64.72	37.91	46.97
Overall-FG	47.768	23.490	26.995

all lower-level nodes in the ontology including the RCC8 types (e.g., EC) and directional relative types (e.g., above, below).

Table 2 shows the results of our baseline system that was described in the previous section. Though the results of the roles and relation extraction are fairly comparable to the state of the art [9,14], the results of the relations type classifiers are less matured because a simple pipeline, described in Sect. 4, was used. Table 3 shows the results of the participant systems.

As mentioned before, LIP6 uses the model suggested in [14] and its implementation in Saul [10] for sub-task 1 and sub-task 2. It has a focus in designing a model for sub-task 3. The experimental results using textual embeddings alone are shown under *text only* in the table, and a set of results are reported by exploiting the accompanying images and training the visual embeddings from the corpora. The LIP6's system significantly outperforms the provided baseline for relation type classifiers. Despite our expectations, the results that use the visual embeddings perform worse than the one that ignores images. In addition to the submitted systems, the LIP6 team improved their results slightly by using a larger feature size in their dimensionality reduction procedure with their text-only features. This model outperforms their submitted systems and is listed in Table 3 as *Best model*.

Discussion. The previous research and the results of LIP6 team show this task is challenging, particularly, using this small set of training data. LIP6 was able to outperform the provided baseline using the textual embeddings for relation types but the results of combining the images, in the contrary, dropped the performance. This result indicates that integrating the visual information needs more investigation otherwise it can only add noise to the learning system. One very basic question to be answered is whether the images of this specific dataset can potentially provide complementary information or help resolving ambiguities in the text at all; this investigation might need a human analysis. Although the visual embeddings did not help the best participant system with the current experiments, using other alternative embeddings trained from large corpora might help improving this task. Given the current interest of the vision and language communities in combining the two modalities and the benefits that

Table 3. LIP6 performance with various models for Sub-task 3; LIP6 uses Roberts and Harabagiu [14] for Sub-tasks 1 and 2.

Label		P	R	F1
SP		97.59	61.13	75.17
TR		79.29	53.43	63.84
LM		94.05	60.73	73.81
Overall		89.55	58.03	70.41
Triplets		68.33	48.03	56.41
Text only	Overall-CG	63.829	44.835	52.419
	Overall-FG	56.488	39.038	43.536
Text+Image	Overall-CG	66.366	46.539	54.635
	Overall-FG	58.744	40.716	45.644
Best model	Overall-CG	66.76	46.96	55.02
	Overall-FG	58.20	41.05	45.93

this trend will have for the information retrieval, there are many new corpora becoming available (e.g. [11]) which can be valuable sources of information for obtaining appropriate joint features. There is a separate annotation on the same benchmark that includes the ground-truth of the co-references in the text and image [2]. This annotation has been generated for co-reference resolution task but it seems to be very useful to be used on top of our spatial annotations for finding better alignment between spatial roles and image segments. In general, current related language and vision resources do not consider formal spatial meaning representation but can be used indirectly to train informative representations or be used as source for indirect supervision for extraction of formal spatial meaning.

6   Conclusion

The goal of the multimodal spatial role labeling lab was to provide a benchmark to investigate how adding grounded visual information can help understanding the spatial semantics of natural language text and mapping language to a formal spatial meaning representation. The prior hypothesis has been that the visual information should help the extraction of such semantics because spatial semantics are the most relevant semantics for visualization and the geometrical information conveyed in the vision media should be able to easily help in disambiguation of spatial meaning. Although, there are many recent research works on combining vision and language, none of them consider obtaining a formal spatial meaning representation as a target nor provide supervision for training such representations. However, the experimental results of our mSpRL lab participant show that even given ground truth segmented objects in the images and having the exact geometrical information about their relative positions, adding useful information for understanding the spatial meaning of the text is very challenging. The experimental results indicate that using the visual embeddings and using the similarity between the objects in the image and spatial entities in the text can turn to adding noise to the learning system reducing the performance. However, we believe our prior hypothesis is still valid, but finding an effective way to exploit vision for spatial language understanding, particularly obtaining a formal spatial representation appropriate for explicit reasoning, remains as an important research question.

References

1. Grubinger, M., Clough, P., Müller, H., Deselaers, T.: The IAPR benchmark: a new evaluation resource for visual information systems. In: Proceedings of the International Conference on Language Resources and Evaluation (LREC), pp. 13–23 (2006)
2. Kazemzadeh, S., Ordonez, V., Matten, M., Berg, T.L.: Referit game: referring to objects in photographs of natural scenes. In: EMNLP (2014)
3. Kiros, R., Salakhutdinov, R., Zemel, R.S.: Unifying visual-semantic embeddings with multimodal neural language models. CoRR abs/1411.2539 (2014). http://arxiv.org/abs/1411.2539
4. Kolomiyets, O., Kordjamshidi, P., Moens, M., Bethard, S.: Semeval-2013 task 3: Spatial role labeling. In: Second Joint Conference on Lexical and Computational Semantics (*SEM): Proceedings of the Seventh International Workshop on Semantic Evaluation (SemEval 2013), Atlanta, Georgia, USA, vol. 2, pp. 255–262, June 2013
5. Kordjamshidi, P., Bethard, S., Moens, M.F.: SemEval-2012 task 3: Spatial role labeling. In: Proceedings of the First Joint Conference on Lexical and Computational Semantics: Proceedings of the Sixth International Workshop on Semantic Evaluation (SemEval), vol. 2, pp. 365–373 (2012)
6. Kordjamshidi, P., van Otterlo, M., Moens, M.: Spatial role labeling annotation scheme. In: Pustejovsky, J., Ide, N. (eds.) Handbook of Linguistic Annotation. Springer, Dordrecht (2015)
7. Kordjamshidi, P., van Otterlo, M., Moens, M.F.: Spatial role labeling: task definition and annotation scheme. In: Calzolari, N., Khalid, C., Bente, M. (eds.) Proceedings of the Seventh Conference on International Language Resources and Evaluation (LREC 2010), pp. 413–420 (2010)
8. Kordjamshidi, P., van Otterlo, M., Moens, M.F.: Spatial role labeling: towards extraction of spatial relations from natural language. ACM - Trans. Speech Lang. Process. 8, 1–36 (2011)
9. Kordjamshidi, P., Moens, M.F.: Global machine learning for spatial ontology population. Web Semant. 30(C), 3–21 (2015)
10. Kordjamshidi, P., Wu, H., Roth, D.: Saul: towards declarative learning based programming. In: Proceedings of the International Joint Conference on Artificial Intelligence (IJCAI), July 2015
11. Krishna, R., Zhu, Y., Groth, O., Johnson, J., Hata, K., Kravitz, J., Chen, S., Kalantidis, Y., Li, L., Shamma, D.A., Bernstein, M.S., Li, F.: Visual genome: Connecting language and vision using crowdsourced dense image annotations. Int. J. Comput. Vis. 123(1), 32–73 (2017)

12. Pennington, J., Socher, R., Manning, C.D.: Glove: global vectors for word representation. In: EMNLP, vol. 14, pp. 1532–1543 (2014)
13. Pustejovsky, J., Kordjamshidi, P., Moens, M.F., Levine, A., Dworman, S., Yocum, Z.: SemEval-2015 task 8: SpaceEval. In: Proceedings of the 9th International Workshop on Semantic Evaluation (SemEval 2015), 9th International Workshop on Semantic Evaluation (SemEval 2015), Denver, Colorado, 4–5 June 2015, pp. 884–894. ACL (2015). https://lirias.kuleuven.be/handle/123456789/500427
14. Roberts, K., Harabagiu, S.: UTD-SpRL: a joint approach to spatial role labeling. In: *SEM 2012: The First Joint Conference on Lexical and Computational Semantics: Proceedings of the Sixth International Workshop on Semantic Evaluation (SemEval 2012), vol. 2, pp. 419–424 (2012)

Author Index

Printed in the United States
By Bookmasters